Nanotechnology

Nanotechnology

Advances and Real-Life Applications

Edited by

Cherry Bhargava and Amit Sachdeva

CRC Press
Taylor & Francis Group
Boca Raton London New York

CRC Press is an imprint of the
Taylor & Francis Group, an **informa** business

First edition published 2020
by CRC Press
6000 Broken Sound Parkway NW, Suite 300, Boca Raton, FL 33487-2742

and by CRC Press
2 Park Square, Milton Park, Abingdon, Oxon, OX14 4RN

© 2021 Taylor & Francis Group, LLC
CRC Press is an imprint of Taylor & Francis Group, LLC

ISBN: 9780367536732 (hbk)
ISBN: 9781003082859 (ebk)

Typeset in Times
by Deanta Global Publishing Services, Chennai, India

Contents

Contents

Preface

Nanotechnology is the study and application of extremely small things and can be multidisciplinary. Nanotechnology encompasses the understanding of the fundamental physics, chemistry, biology, and technology of nanometer-scale objects. An example is nanotechnology and the development of new tools to analyze and manipulate matter on a molecular level. The pace of technology growth and its rate of proliferation across the world also present major new challenges. Nanotechnology is an emerging science that is expected to have rapid and strong future developments. This book targets the future of nanotechnology in various industries and disciplines. Nanoparticles can be designed as biosensors for plant disease diagnostics and as delivery vehicles for genetic material, probes, and agrichemicals. In the past decade, reports of nanotechnology in phytopathology have grown exponentially. Nanomaterials have been integrated into disease management strategies and diagnostics and as molecular tools.

Developments in carbon nanotubes and scanning tunneling microscopy have driven a need for ever-more-powerful techniques for structural and chemical characterization on the "nanoscale" and "nano-characterization" – the theme of this book now encompasses a diverse range of techniques that are the essential tools underpinning much of nanotechnology. As nanostructures become smaller, their three-dimensional shape assumes much greater importance. This is particularly the case for supported catalysts, where active sites may be determined by such entities. Although electron tomography had been previously applied successfully to biological structures, it has only recently been extended to applications in nanoscience. From the mechanical to the medicinal, nanotechnology has emerged as a multidisciplinary branch that deals with making things the size of atoms and molecules. The interest of nanotechnology is increasing related to the medical field. It has also triggered the emergence of a new field known as nanomedicine.

This book offers an introduction to the increasingly developing nanotechnology field by highlighting the key fundamentals and the application of advanced nanotechnology in real life. It features a robust framework of the subject. This work covers the basics of nanotechnology and provides a solid understanding of the subject. Starting with a review of basic quantum mechanics and materials science, the book helps to gradually build an understanding of the various effects of quantum confinement, optical-electronic properties of nanoparticles, and major nanomaterials. The book covers the various physical, chemical, and hybrid methods of nanomaterial synthesis and nanofabrication.

This book is a collection of essays about researchers involved in all facets of nanotechnologies. It will act as a superb introduction to nanotechnology and nanoscience both for students and those intrigued by the nano world and its applications.

About the Editors

Cherry Bhargava is Associate Professor and Head, VLSI domain, School of Electrical and Electronics Engineering at Lovely Professional University, Jalandhar, India. She has more than 15 years of teaching and research experience. She has a PhD (ECE) from IKG Punjab Technical University, State Govt. University, Punjab, M.Tech (VLSI Design & CAD) from Thapar University, and B. Tech (EIE) from Kurukshetra University. She is GATE qualified with All India Rank 428. She has authored about fifty technical research papers in SCI, Scopus, indexed quality journals, and national and international conferences. She has sixteen books to her credit. She has registered two copyrights and filed twenty-one patents. She is a recipient of various national and international awards for being outstanding faculty in engineering and an excellent researcher. She is an active reviewer and editorial member of numerous prominent SCI and Scopus indexed journals. Her research area is Nanotechnology and Artificial Intelligence.

Amit Sachdeva is Associate Professor at Lovely Professional University, Jalandhar, India. He has teaching experience of more than 7 years and his field of specialization is Material Technology. Dr. Sachdeva has authored around 20 technical research papers in SCI, Scopus, indexed quality journals, and national and international conferences. Dr. Sachdeva is an editorial member of various scientific indexed journals and is a lifetime member of IAENG and IFERP. Dr. Sachdeva has received the Young Scientist Award from the University of Malaya at ICFPAM 2019, organized at Penang Island, Malaysia. He chaired a session and was selected as a judge for evaluating Poster sessions. He has participated in around 15 international conferences and has also been part of organizing committees at 6 international conferences.

Contributors

Vibha Aggarwal
Punjabi University Neighbourhood
   Campus
Rampura Phul, India

Iqbal Altaf
Department of Chemistry
University of Karachi
Karachi, Pakistan

Anil Anora
Thapar University
Patiala, India

Anupriya
School of Pharmaceutical Sciences
Lovely Professional University
Phagwara, India

Shivani Arora Abrol
School of Electronics and Electrical
   Engineering
Lovely Professional University
Phagwara, India

Arti
Department of Electronics
Sanatan Dharma College
Ambala Cantt, India

Ankit Awasthi
School of Pharmaceutical Sciences
Lovely Professional University
Phagwara, India

Rafia Azmat
Department of Chemistry
University of Karachi
Karachi, Pakistan

**Varimadugu Bhanukirankumar
Reddy**
School of Pharmaceutical Sciences
Lovely Professional University
Phagwara, India

Cherry Bhargava
School of Electronics and Electrical
   Engineering
Lovely Professional University
Phagwara, India

Neeraj Choudhary
School of Pharmaceutical Sciences
Lovely Professional University
Phagwara, India

Leander Corrie
School of Pharmaceutical Sciences
Lovely Professional University
Phagwara, India

Linu Dash
School of Pharmaceutical Sciences
Lovely Professional University
Phagwara, India

Anamika Gautam
School of Pharmaceutical Sciences
Lovely Professional University
Phagwara, India

Monica Gulati
School of Pharmaceutical Sciences
Lovely Professional University
Phagwara, India

Sandeep Gupta
Punjabi University Neighbourhood
   Campus
Rampura Phul, India

Shriya Iyer
Department of Engineering
Amity University Dubai
Dubai, United Arab Emirates

Athira J
Department of Engineering
Amity University Dubai
Dubai, United Arab Emirates

Rubiya Khursheed
School of Pharmaceutical Sciences
Lovely Professional University
Phagwara, India

Abhishek Kumar
School of Electronics and Electrical
    Engineering
Lovely Professional University
Phagwara, India

Bimlesh Kumar
School of Pharmaceutical Sciences
Lovely Professional University
Phagwara, India

Narendra Kumar Pandey
School of Pharmaceutical Sciences
Lovely Professional University
Phagwara, India

Rajan Kumar
School of Pharmaceutical Sciences
Lovely Professional University
Phagwara, India

Ankit Kumar
School of Pharmaceutical Sciences
Lovely Professional University
Phagwara, India

Rajneesh Kumar Gupta
School of Pharmaceutical Sciences
Lovely Professional University
Phagwara, India

Pardeep Kumar Sharma
School of Pharmaceutical Sciences
Lovely Professional University
Phagwara, India

Sachin Kumar Singh
School of Pharmaceutical Sciences
Lovely Professional University
Phagwara, India

Virinder Kumar Singla
Punjabi University Neighbourhood
    Campus
Rampura Phul, India

Asha Anish Madhavan
Department of Engineering
Amity University Dubai
Dubai, United Arab Emirates

Summyia Masood
Department of Chemistry
University of Karachi
Karachi, Pakistan

Indu Melkani
School of Pharmaceutical Sciences
Lovely Professional University
Phagwara, India

Reshmi S Nair
Department of Engineering
Amity University Dubai
Dubai, United Arab Emirates

Sakshi Panchal
School of Pharmaceutical Sciences
Lovely Professional University
Phagwara, India

Amica Panja
School of Pharmaceutical Sciences
Lovely Professional University
Phagwara, India

Pooja Patni
School of Pharmaceutical Sciences
Lovely Professional University
Phagwara, India

Amina Pervaiz
Department of Chemistry
University of Karachi
Karachi, Pakistan

Shilpa Prasad
Department of Biotechnology
Amity University Jharkhand
Ranchi, India

Pankaj Prashar
School of Pharmaceutical Sciences
Lovely Professional University
Phagwara, India

Parteek Prasher
University of Petroleum and Energy
Studies
Dehradun, India

Priyanka
Punjabi University Neighborhood
Campus
Rampura Phul, India

Balwant Raj
Dept. of ECE
Punjab University SS Giri Regional
Centre
Hoshiapur, India

Balwinder Raj
Dept. of Electronics and
Communication Engineering
NITTTR
Chandigarh, India

Dania S
Department of Engineering
Amity University Dubai
Dubai, United Arab Emirates

Sharon Santhosh
Department of Engineering
Amity University Dubai
Dubai, United Arab Emirates

Mousmee Sharma
Department of Chemistry
Uttaranchal University
Dehradun, India

Rajani Sharma
Department of Biotechnology
Amity University Jharkhand
Ranchi, India

Dileep Singh Baghel
School of Pharmaceutical Sciences
Lovely Professional University
Phagwara, India

Harmeet Singh
Department of Mechanical Engineering
Guru Nanak Dev Engineering College
Ludhiana, India

Jeetendra Singh
Dept. of Electronics and
Communication Engineering
NIT Sikkim
Sikkim, India

Saurabh Singh
School of Pharmaceutical Sciences
Lovely Professional University
Phagwara, India

Ankita Sood
School of Pharmaceutical Sciences
Lovely Professional University
Phagwara, India

Pavani Sriram
Vaagdevi College of Pharmacy
Kakatiya University
Warangal, India

Nimmy Srivastava
Amity University Jharkhand
Ranchi, India

Ashish Suttee
School of Pharmaceutical Sciences
Lovely Professional University
Phagwara, India

Shipra
Punjabi University Neighborhood
    Campus
Rampura Phul, India

Lalit Thakur
Department of Mechanical Engineering
National Institute of Technology
Kurukshetra, India

Hitesh Vasudev
School of Mechanical Engineering,
Lovely Professional University
Phagwara, India

Nada W
Department of Engineering
Amity University Dubai
Dubai, United Arab Emirates

1 Introduction to Nanomaterials
History, Classification, Properties, and Applications

Arti
Sanatan Dharma College, Ambala Cantt, India

CONTENTS

1.1   INTRODUCTION: BACKGROUND AND DRIVING FORCES

Science has expanded very fast since the ages of Newton and Galileo. The basic concepts in chemistry and physics have had a strong influence on the ability to produce new materials. There are many scientists such as Faraday, who gave laws of electrolysis used to prepare and purify the materials. The ability to determine the characteristics and various properties of the materials gave different processes to develop better materials. For a long time, different natural and manmade materials

1

have been used in our lives in one way or another. The expertise gained by people in the last few decades has added lots of new approaches to material development which involves designing materials as per the demand of people. As the field has expanded, material scientists have created totally different types of materials such as nanotubes, buckyballs, etc. which are very small spheres or cylinders made up of carbon atoms. Now researchers have discovered the materials have new dimensions which are very small in size, bonded together atom by atom and having properties altogether different than the same material at a larger scale. The enhancement of the material science is nowadays known as "nanoscience", which is a multidisciplinary subject covering solid-state physics, chemistry, atomic and molecular physics and many other disciplines. All this increase in knowledge has arisen with nanostructures which may be manipulated to explore the physical, chemical and biological properties of the system that are intermediate in size, somewhere between single atoms, molecules and bulk materials. Richard Feynman, a famous Nobel laureate said in his lecture "There's Plenty of Room at the Bottom" that "nothing in the laws of physics prevented us from arranging the atoms the way we want". His lecture proved to be a landmark in the field of nanoscience, which is entirely based on the arrangements of the atoms inside materials. Atoms are basic building blocks of all matters available not only on earth but in the universe too. The properties of these materials depend on the arrangement and settled structure of atoms. The nanosized particles and materials are not an invention or discovery as these materials are defined by their size. Nanoparticles differ from the coarser particles of the same materials by their capability to form agglomerates.

1.2 NANOSCIENCE, NANOTECHNOLOGY AND NANOMATERIALS

Nanoscience is the branch that deals with the efficient use of various materials at atomic, macromolecular and molecular levels where properties of materials are totally different from the materials at larger scale. Thus nanoscience is the science of small things that are less than 100 nm in size and known as 'nanomaterials'.

Nanotechnology is the design, production, characterization and application of the devices and systems by controlling the size and shape at nanometer scale. At such a small scale, boundaries between physics, chemistry and biology are quite difficult to separate. Despite this small scale, nanotechnology plays a very important role in all these disciplines. This field is interdisciplinary with combinations of various ideas becoming integrated. Researchers from various specializations such as physics, chemistry, biology, engineering, information technology and many more are contributing in this interdisciplinary field at their level best. Some industries are also contributing by simply applying the knowledge gained by the various researchers in the field of nanotechnology. Nanotechnology is the only science which has the potential and caliber to impact all devices available now and in the future. All industries are dependent on the materials and various devices made by atom and molecules which may be improved by the applications of nanotechnology. The advances and innovations achieved in the size, shape and uses of different materials will directly benefit our lives. Nanotechnology is certain to have a revolutionary impact on the methods of designing and manufacturing things.

1.3 HISTORY OF NANOMATERIALS

Human needs and desires always give birth to new inventions and discoveries. Nanotechnology, at the frontier of the 21st century, is a product of such desires. Nanotechnology, nanomaterials and nanoparticles are common words, not only in research areas but in everyday life too. Nanostructures are not a new discovery; the oldest proof of the existence of nanoparticles and nanomaterials is in the form of early meteorites. Nature later evolved many other nanostructures like skeletons, seashells and many other structures (Hornyak, Dutta and Tibbels, 2008). Many nano-scaled smoke particles were formed and identified during the use of fire by early humans. However, the scientific journey of these materials started much later. In 1857, the properties of nanomaterials were proven by Faraday in his research paper "Experimental relations of gold (and other metals) to light". This was the first scientific report presented about nanomaterials. Later, after so many years of research in the early 1940s, silica nanoparticles were manufactured by researchers in the USA to create ultrafine carbon black for rubber reinforcement. In the 1960s and 1970s, metallic nanopowders were developed and used for magnetic recording tapes. In 1976, Graqvist and Buhrman used gas evaporation techniques on nanocrystals for the first time. The golden era of nanotechnology started in the 1980s when Kroto and Curl discovered fullerenes and Eric Drexler of Massachusetts Institute of Technology (MIT) used ideas from Feynman's "There's Plenty of Room at the Bottom" and Taniguchi's used the term "nanotechnology" in his 1986 book titled "Engines of Creation: The Coming Era of Nanotechnology". This was the beginning of the use of nanomaterials.

1.4 CLASSIFICATION OF NANOMATERIALS

Nanomaterials are entities having a very small size in the range from 1 nm to 100 nm i.e. in the nanoscopic range. These materials show distinct properties which are absent in macroscopic as well as in microscopic states of matter. Nanomaterials are of different shapes like nanobuds, nanotubes, nanowires, nanorods, nanoflowers, nanotowers, nanoneedles, nanocombs, etc. Nanomaterials are divided into various categories described as follows:

1.4.1 NANOMATERIALS BASED ON THE QUANTUM CONFINEMENT

Quantum confinement is the state of change of electronic and optical properties of the materials which are very small in size, especially 10 nm or less. The nanomaterials are classified into three groups depending on the number of directions of quantum confinement. The schematic of the various nanomaterials based on quantum confinement is shown in Figure 1.1.

1. Zero-dimensional (0D) nanomaterials: In these nanomaterials, all three directions are in nanoregime i.e. quantum confinement is in all three directions. Such types of materials are known as nanoparticles. Here, no dimension in any direction is more than 100 nm. Nanocubes, nanoclusters,

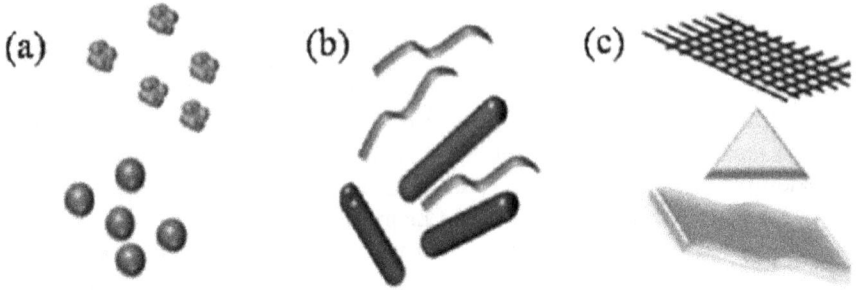

FIGURE 1.1   (a) 0D spheres and clusters; (b) 1D nanofibers: wires and roads; (c) 2D film, plates and networks.

quantum dots and nanoparticles are examples of 0D materials. The nanoparticles may be of amorphous, single crystalline or polycrystalline structure having various shapes and forms (Source: Shah and Ahmad, 2011).

2. One-dimensional (1D) nanomaterials: In such nanomaterials, quantum confinement is in two directions i.e. only one direction is outside the nanoregime. Examples of such materials are nanotubes, nanowires, nanofibers, nanotapes, nanofilaments and many more. These materials are amorphous or crystalline in nature and may exist in pure or impure form, embedded within other mediums.

3. Two-dimensional (2D) nanomaterials: In these materials, the quantum confinement is in one direction only. That means two directions are outside the nanoscale regime and only one direction is in nanoregime. This class of materials includes plate-like shapes such as nanofilms, nano-networks, ultra-thin films, nanocoating, nanolayers and graphene, etc. These materials are made up of various chemical compositions to form amorphous or crystalline structures.

1.4.2 SUPERLATTICES

Superlattices are those nanostructures having repeated 0D, 1D or 2D materials appearing in a periodic structure composed of two or more materials with the thickness of a few nanometers to tens of nanometers. The types of superlattices are shown in Figure 1.2.

1.4.3 3D NANOMATERIALS OR BULK NANOMATERIALS

The materials which are having all the three dimensions more than 100 nm are known as 3D nanomaterials or Bulk nanomaterials. Bulk nanomaterials are having multiple arrangements of nanosize crystals known as crystallites. These materials have all three parameters of length, breadth and height. Different types of materials like nanocrystalline materials, nanoporous materials and nanocomposites are bulk nanomaterials (George, 2008).

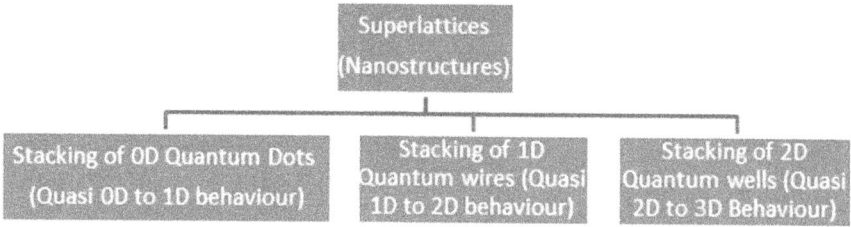

FIGURE 1.2 Superlattice types based on their periodic structure.

Nanocrystalline materials are basically single phase or multiphase polycrystalline structures having a grain size of less than 100 nm. The state of grain boundaries plays an important role in controlling the properties of these materials. Similarly, nanocomposites are the synthesized materials composed of scattering or dispersion of nanomaterials in the form of a matrix. These nanomaterials act as fillers in a matrix. Nanocomposite materials represent a matrix fillers' combination where 0D, 1D, 2D nanomaterials are surrounded by a matrix, and bound together as a separate entity to produce a 3D network. The matrix can be of any type, such as metallic, ceramic or polymeric having dimensions in bulk range, whereas fillers are in the nano range. The dimensions of nanofillers decide the type of nanocomposite. Basically, these are of three types: *Zero-dimensional (0D) type* nanocomposites where there is dispersion of zero type nanoparticles in matrix phase, *one-dimensional (1D) type* nanocomposites have dispersion of one-dimensional nanoparticles in matrix phase and *two-dimensional (2D) type* nanocomposites have dispersion of two-dimensional nanoparticles in matrix phase.

1.4.3.1 Importance of Nanomaterials

Nanomaterials are of special interest due to their size, even smaller than the thickness of human hair. Also, nanomaterials have been an area of interest for researchers in the past decade due to their specific mechanical, electrical, magnetic and optical properties. A few examples are as follows:

- Nanophase ceramics are important for researchers due to their ductile nature at elevated temperatures as compared to coarse-grained ceramics.
- Nanostructured semiconductor materials show various non-linear optical properties. These particles show quantum confinement effects which cause luminescence in silicon powders and also lead to silicon germanium quantum dots as infrared optoelectronic devices. These semiconductors are also used in solar cells as window layers.
- Nanosized metallic powders have been used for the production of gas tight materials, porous coating and dense parts. Cold welding properties of these materials when combined with ductility make them suitable for metal-metal bonding, to be used especially in the electronic industry.

- Nanostructured metal-oxide thin films are used for making gas sensors (NOx, CO_2, CO, CH_4 etc.) with enhanced selectivity and sensitivity. Nowadays, rechargeable batteries for consumer goods are also made with nanostructured metal-oxide (MnO_2). Nanocrystalline silicon films are used for highly transparent contacts in thin film solar cell and nanostructured titanium oxide porous films due to its high transmission and significant surface area enhancement, leading to strong absorption in dye-sensitized solar cells.
- Polymer-based composites with a high content of inorganic particles leading to a high dielectric constant are interesting materials for photonic band gap structure.

1.5　PROPERTIES OF NANOMATERIALS

Nanotechnology's fast-growing popularity is due to the novel properties of the nanomaterials which have been discovered, and those yet to be discovered. For example, nanowires are used in laser, nanoelectronics, solar cells, nanophotonics, resonators and different type of sensitive sensors. Similarly, nanoparticles may be used for coating in catalysts, energy storage, nanoelectronics, drug delivery systems and in biomedical applications (Shah, 2019). Nanostructures are being used in display systems, light-emitting diodes and high-efficiency photovoltaic cell. There exists a huge variety of nanomaterials, and thus their properties and applications are also varied. Due to these variants, researchers are on the peak threshold of understanding the properties of these materials. In this chapter, the emphasis is on the properties which are extensively used by present researchers. The important properties of nanomaterials are summarized below:

1.5.1　HIGH SURFACE AREA

The surface atoms in nanosized materials are too high in percentage. These surface atoms are coordinately unsaturated and pose the chemically unsatisfied valence orbitals. The energy of these surface atoms is too high which makes them more reactive in comparison to the interior bulk atoms. This increased surface area is more responsible for the enhanced hydrogen storage property (energy storage), better catalytic properties, enhanced surface scattering reducing electrical conductivity, increased solubility, tendency to agglomerate (accumulation) and reduced melting points.

1.5.2　QUANTUM CONFINEMENT EFFECT

It increases the energy gap between energy levels which may lead to transitions from metal to semiconductor to insulator known as metal to insulator transition, which results in different electrical and optical properties of nanosized materials from those of bulk materials.

1.5.3　LESS STRUCTURAL DEFECTS

The structural defects such as point defects and dislocations are very small in nanosized materials as compared to large grain size and bulk materials which lead to improvements in their mechanical properties.

1.5.4 MECHANICAL PROPERTIES

There are various scientific challenges in nanotechnology and nanoscience which include the development of nanomaterials with novel and specific mechanical properties (Das and Das, 2016). The interest in the mechanical behavior of nanostructured materials develops from the distinct properties observed from the materials prepared by the gas condensation method. The early observations in these materials were lower elastic moduli than the conventional grain size materials, their strength and high hardness in metals approximately 10 times higher than those of high grain metals.

The early measurements in elastic constants of nanocrystalline materials prepared by inert gas condensation method gave a lower value of Young's Modulus (E) than the conventional grain size materials. This low value of E was suggested due to the presence of various cracks and extrinsic defect pores. The researchers gave various results which show that porosity is the important factor in producing low E values. Hardness and strength of nanosized materials are better due to reduced structural defects.

1.5.5 OPTICAL PROPERTIES

The significant reduction in the size of nanomaterials affects the optical properties. Quantum confinement and surface plasma resonance are key factors of change in optical properties. Surface Plasmons (SP) are the cause of the color of nanomaterials. It is a natural oscillation of electron gas inside the given nanosphere. If this sphere is small compared to the wavelength of light, and light has a frequency close to Surface Plasmon (SP), then SP will absorb the energy. The frequency of SP depends on the shape and dielectric function of the nanomaterials (Darwesh, 2017).

1.5.6 MAGNETIC PROPERTIES

The magnetic properties of the materials are the result of net magnetic moments of the electrons in atoms which compose the material. These properties are measured from certain defined points achieved from the variation of magnetization with the magnetic field. The magnetic properties are categorized as structure-sensitive and structure-insensitive properties. The structure-sensitive properties are affected by impurities such as nitrogen, carbon, oxygen and sulfur. These impurities tend to locate at interstitial sites in the crystalline lattice which result in the strained lattice. As a result, a small concentration of these impurities has a large effect on the magnetic properties of the nanomaterials. Hysteresis losses, remanence, magnetic stability, permeability and coercivity are all structure-sensitive. They can be controlled through mechanical and thermal treatment processing. Structure-insensitive properties are the properties which are not affected by the changes in the material structural properties such as resistivity and saturation magnetization. These are dependent on the decomposition of the particular alloy.

1.5.7 Electrical Properties

Nanomaterials have an increased surface area causing more scattering resulting in reduced conductivity. But the electrical conductivity increases in these nanomaterials because of increased perfection and ballistic transport. This increases their capacity to hold more energy than conventional materials. This capability makes nanomaterials better to replace conventional and rechargeable batteries, as both regular and rechargeable batteries require frequent recharging. The solution to this problem is nickel-metal hydride batteries made of nanocrystalline nickel and metal hydrides which require far less frequent charging and last for a much longer time. The effects of size on the electrical conductivity of nanomaterials and nanostructures are based on the mechanism of coulomb charging, tunneling and widening, surface scattering, etc. Also, defects in structure, reduced impurities and dislocations would affect the electrical conductivity of the nanomaterials and nanostructures.

1.5.8 Reduced Lattice Constants

The coordinatively unsaturated surface atoms tend to form strong bonds with the interior atoms and also with the nearer surface atoms, which results in a reduction of the spacing between the atoms i.e. lattice constant is reduced.

1.6 APPLICATIONS OF NANOMATERIALS

The last decade has seen a huge impact of nanotechnology on different fields such as medicine, electronics, communication, health care, cosmetics, food agriculture, textile, chemical industries, water purification, energy storage and many more. The important fields of applications of nanomaterials and nanotechnology are summarized in Table 1.1.

TABLE 1.1

Summary of the Applications of Nanomaterials and Nanotechnology

Applications of Nanomaterials and Nanotechnology	Biomedical: Detection and diagnosis, cancer therapy including hyperthermic treatment, drug delivery, wound dressing with antibacterial activities, biomarking and imaging, biocomposites
	Cosmetics: Sunscreens, antioxidants
	Electronics: Nanosized transistors, high density data storages, CNT FETs, flat display, high sensitive sensors
	Textiles: Polymer hybrid fibers, anti-stain textile, heat retaining textiles, UV-blocking textiles
	Environment: Water purification, waste water treatment, pollution monitoring
	Defense and security: Smart materials, highly sensitive sensors

1.7 CONCLUSION

This chapter highlighted the brief history of nanomaterials and nanotechnology and impresses upon us that nanomaterials are very useful materials in a current context. Various researchers are trying to gain an in-depth knowledge of these materials; nanotechnology still has much to do in this field. Nanotechnology is a critical component for strong competitive technology on a global scale. Almost all developed countries are putting their effort into the establishment of nanotechnology and the study of nanomaterials. Nanotechnology incorporates all disciplines including physics, chemistry, electronics and even biology and the mathematical sciences. Nanomaterials and nanostructures are finding a solution to every problem, having applications in various fields such as biomedical, defense, electronics and many more. Significant short-term and long-term applications in different sectors are underway and the last decade has borne witness to all the efforts undertaken by various researchers in this field. Thus nanomaterials and nanostructures as a part of the vast field of nanotechnology have a bright future to offer both wellness and development to society and the world.

REFERENCES

H.H.M. Darweesh, 2017, Nanomaterials: Classification and Properties-Part I, *Nanoscience*, Volume 1, pp. 1–11.

A.K. Das, M. Das, 2016, *An Introduction to Nanomaterials and Nanoscience*, CBS Publishers & Distributors Pvt. Ltd.

C. Hadjipanyas George, 2008, *Synthesis, Properties and Applications*, Kulwer Academic, Netherlands.

G.L. Hornyak, J. Dutta, H.F. Tibbals, A. Rao, 2008, *Introduction to Nanoscience*, CRC Press, Boca Raton.

M.A. Shah, T. Ahmad, 2011, *Principles of Nanoscience and Nanotechnology*, Narosa Publications.

M.A. Shah, K.A. Shah, 2019, *Nanotechnology: The Science of Small*, Wiley Publications.

2 Nanomaterials
Methods of Generation

Anil Arora
Thapar University, Patiala, India

CONTENTS

2.1   INTRODUCTION

One of the most revolutionary areas of microelectronics and nanotechnology research is techniques of micro-/nano-lithography and etching. This approach is sometimes known as "top-down nanotechnology" where small structures are made by starting with large materials. These materials are patterned and carved down to make nanoscale structures in the desired precise patterns. The other approach is to start with the atomic scale and build the structures and materials atom-by-atom, and this is called "bottom-up nanotechnology". Here, the forces of nature are used to assemble various nanostructures – for this, the term "self-assembly of nanostructures" is used more often. In this chapter, these two generalized processes for nanomaterial fabrication are described. The various techniques falling under both approaches are discussed. The mechanisms of atoms and molecules reacting under chemical and physical circumstances are addressed briefly. There are advantages and disadvantages to both the top-down and bottom-up approach, which are also listed in this work.

2.2   METHODS OF NANOMATERIAL SYNTHESIS

Materials are made of grains which in turn are composed of atoms. Depending upon the grain size, the materials are visible or invisible to the naked eye. When not compressed, the nanomaterials are known as nanopowders having a grain size varying from 1–100 nm in at least any one of three coordinates. The nanomaterials are not a new entity but understanding of the preparation of metals, oxides, polymers and many more substances may be recent as a matter of research. The preparation of these nanomaterials with a well-defined structure and size is a challenge for researchers. A variety of techniques are used for the generation and synthesis of nanomaterials depending upon the material of interest and required size. Several points are to be considered with regard to the use and utility of the particular process, as listed below:

(a) Applicability range for different classes of materials
(b) Reproducibility of average size and shape of particles
(c) Control over particle size and range of size
(d) Homogeneity of the phase of product

The basic fundamental behind the generation of the nanoparticles is to produce a large number of nuclei and to prevent the growth and accumulation of the grains. The approaches which are used for the synthesis and generation of the nanomaterials are based on well-developed physical and chemical techniques. Top-down approaches are based on the continuous breaking of the bulk materials and generally are an extension of the lithography process. Lithography is the technique used in the microfabrication of integrated circuits (ICs). Bottom-up approaches are related to the process of atom-to-atom accumulation, molecule to molecule, or cluster by cluster fabrication. It plays an important role in the processing and fabrication of the nanomaterials. This technique is quite economical and easy to use but takes a considerable amount of time to complete the process. The various techniques used in these two approaches are given in Table 2.1.

   All of the above-mentioned techniques are described in the following sections in brief.

2.3   TOP-DOWN FABRICATION METHODS

The top-down approach deals with the subtractive methods of the making of nanomaterials. One can visualize these methods with the process of grinding, ball milling, crushing, mechanical grinding, etc. Grinding and crushing are the rough methods used industrially, but nanotechnology has the advantage of highly precise and controlled methods of achieving monodispersed nanoparticles.

   The conventional top-down approach has the problem of damaging the crystallographic structure and some additional defects may also be introduced during this approach. Still, this approach plays an important role in the fabrication and synthesis of various nanostructures.

   The terms vacuum "deposition", "vaporization" and "vacuum evaporation" are very important in the techniques used for the synthesis of the nanomaterials, so

TABLE 2.1

Techniques for Generation of Nanomaterials

Approach Used	Techniques Used	Nanomaterials Used
Top-Down Approach	Arc Discharge Method	Carbon Nanotubes (CNTs) and Fullerenes
	Laser Ablation	Broad Range of Nanoparticles Including CNTs
	Ball Milling	Mixtures and Composites of Elemental Powders
	Inert Gas Condensation	Oxides, Semiconductors and Alloys
	Physical Vapor Deposition	Alloys, Semiconductors and Sulfides
Bottom-Up Approach	Chemical Vapor Deposition	CNTs, Boron Nanotubes, Fullerene
	Molecular Beam Epitaxy	Thin Films and Compound Semiconductors
	Sol-Gel Synthesis	Oxide and Colloidal Nanoparticles
	Hydrothermal Synthesis	Elemental Nanopowders and Oxides
	Microwave Method	Oxide Nanoparticles including TiO_2

they are important to understand. In the vacuum-deposition process, compounds, elements and alloys are vaporized and deposited in vacuum, while in vaporization the material is vaporized with the thermal process or heating. The process is accomplished at a pressure less than 0.1 Pa and vacuum level of 10 to 0.1 MPa at a temperature ranging from ambient temperature to 500° C. Vacuum deposition processes are of high deposition rates and quite economical, with a limitation of difficult deposition of many compounds. The various techniques used for the synthesis in top-down approach are described in the following sections.

2.3.1 Arc Discharge Method

The common technique for the production of carbon nanotubes (CNTs) is the arc discharge method where carbon is vaporized between two carbon electrodes. The formation of CNTs in this process is dependent on the pressure of He or Ar gas, applied current and the temperature.

A typical arc discharge chamber is shown in Figure 2.1.

The anode and cathode which are made of graphite rods are placed millimeters apart. The anode graphite rod is generally doped with a metal catalyst particle such as iron. At 100 Amp current, the carbon vaporizes in hot plasma. The discharge causes a high temperature environment resulting in the sublimation of the carbon contained in the negative electrode. This results in the production of a dazzling stream of ionized air or plasma at a temperature of about 6000° C which is near the surface temperature of the Sun. This high temperature gas rises like a hot air balloon while it remains anchored to the current feeding electrodes at its ends. It acquires the upward carving shape hence named an 'arc'. The single- and multi-walled nanotubes

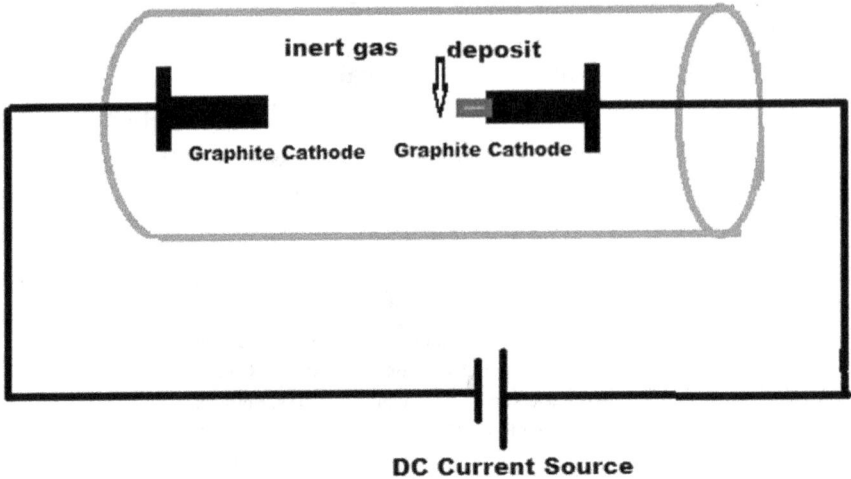

FIGURE 2.1 Arc discharge method.

of varying diameters ranging from 2 to 30 nm in length are produced by the arc discharge method with a 30% yield by weight. The formation of CNTs depends on the following conditions:

1. Helium pressure ranges from 50 to 760 Torr
2. Process temperature ~ 2000° C
3. Applied current from 50 to 100 A
4. Applied Voltage 20 V

Nanotubes of dichalcogenides such as MoS_2, WS_2 and $MoSe_2$ are also obtained by the arc discharge method. MoS_2 and WS_2 can be prepared by starting with stable oxides MoO_3 and WO_3. These oxides are heated at high temperatures in a reducing atmosphere and then reached with H_2S. Selenides are obtained using H_2S. Nanoparticles of metal oxides, nitrides and carbides can also be prepared by carrying out the discharge in a suitable gas medium or by loading the electrodes with a suitable precursor.

2.3.2 LASER ABLATION

In 1995, CNTs were first synthesized by using the laser ablation method with a yield of 70% by weight and producing single-wall CNTs. The graphite rod containing 50:50 mixtures of Co and Ni Catalyst were heated at a temperature of about 1200° C in flowing argon and exposed to a laser pulse. The laser pulse vaporizes the graphite target in a high temperature reactor. There are two consequent pulses which are used to vaporize the target more uniformly and to minimize the carbon black deposition. The setup of simple laser ablation system is shown in Figure 2.2.

The graphite is vaporized by the pulse laser in a high temperature reactor while the inert gas such as argon is blended inside the chamber. Different types of CNTs are

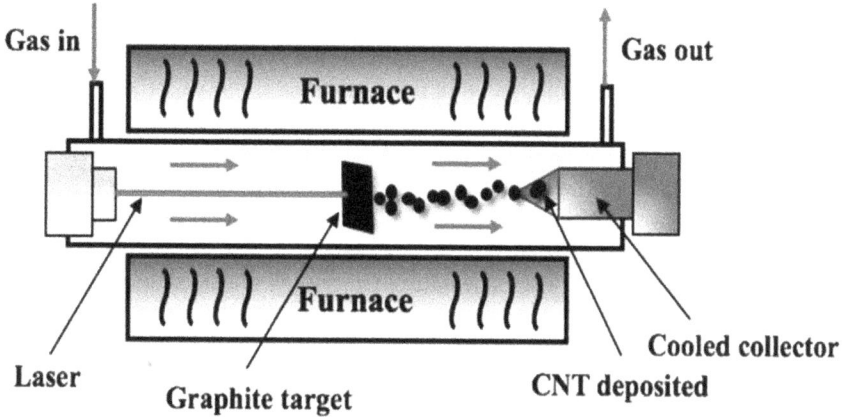

FIGURE 2.2 Laser ablation method of CNT formation (credit: Zan Lu, Raad Radd, Farzad Safaei, Jiangtao Xi, Zhoufeng Liu and Javad Foroughi. 2019. Carbon nanotube based fiber supercapacitor as wearable energy storage. *Frontier in Materials*, Vol. 6, 10.3389/fmats.2019.00138).

formed in a big clump due to the effect of plume leaving the furnace and these CNTs are then separated for their usefulness. The plume behaves like a rocket exhaust and expands from the target with a strong forward-directed velocity distribution of different particles. The nanotubes are generated and developed on the coolant surfaces of the reactor as the condensation of carbon vapors takes place. Gold, palladium, iron and compound of sulfides are also prepared by this method. Laser ablation is suitable for in-situ studies of the toxic materials and costlier than the arc discharge method.

2.3.3 Ball Milling

Ball milling is the most common method used for the generation of nanoscale materials. It is also named as mechanical milling. The objectives of the ball milling method are mixing, size reduction and synthesis of the nanocomposites. In this method, stainless steel or tungsten carbide balls are used and transferred into coarse-grained material with the intention of reducing the size. To prevent the oxidation, the whole process is carried out in controlled atmospheric conditions. In the ball milling process, the milling is carried out by stirring action of the agitator that has a vertical rotating shaft with horizontal arms. This motion causes differential movements between the balls. The nanopowder is achieved by the horizontal arms attached to the rotating vertical shaft and provides a much higher degree of surface contact. The following conditions must be satisfied for the formation of nanostructures using ball milling:

- Frequency > 1000 cycles/min
- Ball Velocity > 5 m/s

2.3.4   Inert Gas Condensation

Gas Condensation is the technique firstly used for the synthesis of metallic or inorganic materials. In this method, metal or inorganic material is vaporized using thermal evaporation sources such as an electron beam device or joule-heated refractory crucibles at an atmospheric pressure varying from 1 to 50 mbar. A high residual gas pressure (>3mPa) causes the formation of very small particles (100 nm). These particles are formed by collision of evaporated atoms with residual gas molecules. The sources of vaporization used may be resistive heating, low and high energy electron beam or inductive heating. The schematic of inert gas condensation technique is shown in Figure 2.3.

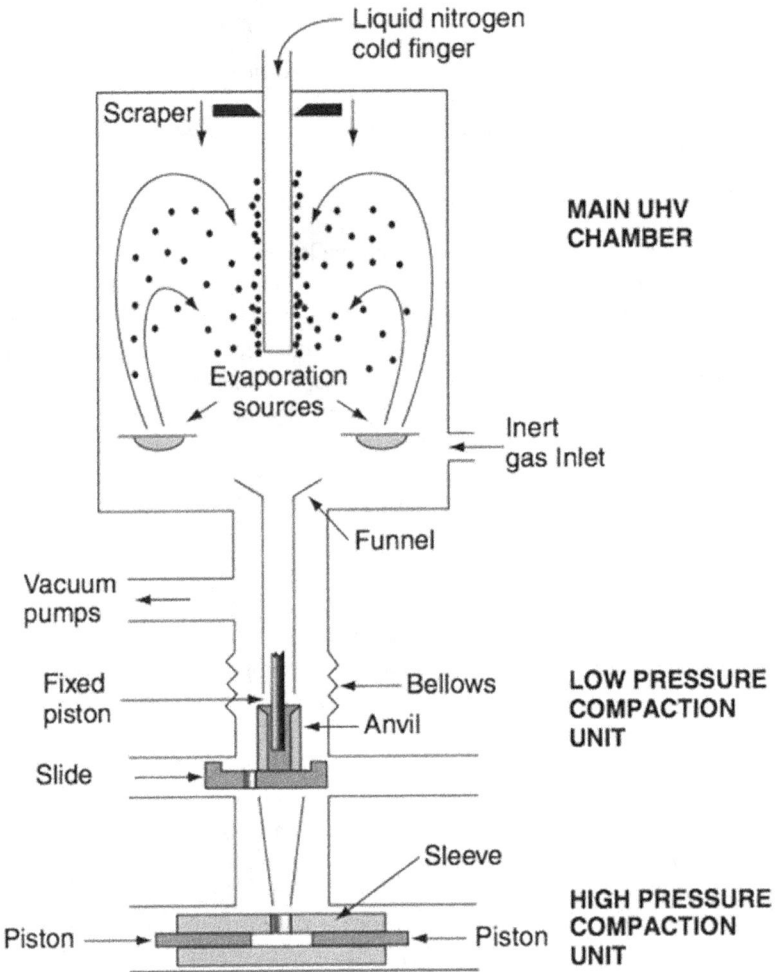

FIGURE 2.3   Inert as condensation method for production of nanoscale powder (credit: M.A. Shah 2019).

The inert gas condensation equipment is comprised of an ultra-high vacuum having an evaporation source, cluster collection device filled with liquid nitrogen, compaction device and cold finger scrapper. When condensation of the heated vapors takes place near the joule-heating device, it is removed by the scraper in the form of a metallic plate. Evaporation is to be performed from Ta, Mo or W refractory metal crucibles.

The nanoparticles are generated with well-controlled size distributions. But the method has certain limitations such as source precursor incompatibility, dissimilar evaporation rates, etc.

2.3.5 PHYSICAL VAPOR DEPOSITION

The physical vapor deposition is the most common technique used with the deposition of alloys, semiconductors and sulfide nanoparticles. The target material is heated at a high temperature at a pressure of the order of 10^{-6} to 10^{-12} Torr. The vapors are condensed on a cool substrate so as to produce the thin films. The schematic diagram of the physical vapor deposition system is shown in Figure 2.4. The characteristics and quality of the deposit substrate depend on the temperature, rate of deposition, ambient pressure and the uniformity of the film. Very low pressure is an essential requirement of the physical vapor deposition because a mean free path between the collisions becomes large enough so that a vapor beam arrives at the substrate unscattered, and reduces the probability of the contamination of the deposited film.

2.4 BOTTOM-UP FABRICATION METHODS

The most emphasized approach used for the generation of nanoparticles is that known as bottom-up fabrication. In this approach, atom-by-atom assembly is used to obtain large structures. An example of such an approach is the production of salt and nitrate

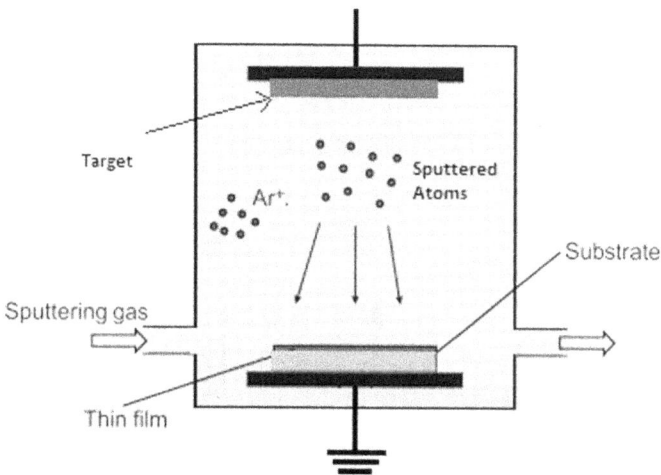

FIGURE 2.4 Schematic diagram of physical vapor deposition.

by the chemical industry and the growth of single crystal and thin film deposition by the semiconductor industry. Variation of the method used for the generation gives rise to differences in chemical composition, crystallinity and their microstructure of the different materials. Also, the material produced by a different method may also have considerable differences in its physical properties. The various techniques used in the bottom-up approach are explained in the following sections.

2.4.1 CHEMICAL VAPOR DEPOSITION

Chemical vapor deposition is the most common method used for the generation of nanoparticles. The approach is to deposit a solid material from the gaseous phase which is somewhat the same as that of physical vapor deposition. In thermal CVD, the reaction activation temperature is above 900°C. A schematic diagram of a typical CVD furnace is shown in Figure 2.5. In this furnace, the reacting gases are introduced inside the tube and the wafers on which materials are to be deposited are placed in the high temperature zone of the furnace. The CVD of the hydrocarbon over a metal catalyst is used to produce various carbon nanomaterials such as carbon filament and fibers. Decomposing CH_4 at 1100°C can produce CNTs. As the gas decomposes, the carbon atoms are condensed on a cooler substrate which contains various catalysts such as iron. To a great extent, some CNTs are produced by catalyst CVD of acetylene over iron and cobalt catalyst supported by zeolite or silica.

The advantages of CVD include its simplicity, the fact that it is economical to use, and its mass production possibility. Hydrocarbons can be used in CVD in any form i.e. solid, liquid or gas. Any form of CNTs can be produced using the CVD technique, like aligned, straight or coil nanotubes, or any other desired shape.

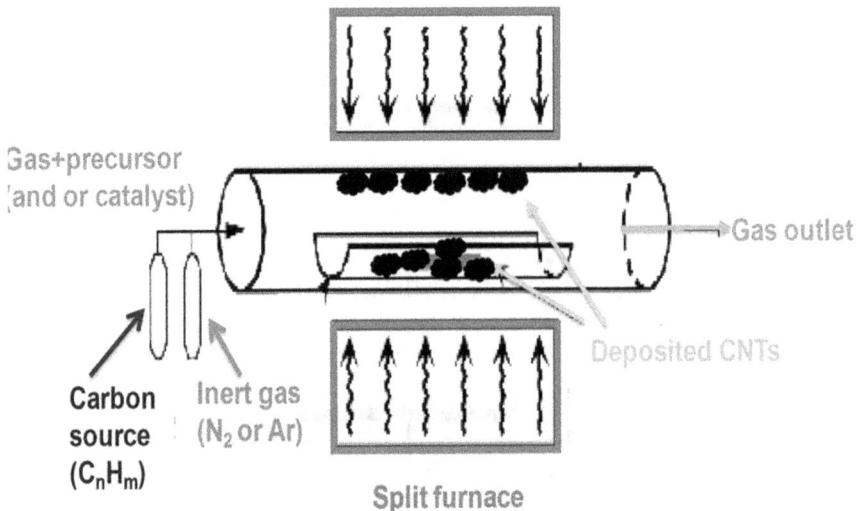

FIGURE 2.5 Schematic diagram of chemical vapor deposition (credit: Waseem Khan).

2.4.2   MOLECULAR BEAM EPITAXY (MBE)

Molecular Beam Epitaxy is an ultra-high precision and ultra-clean chamber with in-situ tools such as Reflection High Electron Diffraction (RHEED) and Auger Electron Spectroscopy (AES) for the characterization of the deposited materials during the growth mechanism.

In the molecular beam epitaxial system, ultra-pure elements such as arsenic and gallium are heated in separate quasi-Knudsen effusion cells until they begin to evaporate slowly. The reaction between the two takes place after there is condensation on the wafer. In the case of gallium and arsenic, a single crystal gallium arsenide is deposited on the silicon wafer. The word 'beam' here means that the evaporated atoms neither react with each other nor with any gases in the vacuum chamber until they reach the wafer. There is continuous rotation of the substrate so as to get the uniform deposition of the thin film. There is also the facility of the mechanical shutter to control which material is to be deposited, and for how much time. For the growth of gallium arsenide, the requirements of the chamber environment are as follows: temperature of 1400° C, ultra-high vacuum of about 10^{-10} Torr and growth rate ~1μm/h. The typical system of the MBE is as shown in Figure 2.6.

2.4.3   SOL-GEL SYNTHESIS

The Sol-gel method is a quite versatile method for the generation of nanomaterials and nanostructures. This method has received tremendous interest for the deposition of the various materials due to its small and economical setup and low processing temperature. The area of technology covered by the sol-gel method is very wide-ranging, from physics to biology. This method is used for making various colloidal

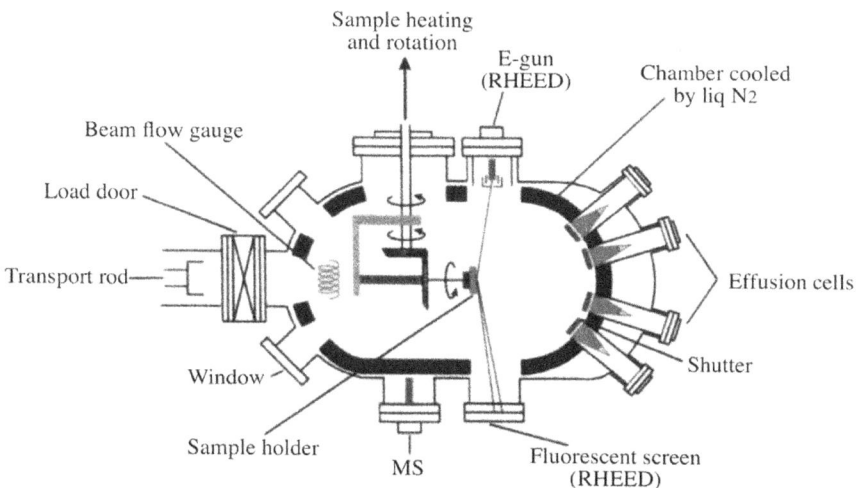

FIGURE 2.6   Molecular beam epitaxy system (credit: Andrew R. Barron).

dispersions of organic and inorganic materials and also to produce their hybrids, especially oxide and oxide-based hybrids. Fibers, thin films and powders may be prepared from such colloidal dispersions. Some specific considerations are required for the fabrication of different forms of product, but the general approach and fundamentals used in the synthesis of colloidal dispersion remain the same.

The colloidal particles are larger in size than nanoparticles or normal molecules. These particles appear bulky after mixing with liquid colloidal. Besides this, the nanosized molecules always appear clearer. This involves the evolution of the network through the formation of colloidal suspension known as sol and gelation to form a network in the liquid phase known as gel as shown in Figure 2.7.

The Sol-gel method is performed at a temperature ranging from 25° to 300°C. This process involves a homogeneous solution of one or more alkoxides depending on the choice. These are the organic precursors used for titania, zirconia, silica and alumina. The most critical stage in the whole process of sol-gel synthesis is the drying process. If drying is done by simple evaporation method, xerosol is obtained, while if supercritical CO_2 drying is used then the result is aerosols. This method is particularly used for making temperature-sensitive organic–inorganic hybrid material, metal oxides and metastable materials.

2.4.4 HYDROTHERMAL SYNTHESIS

Hydrothermal technique has become an important method to generate advanced materials used in various technical applications such as optoelectronics, electronics, catalysts, ceramics, magnetic storage devices, bio-photonics and biomedical. This technique is not only important in the processing of monodispersed and highly

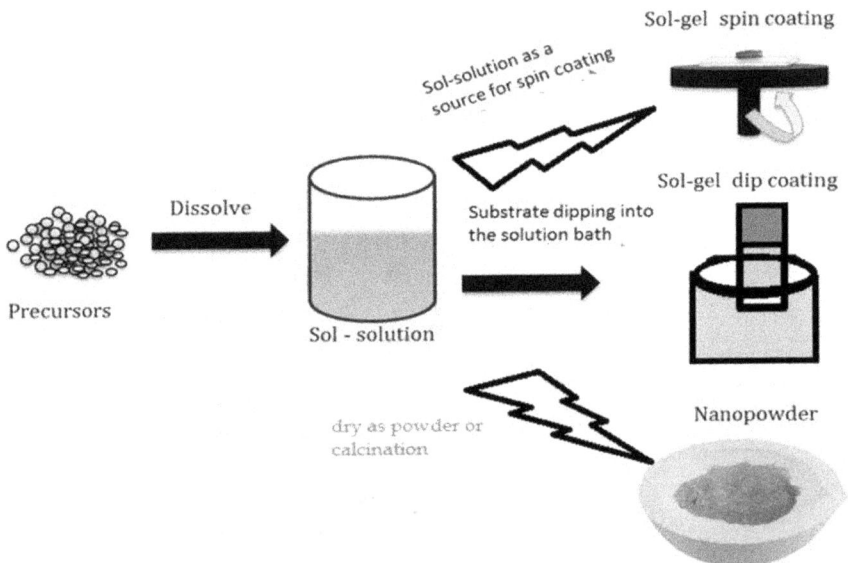

FIGURE 2.7 Sol-gel synthesis for the nano-powder or nanomaterials.

homogeneous particles but also plays an important role in the processing of nano-composites or nanohybrid materials. It uses a unique method of coating compound materials on metal, ceramics, polymers, etc.

Hydrothermal synthesis process involves several techniques to crystallize the substances. It is done at a very high vapor pressure and using a high-temperature aqueous solution. Thus, it is termed using the words "Hydro" + "Thermal" = Hydrothermal method. It is essentially an artificial method to crystalize the material or nanoparticles, depending upon the solubility of the aqueous solution in hot water and high temperature levels. A very strong container known as an autoclave is used inside the hydrothermal reactor which is filled with the solution. Constant mainte-nance of the difference of temperature is necessary between the opposite ends of the compartment where the crystallization takes place. The solvent is dissolved towards the end at the higher temperature while the nanoparticle growth is towards the com-paratively cooler end. This method has the advantage of large-sized nanoparticles and crystals, and at the same time it also provides more control over the composition and content of the material.

2.4.5 MICROWAVE METHOD

The microwave method for generation of the nanoparticles and nanomaterials may be useful and attractive due to uniform heating, high heating rates as compared to thermal heating, very short thermal induction period, wide range of reaction condi-tions and high yields of reaction.

Microwave radiations are applied to prepare various oxide nanomaterials and nanoparticles. This method is basically inclined towards a green chemistry approach where utilization of the principles reduces the use of hazard chemicals used in the design of the material. It also reduces wastage and is designed to increase energy efficiency. Microwave radiations in the range 30 to 300 GHz are used in this method.

2.5 CONCLUSION

Since the microwaves are able to penetrate the materials, heat can be generated throughout the volume which results in volumetric heating. Many researchers have coated the preparation of titanium nanoparticle suspension which is prepared within 5 min to 1 hour time with microwave heating while the same materials are gener-ated using conventional heating in the time period of 32 hours which was too high in comparison to the microwave method of generating the nanoparticles.

BIBLIOGRAPHY

A.K. Das. 2017. *Introduction to Nanomaterials and Nanoscience*. CBS Publications.
S.K. Gandhi. 1994. *VLSI Fabrication Principles: Silicon and Gallium Arsenide*. Wiley & Sons Publications.
M.A. Shah. 2019. *Nanotechnology: The Science of Small*. Wiley Publications.
M.A. Shah and T. Ahmad. 2011. *Principles of Nanoscience & Nanotechnology*. Narosa Publications.

Shrividhya Thiagarajan, Anandhavelu Sanmugan and Dhanasekaran Vikraman. 2017. Facile methodology of sol-gel synthesis for metal oxide nanostructures. In: *Recent Applications in Sol-Gel Synthesis*. ed. Usha Chandra. IntechOpen.

J.N. Tiwari, R.N. Tiwari and K.S. Kim. 2012. Zero-dimensional, one-dimensional, two-dimensional and three-dimensional nanostructured materials for advanced electro-chemical energy devices. *Progress in Materials Science* 57(4), 724–803.

3 Nanotechnology Advances, Benefits, and Applications in Daily Life

Pavani Sriram and Ashish Suttee
Kakatiya University, Warangal, India; and
Lovely Professional University, Phagwara, India

CONTENTS

3.1 INTRODUCTION

Nanotechnology is amongst the most highly innovative fields of science, engineering, and technology, and comprises the study and extremely minute manipulation of nanostructure properties (Abiodun and Ajayi 2014, Priestly, Hartford, and Malcom 2007). It is a promising new advancement that has triggered a lot of new ideas and innovation (Feynman 1960). The word "nanotechnology" was primarily used in 1974 by the researcher Norio Taniguchi, at the University of Tokyo, who uses it to discuss the capability of engineer resources at nanoscale (Miyazaki and Islam 2007).

As per OSHA, "Nanotechnology includes the manipulation, understanding, and engineering of atoms also molecules, imaging, quantifying, modeling at the nanoscale dimensions of approximately 1 to 100 nm" (Priestly, Hartford and Malcom 2007). The prefix "nano" is derived from the Greek word for "dwarf" and can be defined as the measure of one-billionth of a meter (Miyazaki and Islam, 2007).

According to a description endorsed by the EU in 2011, nanomaterial are natural or synthetic objects with their external dimensions in the nanoscale and have a specific surface area through volume, which represents direct control of materials, or a determined set of processes that allow for almost entire control over the structure of matter (Commission Recommendation 2011, Wohlleben et al. 2017) to work at atomic, molecular and supramolecular levels. The characterization, fabrication, and employment of materials, structures, devices, procedures, processes, and schemes by regulating shape and size at nanometer scale with predominately new properties are proposed (Roco 2003).

The manipulated materials are specially designed for definite purposes that make them unique. This allows for various advantages that are unattainable by any other form or state of material (Loureiro et al. 2018).

Scientifically, nanotechnology is utilized in exhibiting new and important physical, chemical, and biological properties as well as the methods empowered with the capability to control properties at nanoscale. The variations in properties are due to a rise in surface area and the supremacy of quantum effects which is linked to very small sizes and the large surface area- volume ratio (Miyazaki and Islam 2007).

Like each advanced field, it is filled with both benefits and risks. Several dangers associated with nanotechnology have been noted. In spite of these hazards, it remains a potentially attractive area and a leading area of development in the field of science and technology. Various government and private organizations are capitalizing on research into nanotechnology as there is strong potential for its use in diverse areas such as medicine, energy, industry, IT, information, communication, agriculture, the environment, along with many more fields of activity (Mansoori GA).

Currently, a life without nanotechnology is hard to visualize. Nanotechnologies have been used in many products and industrial applications. Some of the particles in the nano-range are: Nanorods, nanogels, nanopores, nanotubes, nanobarcodes, nanoshells, nanospheres, nanoscales, liposomes, quantum dots, nanofluidics, nanowires, dendrimers, nanocapsules, carbon nanotubes, nanomedicine (Silva 2004, Cobb and Macoubrice, 2004).

3.2   NANOTECHNOLOGY: ADVANTAGES AND DISADVANTAGES

Several potential *advantages* are:

- Advances in medicine and life-saving treatments for illnesses such as cancer, clearing the blockages by sending nanobots into a patient's arteries.
- Development of strong, durable, and lighter products with nanotechnology.
- Enhanced imaging and diagnostic purposes.
- Nanotechnology can be used to improve drug fabrication by altering medicines at a molecular level to produce drugs with more efficacy and lessened side effects.
- Nanotechnology will allow circuits to be built very precisely at an atomic level.
- Potential to rectify genetic problems by fixing impaired genes.
- Help with progress in the manufacturing of lightweight, durable and efficient tools of production.
- Reduce dependency on non-renewable energy sources.
- Transformation and production of energy-efficient fuel and solar cells, thereby decreasing the cost of construction and equipment.
- Upgraded electronic devices and computing, comprising nanotransistors, nanodiodes, faster kinds of computers, quantum computers, for illumination, for purposes such as plasma and LED displays.
- Use of nanorobots to rebuild the ozone layer.
- Useful in cleaning polluted areas.
- With this knowledge, there is the possibility of developing effective energy-producing, absorbing, and storing products in smaller, effective, and competent devices (Feynman 1960, Commission Recommendation 2011, Wohlleben et al. 2017).

Several *disadvantages* are:

- Because of their small size, these particles can cause respiratory problems from the inhalation of minute asbestos particles.
- Economic market disruption associated with a probable decrease in the value of oil, due to the high degree of reproduction of efficient energy sources, and other materials like gold or diamonds with molecular manipulation.
- Loss of employment in the manufacturing sector and traditional agriculture.
- Nanotechnology develops negative effects on the environment due to the generation of pollutants as potential new toxins.
- Possible threats and risks to humans, security, privacy, and health.
- Upgraded atomic weaponry and attainability and more accessibility of weapons of mass destruction.

- Very expensive and also difficult to manufacture (Mansoori 2005, Silva 2004, Kovvuru, Mahita and Babu 2012).

3.3 BENEFITS OF NANOTECHNOLOGY

After many years of research, applications of nanotechnology are bringing in equally predictable and unpredictable possibilities to help the world.

Nanotechnology is assisting significantly in developing, even transforming, many technological and industrial areas including IT, security, environment, biology, chemistry, medicine, agriculture, food, construction, transportation, safety, household, cosmetics, energy, textiles, electronics, and many others (Cobb and Macoubrice 2004).

3.3.1 Is Nanotechnology Harmless?

The benefits of nanotechnology are numerous in a broad context, and so we should make the best use of them. Equally, efforts to reduce the risks should be made (Cobb and Macoubrice 2004, Roco 2003, Taniguchi 1974).

- Advances in transportation systems, cheaper and cleaner energy.
- Entrapment of toxins and organisms by creating filters.
- Helpful in monitoring, analyzing, and managing chronic diseases.
- Increases oral bioavailability of drugs through specialized uptake mechanisms (Cobb and Macoubrice 2004, Roco 2003).
- Relevant to bone grafting, tissue engineering, and regeneration (Taniguchi 1974).
- Helps in generating novel therapeutic agents and carriers for the management of life-threatening diseases.
- Nanoparticles are employed progressively in catalysis to enhance chemical reactions by reducing the number of ingredients to produce anticipated outcomes (Cobb and Macoubrice 2004).
- Nanotechnology benefits enrich nanodentistry by reducing the polymerization contraction and thermal enlargement and improving polishing ability (Taniguchi 1974).
- Nanotechnology can lead to significant fuel savings.
- Nanotechnology identifies food spoilage and pathogens.
- Nanotechnology improves the production of drugs that are not able to go through clinical trials.
- Nanotechnology is useful in the manufacturing of devices and reformulation of existing drugs for effectiveness, safety, and better healthcare, thereby reducing the costs.
- Obtainability of tougher, stronger, and lighter resources for engineering.
- Protect drugs from enzymatic degradation in the GIT by encapsulation.
- Removal of pollutants from the environment (Roco 2003).
- Surface treatments of textiles aid them to resist wrinkling, staining, and bacterial growth.

3.4   ADVANCES AND APPLICATIONS OF NANOTECHNOLOGY

Nanotechnology is interdisciplinary and it assimilates all disciplines, especially bio-medicine, engineering, and technology. The list of applications (Figure 3.1) of nano-technology has grown rapidly and it has the potential to modernize several areas. Nowadays technology is finding emergent applications in industry, medicine, nano-biotechnology, nanoelectronics, nanocoatings, nanosensors, cosmetics, construc-tion, displays, transportation, food, etc. (Stylios, Giannoudis, Wan 2005, The Royal Society 2004).

3.4.1   AGRICULTURE

The adoption of nanotechnology can likely transform the entire agricultural sector and food companies from food generation to conservation, as well as approaches to bundling, transit, and waste management. Nanoscience concepts and their applica-tions (Figure 3.2, 3.3) can probably renew the generation cycle, increase production

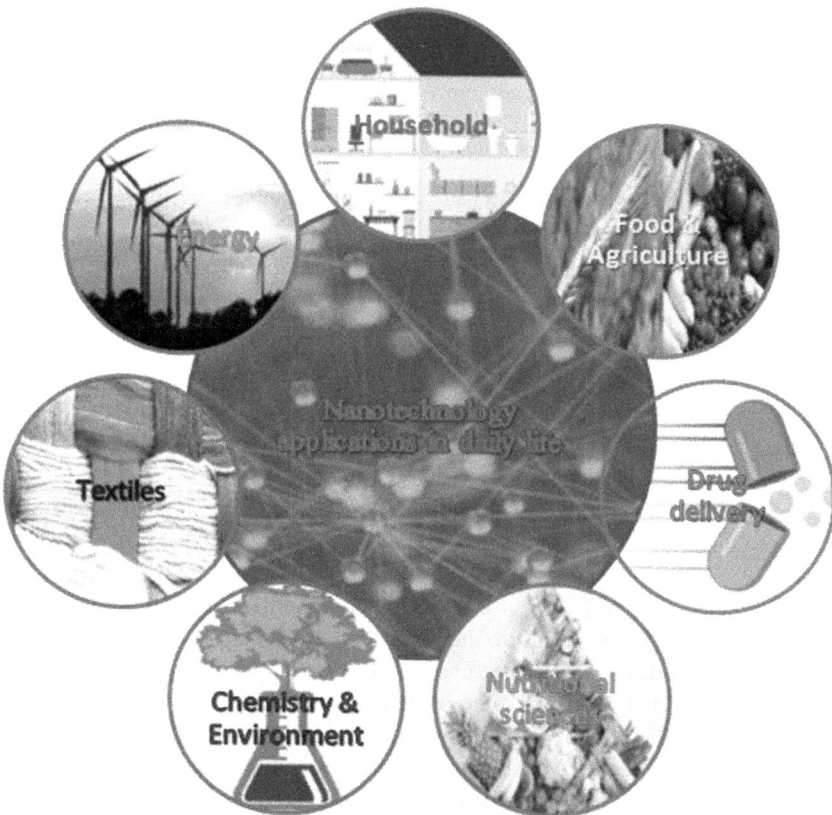

FIGURE 3.1   Applications of nanotechnology.

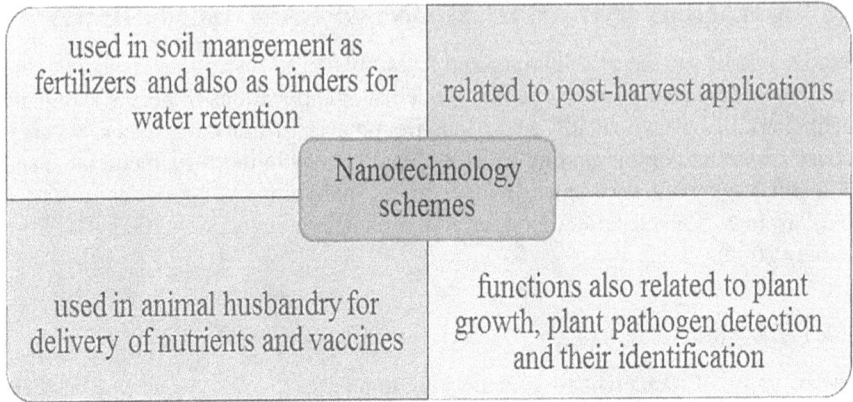

used in soil mangement as fertilizers and also as binders for water retention

related to post-harvest applications

Nanotechnology schemes

used in animal husbandry for delivery of nutrients and vaccines

functions also related to plant growth, plant pathogen detection and their identification

FIGURE 3.2 Broad nanotechnology applications in the agricultural sector.

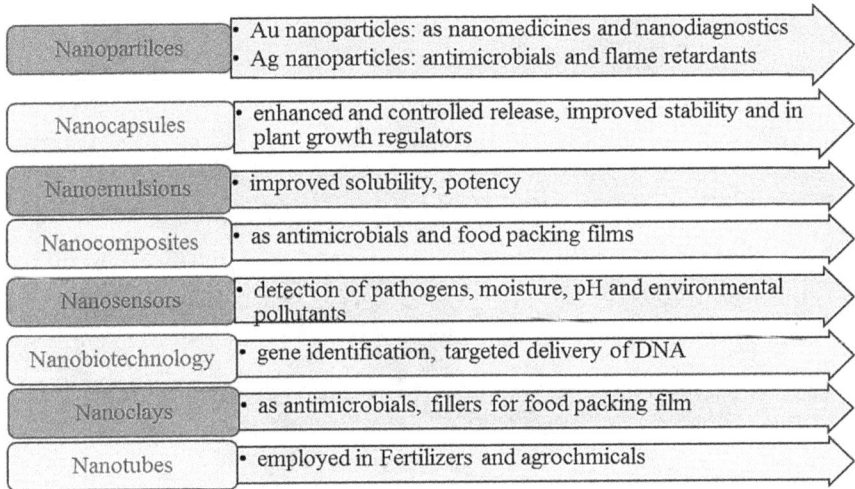

Nanopartilces
- Au nanoparticles: as nanomedicines and nanodiagnostics
- Ag nanoparticles: antimicrobials and flame retardants

Nanocapsules
- enhanced and controlled release, improved stability and in plant growth regulators

Nanoemulsions
- improved solubility, potency

Nanocomposites
- as antimicrobials and food packing films

Nanosensors
- detection of pathogens, moisture, pH and environmental pollutants

Nanobiotechnology
- gene identification, targeted delivery of DNA

Nanoclays
- as antimicrobials, fillers for food packing film

Nanotubes
- employed in Fertilizers and agrochmicals

FIGURE 3.3 Nanotechnology area/types and their applications in agriculture.

and stability forms (Rashidi and Khosravi 2011). Some of the applications of nano-technology in agriculture include:

- Energy Harvesting: There are devices other than the notable solar cell and thermo-electrics that acquire the ability to power nanosystems. An advantage of nanodevices and structures is they generally run at remarkably low power. The energy accumulated from the surroundings may be satisfactory to power the technique (Habibi, Lucia, and Rojas 2010).
- Improved delivery of medicaments to plants: It can also be utilized to enhance the delivery of drugs or other chemicals to plants or soil either by microencapsulation or emulsifying (Sasson, Levy-Ruso and Ishaaya 2007, Das, Saxena and Dwivedi 2009).

- Water retention or purification: Retaining water is a required goal in agriculture. For instance, a natural wetting agent (zeolite) assists in absorbing water and improves porosity in sandy soils and clay soils respectively. Water purification can be attained through nanomaterials, which can eliminate toxins and other unwanted chemicals (Nandita, Shivendu, and Ramalingam 2017).

3.4.2 CONSTRUCTION

Nanotechnology can feasibly make progress faster, economical, safer, and help transform construction. The automation of nanotechnology advances can be incorporated in the development of construction from homes to vast high-rises noticeably more quickly and at a considerably lower price. For example, silica nanoparticles (Aerogel) which are separated by nanopores provide excellent insulation in building materials (Sawhney et al. 2008).

3.4.3 CHEMICAL SENSORS

It can allow sensors to identify insignificant quantities of chemical vapor. Numerous types of sensing elements, like carbon nanotubes, ZnO nanowires, or Pd nanoparticles can be made use of in this technology-dependent sensor. As the size of nanomaterials or nanoparticles is small, a little concentration of gas/chemical molecules is enough to alter the electrical properties of the sensing material for detection (Rai, Acharya and Dey 2012).

3.4.4 COSMETICS

Nanoscale constituents are also integrated into a wide range of personal care products to improve performance. Nanoscale TiO_2 nanoparticles and ZnO have an identical UV security property and have been used in sunscreen for years to offer protection from the sun. It is invisible on the skin (Smith, Simon and Baker 2013).

3.4.5 CHEMISTRY

Nanotechnology has the ability to advance catalysis and improves energy fuel management. It permits higher active combustion and diminished friction which has an effect of lowering energy utilization in vehicles and power stations. The eventual usage of nanoparticles in catalysis can be as diverse as energy mobiles to catalytic converters and photo-catalytic appliances. Catalysis is critical for the progress of substances. The practice of using nanomaterials in catalysis for enhancing the chemical rate has increased, as they lessen the number of catalytic ingredients required as well as saving capital. Now jewelry nanoparticles are increasingly used in automobile catalytic converters as they are able to reduce the amount of platinum needed (Ramesh Kumar et al. 2014).

3.4.6 DISPLAYS

The need for display designs with low energy consumption could be met with the use of carbon nanotubes. Because of their conductivity and slight distance through a few nm, they can be enforced as field emitters with a pronounced degree of ability for field discharge (FED).

3.4.7 ENVIRONMENT

Nanotechnology has also found some noticeable applications in environmental fields. It helps with a simple and inexpensive way of detecting impurities, thereby removing contaminants from drinking water and desalination. Nanoporous materials are apt for "nanofiltration" with extremely small pores less than 10 nm containing nanotubes. It is principally used for the elimination of ions (Raghavan and Connolly 2014). Engineers established a thin nanopore film as a filter which can modify 2–5 times more than conventional filters. Scientists are also working on developing nanoparticles that can clear industrial toxins in groundwater. These particles can reduce the toxins and render them harmless with a chemical reaction. It is also being studied for cleaning oil during oil spills, by using magnetic nanoparticles.

3.4.8 ELECTRONICS AND IT ITEMS

Nanotechnology answers calls for how we can improve the proficiency of electronic procedures, by decreasing their weight and power utilization. It has significantly contributed to the growth of the IT industry. It has entrusted the expansion of smaller, faster, and more portable systems. The transistors that facilitate all recent computing are becoming smaller and smaller, leading to the storage of superior memory on a small chip. Some transistor developments of nanotechnology are Intel, producing a 14 nm transistor (2014), IBM, the first 7 nm (2015), and the 1 nm transistor at the Lawrence Berkley National Laboratory (2016).

Some of the information technology and electronic commodities which contain nanomaterials are sensors, flash memory chips, USB drives, ultra-responsive hearing aids, quantum dots and nanowire inclusion in cameras and computers, keyboards with antimicrobial/antibacterial coatings, the replacement of cathode rays with nanotubes in televisions, cell phone casings, conductive inks, and flexible displays for e-book readers. Nanoparticle Cu suspensions are developed as a safer, economical, reliable option to Pb which are used to fuse electronics in the assembly process. Other materials like graphene and cellulosic are employed for various flexible types of electronics making them flat, light in weight, non-brittle, and highly efficient electronics to countless smart products.

Another advance is the fabrication of nanoporous self-assembly alumina templates, easy to implement and cost effective, in order to achieve ultra-high-density patterned media (Nelson, Patrick, and Ndidi, 2009).

3.4.9 ENERGY

Nanotechnology can contribute to a huge decrease in the burning of motor toxins by nanoporous medium, which can clear up fumes automatically, by exhaust arrangement in the outlook of nanoscale consistent metal particles or through reactant coatings. The usage of batteries with greater energy material or the employment of rechargeable batteries or supercapacitors with an enormous rate of employing nano-materials could be effective for the battery transfer problem.

Several nano-engineered elements used in the automotive industry can enhance a product's realization and capacity. The world's energy requirements have broadened and in such a situation, nanotechnology is playing an integral role with views to improving the productivity of energy production techniques. Many scientists are try-ing to develop clean and effective processes of energy creation aside from lowering the toxic pressure on the atmosphere and managing the energy utilization of nature (Nelson, Patrick, and Ndidi, 2009).

Researchers are also exploring nanotechnology using CNT scrubbers for separat-ing CO_2 from power plant exhausts. In the future, the fusion of nanotechnology into solar panels will be able to produce solar energy economically. It will also develop efficient and light batteries with a high current density which can perform at a high level with a quick charge. For example, the CNT of epoxy coating is employed for the production of lengthier and more lightweight windmill blades which have the capability to develop higher electricity.

Scientists are emerging with CNT wires that have lower resistance compared to high-tension wires that are presently used in the electric grid, further reducing transmission power loss. It is also incorporated into solar panels for the efficient transformation of sunlight into electricity. Present best solar panels have numerous different semiconductors placed together, but they can handle only 40% of the Sun's energy. From a market viewpoint, they are less effective. Therefore, nanotechnology may help to enhance the performance in such a manner with a slight alteration in using nanostructures' bandgaps. Nanostructured solar cells might be inexpensive to produce and easy to install. It can also be used to produce many new types of batteries that are more effective, lighter in weight, and have a higher power density that can resist longer electrical charges compared with conventional batteries (Elena, Guillermo, Javier 2009).

3.4.9.1 Power

The most progressive nanotechnology trends related to power are storage, conversion manufacturing utilizing reducing materials and focusing on process rates, power-saving techniques, and improved alternative power. Modern techniques like LEDs or QCAs might lead to a strong decline in power consumption for illumination. Some advances are quantum spots and quantum dab utilized for the development of lasers which are of high quality and inexpensive. Some more recent advances are (Kuldeep, Pooja and Rajesh 2012):

- Nanohorns: An irregularly shaped single-walled carbon nanotube (SWCN), a critical element of new-generation fuel cells. The main component when grouped together, an agglomeration of around 100 nm is constituted. A ben-efit is not only its extreme surface area but also the fact that it permeates easily. Besides, nanohorns are simply primed with high purity compared to that of normal nanotubes.
- Nanowires: One-dimensional nanowire structures have unique electrical and optical properties that are used as building blocks in the manufacturing

of nanoscale devices. They act as transistors, light-emitting diodes, and bio-chemical sensors.

- Nanosprings: A nanowire enveloped into a helix could serve as a positioner or magnetic field detectors for nanomachines in the future. Silica nano-springs are used for hydrogen storage.
- Nanomesh and nanofibers are currently used in air and liquid filtration applications with the electrospinning process. They are predominately used to prevent the sticking of body tissues together as they heal.

3.4.9.2 Fuel Cells

Nanotechnology is used to lower the price of catalysts in fuel cells to yield hydrogen ions such as methanol, which may also report an absence of fossil fuels by affecting the production from low-grade raw materials. This makes them economical, by pro-ducing the fuels from simple raw materials, thus rendering its use more effectively (Elena, Guillermo and Javier 2009).

3.4.10 FOOD

Nanotechnology affects numerous features of food science, starting from the food's inception to how it is packed. Firms are producing nanomaterials that make a change not only with the taste, but also in the well-being benefits that food provides. Both food quality and safety can be improved (Table 3.1).

Customers also benefit, as such technology leads to a high-quality product for the same price. An example is with vitamins and minerals. Foods are treated and packed with the aid of nanomaterials formed with nanotechnology (Muthu, Govindaswamy, and Ming 2018). Nano activities in the food system are shown in Figure 3.4 and their benefits and risks are documented in Figure 3.5.

3.4.10.1 Is Nanotechnology Innocuous for Food Application?

Threats to food safety will probably be governed by the specific type of nanotechnol-ogy employed, the kind of application, and other circumstances. Below is a list of benefits, positive achievements, and dangers with the use of nanotechnology in food (Bhattacharya et al. 2007, Ravichandran 2010, Rhim, Park and Ha 2013).

3.4.11 HOUSEHOLD

Nanomaterials are also utilized for creating superior merchandise for domestic use. The most distinctive application of nanotechnology is cleaning individualization or for cleaning the surface of pottery or glass. They are useful in developing superior household products such as air purifiers, filters, stain removers, paints, environmen-tal sensors, antibacterial cleansers, as well as sealing products, camera and computer displays rendered residue-repulsive by applying films of nanoscale. Nanoparticles (nanoceramic) have improved the flatness and quality of regular domestic materials. Conventional wear-impervious coatings are less strong compared to nanoceramic for machine parts (Smith, Simon and Baker 2013, Rakesh, Divya and Shalini 2015).

TABLE 3.1

Applications of Nanomaterials in Food Technology

Nanomaterials	Incorporated In	Usages
Nanosilvers	Cutting boards, cleaning sprays, kitchen utensils, storage containers, and refrigerator shelves	• Possess a distinctive antimicrobial property
Nanoparticles	Amalgamated with Zn	• Used for targeted delivery of drugs • Enhance flow properties, color, and shelf-life of various foods, e.g. Al_2SiO_5 used as an anticaking agent in powdered refined foods, anatase metastable form of TiO_2 as a whitener in sweets and cheeses
Nanoclays	Plastic beverage bottles	• Improve and extend strength and shelf life by acting as a barrier to retain oxygen externally and carbon dioxide internally, e.g. clay and silicate nanoplatelets, SiO_2 nanoparticles, chitosan into the polymer matrix makes it lighter, tougher, fire resistant, and has superior thermal properties
Nanochips		• Identify storage circumstances conducive to decomposition
Nanosensors (in combination with polymers, biosensors)	Food pallets during transportation in refrigerated automobiles	• Identify temperature • Determine the abilities of various foods, comprising wine, coffee, juice, and milk • Recognize pathogens contaminating food, e.g. dimethyl siloxane immune sensor combined with Ab restrained on alumina was established for the rapid discovery of foodborne pathogens • Monitor food pathogens in storage and transportation processes in packaging and smart delivery, e.g. *E. coli*, *Campylobacter* and *Salmonella* • Employed in the agricultural production to aid with rebellious viruses and to enhance the efficiency of agrochemicals at lesser amounts
nano-based inserts (flexible)		Finding of *E. coli*, *Salmonella*, could notify customers about inadequate temperature due to which the product gets spoiled
Nano-sized devices		help to trace food source of origin

(Continued)

TABLE 3.1 (CONTINUED)
Applications of Nanomaterials in Food Technology

Nanomaterials	Incorporated In	Usages
Nanoemulsions		• Improve bioavailability, show better consistency regarding particle gathering and gravitational separation, high clarity deprived of compromising product and flavor
Nanoencapsulation		• Provide shielding hurdles, taste masking, and flavor concealing
		• Control interfaces of active ingredients by means of the food matrix and confirm the availability at a specific rate and time
		• Protect from degradation while processing,
		• storage, utilization, and shows compatibility with other compounds from moisture, heat chemical, or biological
		e.g.: curcumin, the utmost vigorous and minimum stable bioactive constituent of Curcuma longa exhibited condensed antioxidant activity and found to be constant toward pasteurization at diverse ionic strength because of encapsulation
Nanocomposites		• Encapsulation of drugs from protection against ecological issues can be used
		e.g. Guard IN Fresh aids fruits and vegetables to mature by taking ethylene gas
		• bio-nanocomposite
		• Protects food and also progresses the shelf-life
		• Nanocomposites of polymer polyaniline identify foodborne pathogens such as *Bacillus cereus*, *Salmonella*, and have extra mechanical and thermos-resistant packing provisions.

(Muthu, Govindaswamy and Ming 2018, Bhattacharya et al. 2007, Biswal et al. 2012, Gupta et al. 2016, Pandey, Zaidib and Gururani 2013, Tang et al. 2009, Valdés et al. 2009, Wang et al. 2008, Sorrentino, Gorrasi and Vittoria 2007, Sozer and Kokini 2009)

3.4.12 MEDICINE

This developing field is quite significant in assisting with the development of effective and new tools in the field of medicine, as it allows for the possibility of inspecting biological methods that were not feasible earlier. Applications of nanotechnology for management, analysis, monitoring, and control of biological systems are classified as "nanomedicine" by the NIH.

Prior to diagnosis, more personalized treatment options, and higher therapeutic rates are made possible with the use of nanotechnology-enabled imaging

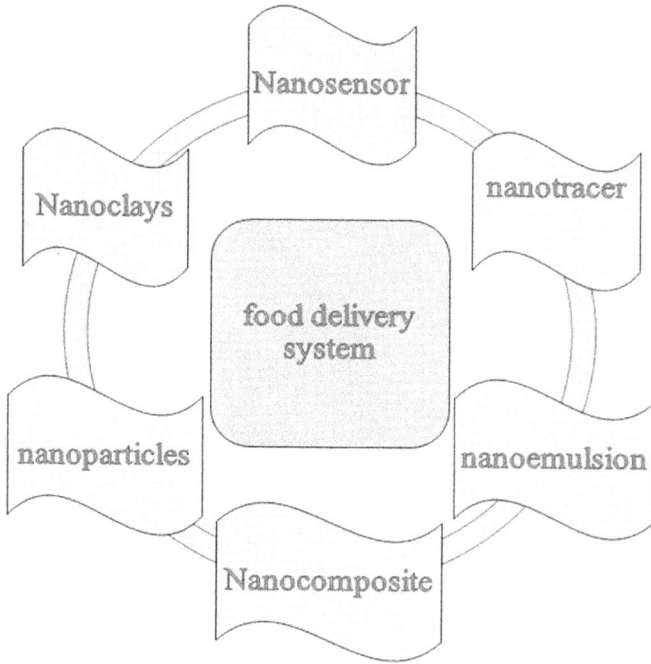

FIGURE 3.4 Nano activity food system.

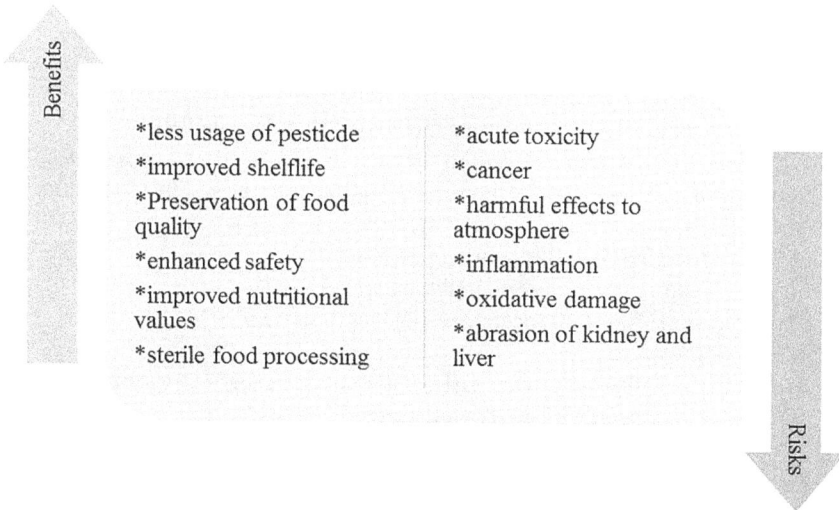

Benefits

*less usage of pesticde
*improved shelflife
*Preservation of food quality
*enhanced safety
*improved nutritional values
*sterile food processing

*acute toxicity
*cancer
*harmful effects to atmosphere
*inflammation
*oxidative damage
*abrasion of kidney and liver

Risks

FIGURE 3.5 Benefits and risks of nano applications in food.

and diagnostic tools, thus increasing life expectancy, detecting and treating diseases, and providing treatments that simulate the body's usual health practices (Figure 3.6).

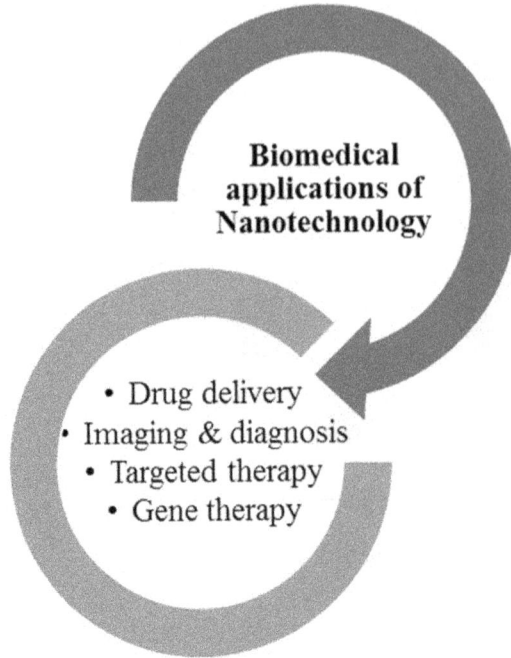

FIGURE 3.6 Biomedical applications of nanotechnology.

Medical use comprises the development of nanoparticles for:

- Artificial cellular proteins.
- Direct drug delivery, regenerative medicine, vaccine drug delivery.
- DNA and protein sequencing with the help of nanopores and nanosprays.
- Either enhancing immune response or avoiding immune system recognition.
- Gene therapy, novel gene sequencing.
- Imaging, diagnostic and screening purposes.
- Lessen bleeding and fasten coagulation.
- Tissue engineering, cell repair, and regeneration.
- Carrier of insulin.
- Treatment of atherosclerosis, early detection of cancer cells, tumors, and reducing the side-effects of chemotherapy.

Technical and scientific analysis has used the special qualities of nanomaterials for various purposes. Table 3.2 outlines some of the clinical uses.

Some additional applications of nanoparticles include: The probable recognition of dysfunction of vascular endothelium, oral administration of insulin, brain targeting for neurodegenerative disorders, transdermal application to improve penetration and distribution across the skin barrier, and to increase oral bioavailability of medicaments like cyclosporine (pH-sensitive nanoparticles), reducing the toxicity of drug (by acting as a carrier, like gentamicin), pulmonary drug delivery

TABLE 3.2

Clinical Applications of Nanotechnology

Type of Nanostructure Nanoparticle	Uses
Aquasomes	Used for drug and antigen delivery
Bio nano-sensors	Monitor health and administer corrective doses of drugs
Carbon nanotubes	Therapeutic applications, used as a vector in the delivery of drugs, proteins, and genes
Dendrimers	Imaging purposes, used as carriers of drug delivery such as polypropylene imine
Fullerenes	Nanodrugs against infections in case of enveloped viruses, drug delivery vehicles
Iron oxide nanoparticles	Contrast agents for MRI
Polymeric nanoparticles	Acts as a carrier for the delivery of insulin
Polymer-coated nanoparticles	Improve oral insulin administration
Inhalable nanoparticles	Useful to administer drugs into lungs
Liposomes	<ul><li>Enhances the drug solubility</li><li>High entrapment capacity of drug molecules as anti-infective agents</li><li>Permeation of drugs across the skin</li><li>Elicit immune response, treatment of cancer, meningitis</li><li>LDC allows the assimilation of both hydrophilic (such as doxorubicin and tobramycin) and lipophilic (such as progesterone and cyclosporine)</li></ul>
Metallic nanoparticles (FeO, Au, Ag, and Ni)	Targeted cellular delivery
Nanocapsules	Prevent and treat infections
Nanoemulsions	vaccine delivery enhance gastrointestinal absorption
Nanorobots	Repair damaged cells, clear blockages in arteries, fixing damaged genes
Nanoshells	*In vitro* imaging
Nanospheres	Entrapment of drugs
Nanovector	Filled with anticancer drugs and detection agents, for targeted gene therapy
Polymeric micelles	For entrapment of drugs or contrast genes
Quantum dots	Imaging agents
Silver nanoparticles	For recognition of inherited series
SLN	Anti-tumor activity
Nanotechnology Tools/ Nanomachines	
Micellar nanocontainers	Hydrophobic drugs are delivered to specific sites
Microchips	For long-term treatment, pulsatile drug delivery
Microneedles	For diagnostic purposes, painless drug infusion, cellular injection
Nanocantilever arrays	Early discovery of precancerous and malignant lesions from biological fluids
Nanopore sequencing	Gene diagnosing of pathogens Detection of single nucleotide polymorphisms
Nanopump	Delivers insulin in small doses for a long duration of time

through nebulization, ocular delivery, and in the management of parasitic infections (Nelson, Patrick and Ndidi 2009, David et al. 2009, Omid and Robert 2009, Moein, Christy and Clifford 2005, Scott 2011, Italia et al. 2007, Sahoo and Labhasetwar 2003; Panyam and Labhasetwar 2003, Peppas 2004, Garcia-Garcia, Andrieux and Couvreur 2005).

3.4.13 REFINERIES

Applying nanotechnology operations, refineries are conveying materials. For example, steel and Al have the ability to eliminate any impurities or contaminants in the components they yield (Rakesh, Divya, and Shalini 2015).

3.4.14 SPORTING GOODS

Present nanotechnology utilization in the sports field comprises enhancing the durability of racquets, filling any deficiencies in and diminishing the speed at which air leaks from tennis balls. Nanomaterials and nanoscale add-ons in polymer complex materials are used to develop a variety of more robust products which are light in weight like baseball bats, rackets, etc.

3.4.15 TRANSPORTATION

Nanomaterial products have the potential to develop automotive achievement. It can allow for the creation of lighter, safer, and more efficient vehicles, which have huge potential in terms of enhancing highway infrastructure and its durability with reduced costs. There are several areas where nanotechnology assists with the art of the automobile by employing high-power battery systems, thermoelectric temperature regulator materials, resistance tires, and continuous audit of the structural reliability by utilization of nanosensors and devices (Rakesh, Divya and Shalini 2015).

It helps with the creation of strong and lightweight materials for most vehicles, hence in spacecraft, decreasing the weight diminishes the energy consumption. Besides that, it also has applications in marine, aircraft, and spacecraft. The purpose of marine transportation broadens from passenger migrating to cargo/weapon carrying, and others.

3.4.16 TEXTILES

An increase in demand for long-lasting and purposeful apparel has created an opportunity for the development of nanomaterials to be unified into textile substrates. Nanocomponents can activate stain-repellent, wrinkle-free, and electrical conductivity without compromising on softness and flexibleness. Nanomaterials also propose a broader application to allow the production of associated garments that are able to sense and react to external response through color or signals. Composite fabrics with nanofibers have led to the development of properties without an increase in weight, thickness, or stiffness and can be washed at a lower temperature in less time. Nanotechnology has been employed to include a film of small carbon particles so that the wearer is guaranteed protection from electrostatic charges. Coating with a thin layer of ZnO nanoparticles on fabrics can generate fabrics that give improved protection from UV (Ali et al. 2016, Shah and Agrawal 2011).

3.4.17 MISCELLANEOUS

- Nanoclays have become essential as construction supplies and are under constant improvement.
- Nanocomposites containing nano-sized flakes of clay and plastic are also used in car bumpers.
- A nanocoating has recent applications comprising of a self-cleaning glass window coated in TiO_2 acts as water repellent (Pilkington Activ is the name of the glass, as produced by the company). It is also scratch-resistant, acting as antibacterial bandages containing silver nanoparticles since Ag ions obstruct microbes, swimming pool cleaners, disinfectants (nano-sized oil particles).
- It also has had an impact on the tennis world by introducing the VS Nanotube Power racket made of graphite and the double-core tennis ball manufactured by Wilson, which is coated with nanoparticles inside which act as a sealant for the air to escape.
- Nanopaints: Performance can be improved by incorporating nanoparticles in the paints making them thinner and lighter, e.g. a thin coating on aircraft reduces their weight and makes it more beneficial to the surroundings.
- Nanolubricants: Inorganic nanosphere nano-sized materials (ball bearings) can also be used as lubricants by controlling their shape making them durable than conventional solid lubricants.
- Space: Advances in nanotechnology assist in the creation of lightweight spacecraft which significantly reduces the required quantity of fuel by lowering the cost of traveling in space and reaching the orbit.
- Superior air quality: Nanotechnology can be adopted to convert vapors emerging from cars or industrial types of machinery into innocuous gases. That is because catalysts made from nanoparticles include a higher surface area among the reacting chemicals than stimulants prepared from huge particles. The large surface area and additional chemicals relate with the catalyst concurrently, which causes the catalyst to be more efficient.
- Better water quality: Nanotechnology is also used to establish solutions to three distinctive obstacles in water quality. One objection is the elimination of industrial waste, such as a purification solvent called TCE, from groundwater. It can also be used to modify the polluting chemical by a chemical reaction to make it powerless. Research has revealed that this process can be adopted well to reach contaminates which are diffused in the underground at an often lower price than practices that need to draw the water out of the soil for analysis.
- Gas and oil extraction: Technology, too, has discovered an application in gas and oil extraction. It has enabled gas lift valves in offshore operations with the help of nanoparticles to perceive the tiniest of pipeline fissures.

Nanotechnology is increasing the productivity of fuel generation from natural petroleum sources by improved catalysis. It is also facilitating lessened fuel utilization in vehicles and power plants by higher-efficiency combustion and reduced friction.

- Optics: The shades handling defensive and anti-reflective ultra-fine polymer coatings are accessible through nanotechnology. Additionally, it extends scratch-proof surface coatings (addition of aluminum silicate) inside nano-composites by making them more effective and improving resistance from scratching and chipping. Nano-optics could see an increase in efficiency in distinct kinds of laser eye surgery. Another advance is the nanoshell metallic sphere, an astonishingly adaptable optical component in nanometer range that can absorb or scatter light at virtually any wavelength providing a new path to optics. (Roco 2003, Bhandare and Narayanam 2014, Kim et al. 2008, Shea 2005, Tratnyek and Johnson 2006, Salamanca-Buentello et al. 2005, Ball 2005, Bittner 2005, Sahoo and Labhasetwar 2003).

3.5 CONCLUSION

The interdisciplinary nature of nanotechnology allows for a comprehensive development of the quality of life. Researchers in various areas such as material science, food, biomedical sciences, engineering, agriculture, energy, transportation, water treatment, imaging, textiles, environmental sciences, and information technology must be up to date with and use nanotechnology, as suitable, for the progression of R&D. Besides, it is a technology that each developing nation ought to invest in in order to progress in various fields. In outline, the research studies have confirmed that nanotechnology offers numerous opportunities, challenges, benefits, such as a single-molecule transistor, gene targeting, enhanced stability, good bio-distribution, low toxicity, targeted delivery, prevention of pathogens, prevention of contamination, increase in the agriculture yield, optimization of fertilizers, delivery of vaccines, diagnostics, patient compliance, improved efficacy, specificity. Tasks for nanotechnology include the establishment of focussed research in order to keep pace with the progress of nanotechnology.

The use of this technology is constantly changing the usage of products, making consumer goods abundant, economical, and long-lasting. Finally, nanotechnology will continue to fabricate new particles that acquire new and remarkable properties. Though several nanotechnologies have been highlighted, there is still on-going research in laboratories. To overcome obstacles, the association between various research organizations, governments, and other sponsors is crucial. Suitable legal and regulatory supervision of nanotechnologies is also essential to effectively safeguard the environment, customers, employees, and culture in general without obstructing technological novelties and encouragements to progress with the marketing of innovative products. With the constant application of nanotechnology, universal lifestyles will be transformed radically and have a strong impact on human behavior and civilization. In conclusion, nanotechnology has gradually increased in importance, and is bound to continue to develop new applications. New and upcoming inventions in nanotechnology improve the safety, health, and quality of many products.

REFERENCES

Abiodun, I.M.F., Ajayi, D.M., Arigbede, A.O. 2014. Nanotechnology and its application in dentistry. *Ann Med Health Sci Res* 4(Suppl 3):S171–S177.

Ali, K., Hang, Q., Amir, M., Butt, H., Dokmeci, M.R., Hinestroza, J.P., Skorobogatiy, M., Khademhosseini, A., Yun, S.H. 2016. Nanotechnology in textiles. *ACS Nano* 10(3):3042–3068.

Ball, P. 2005. Synthetic biology for nanotechnology. *Nanotechnology* 16(1):R1–R8.

Bhandare, N., Narayanam, A. 2014. Applications of nanotechnology in cancer: A literature review of imaging and treatment. *J Nucl Radiather* 5:195.

Bhattacharya, S., Jang, J., Yang, L., Akin, D., Bashir, R. 2007. Biomems and nanotechnology-based approaches for rapid detection of biological entities. *J Rapid Methods Autom Microbiol* 15(1):1–32.

Biswal, S.K., Nayak, A.K., Parida, U.K., Nayak, P.L. 2012. Applications of nanotechnology in agriculture and food sciences. *Int J Inno Sci* 2:21–36.

Bittner, A.M. 2005. Biomolecular rods and tubes in nanotechnology. *Natyrwissenschaften* 92(2):51–64.

Cobb, M.D., Macoubrice, J. 2004. Public perceptions about nanotechnology: Risks, benefits and trust. *J Nanopart Res* 6(4):395–405.

Commission recommendation of 18 October 2011 on the definition of nanomaterial, (2011/696/EU). *Off J EU* L275:38–40.

Das, M., Saxena, N., Dwivedi, P.D. 2009. Emerging trends of nanoparticles application in food technology: Safety paradigms. *Nanotoxicology* 3(1):10–18.

David, P. C., Torjus, S., Zahi, A.F., Willem, J.M. 2009. Nanotechnology in medical imaging probe design and applications. *ATVB Focus Mol Imaging Cardiovasc Dis* 29:992–1000.

Elena, S., Guillermo, R., Javier, G.M. 2009. Nanotechnology for sustainable energy. *Renew Sust Energ Rev* 13(9):2373–2384.

Feynman, R.P. 1960. There's plenty of room at the bottom. *Eng Sci* 23:22–36.

Garcia-Garcia, E., Andrieux, K., Gil, S., Couvreur, P. 2005. Colloidal carriers and blood-brain barrier (BBB) translocation: A way to deliver drugs to the brain? *Int J Pharm* 298(2):274–292.

Gupta, A., Eral, H.B., Hatton, T.A., Doyle, P.S. 2016. Nanoemulsions: Formation, properties, and applications. *Soft Matter* 12(11):2826–2841.

Habibi, Y., Lucia, L.A., Rojas, O.J. 2010. Cellulose nanocrystals: Chemistry, self-assesmbly and applications. *Chem Rev* 110(6):3479–3500.

Italia, J.L., Bhatt, D.K., Bhardwaj, V., Tikoo, K., Ravi Kumar, M.N.V. 2007. PLGA nanoparticles for oral delivery of cyclosporine: Nephrotoxicity and pharmacokinetic studies in comparison to Sandimmune neoral®. *J Control Release* 119(2):197–206.

Kim, L., Anikeeva, P.O., Coe-Sullivan, S.A., Steckel, J.S., Bawendi, M.G., Bulović, V. 2008. Contact printing of quantum dot light emitting devices. *Nano Lett* 8(12):4513–4517.

Kovvuru, S.K., Mahita, V.N., Babu, B.S. 2012. The emerging science in dentistry. *J Orofac Res* 2(1):33–36.

Kuldeep, P., Pooja, K., Rajesh, P. 2012. Recent advances in nanotechnology. *J Sci Eng Res* 3(11):1–11.

Loureiro, S., Tourinho, P.S., Cornelis, G., Van Den Brink, N.W. 2018. *Soil Pollution: From Monitoring to Remediation. Nanomaterials as Soil Pollutants.* Elsevier, Academic Press. 161–190.

Mansoori, G.A. 2005. *Advances in Atomic and Molecular Nanotechnology. Principles of Nanotechnology*, 1st ed. Singapore: World Scientist Publishing Co. Ptc Ltd. 1–10.

Miyazaki, K., Islam, N. 2007. Nanotechnology systems of innovation - An analysis of industry and academia research activities. *Technovation* 27(11):661–671.

Moein, S., Christy, A., Clifford, J.M. 2005. Nanomedicine: Current status and future prospects. *FASEB J. Rev* 19(3):311–330.

Muthu, T., Govindaswamy, R., Ming, C. 2018. Nanotechnology: Current uses and future applications in food industry. *3 Biotech* 8(1):74.

Nandita, D., Shivendu, R., Ramalingam, C. 2017. Applications of nanotechnology in agriculture and water quality management. *Environ Chem.* 15: 591–605.

Nelson, A.O., Patrick, O., Ndidi, C. 2009. Nanotechnology and drug delivery part 1: Background and applications. *Trop J Pharm Res* 8(3):265–274.

Omid, C.F., Robert, L. 2009. Impact of Nanotechnology on Drug Delivery. *American Chemical Society* 3(1):16–20.

Pandey, S., Zaidib, M.G.H., Gururani, S.K. 2013. Recent developments in clay-polymer nanocomposites. *Sci J Rev* 2:296–328.

Panyam, J., Labhasetwar, V. 2003. Biodegradable nanoparticles for drug and gene delivery to cells and tissue. *Adv Drug Deliv Rev* 55(3):329–347.

Peppas, N.A. 2004. Intelligent therapeutics: Biomimetic systems and nanotechnology in drug delivery. *Adv Drug Deliv Rev* 56(11):1529–1531.

Priestly, B.G., Hartford, A.J., Malcom, R. 2007. Nanotechnology: A promising new technology – but how safe. *Med J Aus* 186(4):187–188.

Raghavan, V., Connolly, J.M. 2014. Gold nanosensitizers for mu;timodal optical diagnostic imaging and therapy of cancer. *J Nanomed Nanotech* 5:238.

Rai, V., Acharya, S., Dey, N. 2012. Implications of nano biosensors in agriculture. *J Biomater Nanobiotechnol* 3(2):315–324.

Rakesh, M., Divya, T.N., Shalini, K. 2015. Applications of nanotechnology. *J Nanomed Biother Discov* 5:131.

Ramesh Kumar, K., Nattuthurai, Gopinath, P., Mariappan, T. 2014. Biosynthesis of silver nanoparticles from Morinda tinctoria leaf extract. *J Nanomed Nanotech* 5(6):1–5.

Rashidi, L., Khosravi, K. 2011. The applications of nanotechnology in food industry. *Crit Rev Food Sci* 51(8):723–730.

Ravichandran, R. 2010. Nanotechnology applications in food and food processing: Innovative green approaches, opportunities, and uncertainties for the global market. *Int J Green Nanotechnol* 1(2):72–96.

Rhim, J.W., Park, H.M., Ha, C.S. 2013. Bio-nanocomposites for food packaging applications. *Prog Polym Sci* 38(10–11):1629–1652.

Roco, M.C. 2003a. Nanotechnology: Convergence with modern biology and medicine. *Curr Opin Biotech* 14(3):337–346.

Roco, M.C. 2003b. Broader societal issues of nanotechnology. *J Nanopart Res* 5(3/4):181–189.

Sahoo, S.K., Labhasetwar, V. 2003. Nanotech approaches to drug delivery and imaging. *Drug Discov Today* 8(24):1112–1120.

Salamanca-Buentello, F., Persad, D.L., Court, E.B., Martin, D.K., Daar, A.S., Singer, P.A. 2005. Nanotechnology and the developing world. *PLOS Med* 2(5):e97.

Sasson, Y., Levy-Ruso, G., Ishaaya, I. 2007. Nanosuspensions: Emerging novel agrochemical formulations. *Insectic Des Using Adv Technol*:1–39.

Sawhney, A.P.S., Condon, B., Singh, K.V., Pang, S.S., Li, G., Hui, D. 2008. Modern applications of nanotechnology in Textiles. *Text Res J* 78(8):731–739.

Scott, E. 2011. *Characterization of Nanoparticles Intended for Drug Delivery. Methods in Molecular Biology*, 1st ed. Humana Press, New York.

Shah, R., Agrawal, Y. 2011. Introduction to fiber optics: Sensors for biomedical applications. *Indian J Pharm Sci* 73(1):17.

Shea, C.M. 2005. Future management research directions in nanotechnology: A case study. *J Eng Technol Manag* 22(3):185–200.

Silva, G.A. 2004. Introduction to nanotechnology and its applications to medicine. *Surg Neurol* 61(3):216–220.

Smith, D.M., Simon, J.K., Baker, J.R. 2013. Applications of nanotechnology for immunology. *Nat Rev Immunol* 13(8):592–605.

Sorrentino, A., Gorrasi, G., Vittoria, V. 2007. Potential perspectives of bio nanocomposites for food packaging applications. *Trends Food. Sci Technol* 18(2):84–95.

Sozer, N., Kokini, J.L. 2009. Nanotechnology and its applications in the food sector. *Trends Biotechnol* 27(2):82–89.

Stylios, G.K., Giannoudis, P.V., Wan, T. 2005. Applications of nanotechnologies in medical practice. *Injury* 36(4)(Suppl 4):S6–S13.

Tang, D., Sauceda, J.C., Lin, Z., Ott, S., Basova, E., Goryacheva, I., Biselli, S., Lin, J., Niessner, R., Knopp, D. 2009. Magnetic nanogold microspheres- based lateral-flow immunodipstick for rapid detection of aflatoxin b2 in food. *Biosens Bioelectron* 25(2):514–518.

Taniguchi, N. 1974. *Proceedings of the International Conference on Precision Engineering (ICPE)*, Tokyo, Japan, 18–23.

The Royal Society and the Royal Academy of Engineering. July 2004. Nanoscience and Nanotechnologies. The Royal Society and the Royal Academy of Engineering Report.

Tratnyek, P.G., Johnson, R.L. 2006. Nanotechnologies for environment cleanup. *Nano Today* 1(2):44–48.

Valdés, M.G., González, A.C.V., Calzón, J.A.G., Díaz-García, M.E. 2009. Analytical nanotechnology for food analysis. *Microchim Acta* 166(1–2):1–19.

Wang, X., Jiang, Y., Wang, Y.W., Huang, M.T., Ho, C.T., Huang, Q. 2008. Enhancing anti-inflammation activity of curcumin through O/W nanoemulsions. *Food Chem* 108(2):419–424.

Wohlleben, W., Mielke, J., Bianchin, A., Ghanem, A., Freiberger, H., Rauscher, H., Gemeinert, M., Hodoroaba, V.D. 2017. Reliable nanomaterial classification of powders using the volume-specific surface area method. *J Nanopart Res* 19(2):61.

4 Nanotechnology Applications in Medicine

Ashish Suttee and Neeraj Choudhary
Lovely Professional University, Phagwara, India

CONTENTS

4.1 NANOTECHNOLOGY

Nanotechnology is a field which deals with extremely small structures. The word "nanotechnology" is derived from the Greek word for "dwarf", representing very small size particles. Its major approach is to convert the structure of molecules, individual atoms or compounds to attain the desired product and devices with unique properties. The particle size should be in the range of 0.1 to 100 nm and the materials should possess different properties such as magnetism, optical effects, chemical reactivity, electrical conductance and physical strength. Nanotechnology is helpful in the formation of several types of nanomaterials and nanodevices and also possesses a wide range of applications [1].

Nanotechnology can be broadly classified into various major branches which are listed below:

1. Nanochemistry – deals with the synthesis of nanoparticles
2. Nanophysics – deals with the artificial assembling and fabrication of nanostructures

3. Nanomaterials science – study of the development and production of novel nanostructured components with distinctive properties
4. Nanoelectronics, nanoengineering and optoelectronics – deals with the novel technological processes' development
5. Nano-biocraft – destined for the development of special nanotools, instrumentations, information and computational methodology relevant with the unique biomachine complex, such as nano biorobots, nano biochips, etc.
6. Nanodevice-building, nanometrology and nano-hand systems – help in the expansion of nanotechnology

4.2   SCOPE OF NANOTECHNOLOGY

Nanotechnology is an emerging field and nanomedicines are estimated to reach US$1.5 trillion by 2018. Twenty million dollars was invested in 2006 by the Indian and Australian governments to start the Australia-India Science Research Funding program. According to a BCC Research report, the nanomedicine industry worldwide value was $63.8 billion in 2010 and $72.8 billion in 2011. The anti-cancer product market was valued at $28 billion in 2011, in comparison with the 2010 market value of $25.2 billion. The nanomedicine market is likely to expand at a rate of 12.5% to achieve $180.9 billion by 2017 [2, 3].

4.3   NANOMATERIALS

Nanoscale materials are the molecules which have one aspect of less than around 100 nanometers. Normally, nanometer size is around 100,000 times less than the length of human hair. Nanoscale materials possess unique magnetic, electrical, optical and other properties. These vital features of the nanoparticles play a major role in developing and promoting the use of nanosubstances in medicine, electronics and other fields [4].

Some nanomaterials have been widely used at a commercial level for several years. Nanomaterials are used in the cosmetics industry, stain-resistant clothing, electronics, tires, sporting goods, paints, varnishes as well as many everyday products. These products are also used in the medical field, specifically in drug delivery, diagnosis and imaging. Conventional marketed products have less advantages and novel properties when compared with engineered nanomaterials. The reason for this is that the particles' increased relative surface area and quantum effects are considerably lower than in conventional marketed products. A greater surface area will lead to increased chemical reactivity and strength of nanomaterials. However, quantum effects are an important tool in analyzing the characteristics, properties and behavior of nanomaterials in magnetic, optical and electrical fields. Nanocomposites and nanocoatings are used in several consumer products, such as sports equipment, windows, bicycles and automobiles. Nanomaterials prevent the deterioration of beverages due to exposure to sunlight and are helpful in providing a coating of UV on glass bottles using butyl rubber/nano-clay as composites. Nanoscale silica acts as a filler in a variety of products that include dental fillings and cosmetics [5].

4.4 CLASSIFICATION OF NANOMATERIALS

Nanomaterials are very tiny-sized particles having one dimension of 100 nm or less. Nanomaterials can have one dimension (e.g. surface films), two dimensions (e.g. fibers or strands) or three dimensions (e.g. particles). They exhibit in different forms such as aggregated, single, fused forms with spherical, tubular or asymmetrical shapes. A major category of nanomaterials includes dendrimers, quantum dots and nanotubes [6].

Dendrimers: These are highly branched macromolecules having three-dimensional architecture emanating from a central core which has a controlled monodisperse. The polymer expansion starts from a central core and further extends its surface direction by a chain of polymerization reaction. Furthermore, the desired size is attained through the degree of polymerization reaction, initially from the formation of little nanometers. Cavities and folding occur in the core structure that results in the formation of channels and cages. The dendrimers' exterior groups open to alteration and can be customized for application. The diagnostic and therapeutic agents normally adhere to the exterior of dendrimers via modification at chemical level and possess a small size i.e. 10 nm that is widely used in drug delivery at a controlled rate and can also be used as image contrast agents [7].

Carbon nanotubes and nanowires: With 0.5–3 nm diameter and 20–1000 nm length, these are mainly used in the identification of mutant DNA, recognition of gene expression and as a disease protein biomarker [8].

Nanocrystals: 2–9.5 nm in size, these are used for poorly soluble drugs formulations and for breast cancer marker HeR2 surface labeling.

Nanoparticles: 10–1000 nm in size, these are used in MRI, targeted drug delivery, permeation enhancers, ultrasound contrast agents, apoptosis and in angiogenesis.

Nanoshells: Used in the imaging of tumor cells and in deep tissue thermal ablation.

Quantum dots: nNanoscale crystalline structures are prepared from several compounds, such as cadmium selenide which alter the light color. They have been used since the 1980s. These agents absorb white light and further re-emit in nanoseconds at a particular wavelength. The composition and size of the quantum dots affect the emission wavelength, altering it from blue to near infrared. The 2 nm quantum dots produce a luminescent dark green color, whereas the 5 nm size quantum dots produce a luminescent red. The dots possess larger flexibility in comparison with available fluorescent substances and these characteristics make them appropriate for the creation of nano-scale computing applications that are based on light processing. These quantum dots are helpful for multicolor optical coding in gene expression studies, in vivo imaging, high throughput screening and size range 2–9.5 nm, helping in the optical identification of proteins, lymph node visualization, genes in cell assays and animal models [9].

4.5 PROPERTIES OF NANOMATERIALS

Nanomaterials possess the structural characteristic of atoms and bulk materials as most microstructure materials possess properties related to those of corresponding bulk materials, but materials which have nanometer dimensions have considerably dissimilar characteristics to those of bulk materials and individual atoms. This is primarily due to the size of nanomaterials, which provides: (i) large surface area; (ii) higher surface energy; (iii) reduced imperfections; (iv) spatial confinement as these features are absent in bulk materials. Smaller dimensional metallic nanoparticles are used as catalysts [10].

4.6 NANOPARTICLES IN DRUG DELIVERY

Various research has been conducted to justify the potential role of nanoparticles as a carrier for drug delivery. Nanotechnology drug delivery systems rely upon three major facts: i) drug encapsulation; ii) targeted drug delivery; and iii) drug release.

In this context, Abraxane, an albumin-based paclitaxel nanoformulation is used in the treatment of lung and breast cancer. The study conducted in mice outlined that nanoparticles can act as a carrier for drug delivery, with enhanced effectiveness in the treatment of head and neck cancer. Earlier, Cremophor EL was used for the intravenous delivery of paclitaxel but Cremophor possesses some toxicity. To overcome this toxicity, the carbon nanoparticles were selected for drug delivery thereby carbon nanoparticles reduce the side effects along with improved drug targeting at the site of action and at a low dose of paclitaxel, the desired therapeutic activity is achieved [11, 12]. The study conducted on mice showed that the nanoparticle chain showed better drug delivery of the drug doxorubicin to breast cancer cells. The doxorubicin-loaded liposome was chemically linked with three magnetic, iron-oxide nanoparticle chains ranging in size of 100 nm. After drug loading, the nanochains penetrated cancer cells, and magnetic nanoparticles were allowed to vibrate by producing radiofrequency fields that resulted in the rupture of liposomes, thereby dispersing the drug in its free form throughout the tumor cell, which in turn resulted in inhibition of the cancer cell. The drug concentration was also reduced by nanodelivery when compared with the conventional method of doxorubicin delivery [13].

The antibiotics' nanoparticles derived from polyethylene glycol were used to target in vivo bacterial infection. The nanoparticles contain pH-sensitive chains of histidine amino acid that enhance the delivery and penetration of nanoparticles in antibiotic-resistant bacteria, effectively destroying the bacteria and also maintaining a steady state concentration at the targeted site. This supports the nanoparticles' potential role in the effective treatment of various infectious diseases [14].

Drug-coated nanoparticles bind selectively narrowed regions in blood vessels similar to that of platelets in order to dissolve blood clots. Biodegradable nanoparticle aggregate covered with tissue plasminogen activator and injected intravenously binds and destroys the blood clots. The dissociation of vessels occurs and thereby releases the tissue plasminogen activator from coated nanoparticles and is used to decrease bleeding that occurs during thrombosis treatment. The chemically and thermodynamically stable RNA nanoparticles are X-shaped, carrying four functional modules and are able to remain intact for approximately 8 hours to resist the breakdown of RNAs in the bloodstream. The RNA which have an X-shape regulate cellular function and gene expression and are capable of binding with cancer cells [15]. Minicell nanoparticles are made from mutant bacteria membranes and paclitaxel-loaded cetuximab antibodies used in clinical trial initial phases for the treatment of cancer cells in patients suffering from untreatable cancer; the mechanism is based on engulfing the tumor cells. The minicell drug delivery system has fewer side effects and can be employed in the treatment of different cancers by using different anti-cancer drugs [16]. Nanosponges improve the bioavailability of poorly soluble drugs having a small size and spongy nature. They are used in a targeted drug delivery that inhibits the interactions of the drug and protein, thereby achieving the controlled drug release pattern. Protein and peptides are macromolecules known as biopharmaceuticals that are widely used in the management of several disorders and diseases via different mechanisms on the human body. Nanobiopharmaceuticals generally consist of nanoparticles and dendrimers that are helpful in controlled or targeted drug delivery [17].

4.7   ROLE OF NANOPARTICLES IN MANAGEMENT OF DISEASE

Immune system: Nanoparticles are useful in myelin antigens delivering, as it is reported to induce immune tolerance in mice with relapsing multiple sclerosis. The myelin sheath peptide-coated polystyrene microparticles enrich the immune system thereby preventing the recurrence of disease by shielding the myelin sheath via the nerve fiber coating. This method is generally applied in the treatment of several autoimmune diseases.

The nanodevice buckyballs affects the release of histamine from mast cell into the blood and tissue system in order to change the immune or allergy response. The research illustrates that nanoparticles are a better alternative than the available antioxidant [18].

Cancer: Nanoparticles are widely used in oncology, particularly in imaging. The quantum dots' confinement characteristics are used in combination with imaging associated with magnetic resonance in order to fabricate images of the cancer sites. The nanoparticles produced require only one light source for excitation, in comparison with available organic dyes. The fluorescent quantum dots exhibit an advanced image at a lower price in comparison with organic dyes. But the quantum dots comprise of some toxic elements. Nanoparticles possess the capability to adhere to different functional groups and then bind to various cancer cells. However, due to their small size (10 to 100 nm) they gather at cancer sites, thereby decreasing the effective lymphatic drainage system. Nanowires are widely used in the development of sensor test chips that help in the identification of biomarkers, proteins and also play a major role in the detection and diagnosis of early stage cancer [19, 20].

Gold-coated nanoshells consist of a 120 nm diameter used against tumor cells in mice. The nanoshells act by making a conjugated bond between cancer cells, peptides or antibodies with the nanoshell surface. Afterwards, a laser is applied to the cancer cell that heats the gold coating sufficiently resulting in the destruction of cancer cells [21]. Nanoparticles have been employed in photodynamic therapy, where the nanoparticle is injected in the body and reaches cancer cells. Then the cells are exposed to photo light, resulting in the production of high energy oxygen molecules that chemically interact with tumor cells, leading to the formation of the cross-linking of base pairs that helps in the destruction of cancer cells. In the past few years, photodynamic therapy has drawn attention to its ability to treat tumor cells [22].

4.7.1 Neurodegenerative Disorders

Nanotechnology is widely used in the management of neurodegenerative disorders. The role of a variety of nanocarriers such as nanoemulsions, dendrimers, nanoparticles, liposomes, nanogels, solid lipid nanoparticles, nanosuspensions and polymeric nanoparticles. The effectiveness for various CNS disorders like Alzheimer's disease, brain tumors, acute ischemic stroke and HIV encephalopathy have successfully been analyzed by *in vivo* and *in vitro* BBB models. The reduction of neurotoxicity and enhanced permeability across BBB can also be achieved with nanomedicines.

Parkinson's disease: The second most common CNS disorder, namely Parkinson's disease, is most common after 65 years of age which involves neuroinflammatory responses leading to difficulty in body movement. Current therapy aims at improving a patient's functional capacity rather than modifying the neurodegenerative process. Nanotechnological research, along with advances in cell biology, neurophysiology and neuropathology will significantly benefit the neuroprotection and rejuvenation of CNS. Advances in novel technologies have made it possible to signal cues for guided axon growth and the protection of neurons. The focus of research is to optimize the design and biometric simulation of intracranial nano-enabled scaffold devices for the targeting of dopamine to the brain, so as to minimize the peripheral side effects of conventional therapy. The treatment for CNS disorders now includes newer tools like peptide nanoparticles and peptides.

New therapeutic and diagnostic tools have been developed due to nanotechnology. Various devices of nanotechnology have facilitated the delivery of drugs and small molecules, neuroprotection, reverse neuropathological disorder and the restoration of injured neurons. Several nanocarriers i.e. nanogels, dendrimers, liposomes, nanoemulsions, solid lipid nanoparticles, nanosuspensions and polymeric nanoparticles are effective in the treatment of Parkinson's disease. The effectiveness for various CNS disorders like Alzheimer's disease, brain tumors, acute ischemic stroke and HIV encephalopathy have successfully been analyzed by *in vivo* and *in vitro* BBB models. Novel targeting moieties having a specificity for brain tissues and the enhancement of drug-trafficking performance should be of focus [23].

Alzheimer's disease: Nanotechnology is significant in therapy of the most common cause of dementia i.e. Alzheimer's disease, which affects more than 35 million people worldwide. Nanoparticulate entities are designed and engineered for the capillary endothelial cells of the brain. Higher binding affinity of nanoparticles towards the circulating amyloid-β (Aβ) forms is made useful for the improvement of the Alzheimer's disease condition. Microscopy with the use of the scanning tunneling technique is able to detect the Aβ1–40 and Aβ1–42 and ultrasensitive nanoparticle-based bio-barcodes along with immune sensors which have helped advance in vitro diagnostics [24, 25].

Tuberculosis: Tuberculosis, a lethal communicable disease needs an extended duration of treatment which involves the development of multi drug-resistant strains due to the pill burden. First-line therapy used to treat children is not available commercially. To reduce the period of drug treatment, antiretroviral drug interactions are necessary, and to overcome drug resistance, newer antibiotics are designed. More effective and compliant medicines can be developed by using carbon-based nano-drug delivery systems [26].

Operative dentistry: Various devices and nanomaterials at molecular level, atomic level and supramolecular structures having a particle size of 0.1 nm to 100 nm are created and utilized by nanotechnology. Improved strength, excellent wear resistance and aesthetic appeal due to the retention of luster are also offered by nano-filled composite resin materials. Nanofillers, usually of size 5-40 nm are made by spherical silicon dioxide (SiO2) particles in operative dentistry. Improvement of the inorganic phase load seems an actual advancement due to nanofillers. In operative dentistry, the best option is micro-hybrid composite nanofillers that mimic the desired characteristics of enamel and dentin. These might prove to be the preferred filler material. Nanoproduct companies have developed nanocomposite resins which can be distributed homogenously in coatings or resins. Alumina silicate powder with an average particle size of 80 ran 1:4 M ratio of alumina to silica having a refractive index of 1.508 is used as a nanofiller. Decreased polymerization shrinkage with enhanced flexural strength and a superior hardness are two characteristics of nano-composites [27, 28].

Ophthalmology: To construct, repair, defend, monitor, control and improve the biological systems at molecular stage by means of nanodevices, nanostructures that work particularly at cellular level for achieving medical benefits is the goal of nano-medicines. Biomimicry, along with artificial intelligence and nanotechnology principles, are applied to develop nanomedicine. Nanoparticles have been used in the measurement of intraocular pressure, treatment of oxidative stress, the prevention of scars that occurs after glaucoma surgery, regenerative nanomedicine, the treatment of degenerative disease of the retina using gene therapy, and prosthetics are some of the applications of nanotechnology to ophthalmology. Postoperative scarring, sight-restoring therapy in patients suffering from degenerative retinal disease, and present-day drug delivery challenges can be resolved with the effective use of nanotechnology [29]. A nanoscale-dispersed eye ointment is used in curing severe evaporative dry eye. Excipients of conventional eye ointment, viz. petrolatum and lanolin, were attached to triglycerides as a lipid, followed by dispersion in polyvinyl pyrrolidone solution to develop nanodispersion [30].

Entrapment of ointment matrix in nanoemulsion of medium-chain triglycerides, having an average size of particle, i.e. 100 nm, was observed in the transmission electron micrograph. Improvement of storage stability along with the elimination of cellular toxicity in human corneal epithelial cells by the optimized formulation of nanoscale-dispersed eye ointment have been some of the improved features, in comparison to commercially available polymer-based artificial tears. Therapeutic improvement has been observed, displaying a trend of positive correlation with higher concentrations of ointment matrix in the nanoscale-dispersed eye ointment formulations. The restoration of normal cornea by nanoscale-dispersed eye ointment was demonstrated by histological evaluation. Nanoparticle-developed systems are liposomes, niosomes, nanoparticles, microemulsions, nanosuspensions, dendrimers and cyclodextrins in the field of ocular drug delivery. Various advances in nano-technology like nano-imaging, nanodiagnostics and nanomedicine can be utilized to explore the frontiers of ocular drug delivery and therapy [31].

Visualization: Tracking movement can be used to determine drug metabolism and its distribution throughout the body by cell dying. Quantum dots adhere to the cell protein thereby penetrating the cell membrane and resulting in fluorescence of the quantum dots as they absorb the light at certain wavelengths. The fluorescence depends upon the sizes and the light exposure. Nanoparticles are used for detection of the cell [32].

Tissue engineering: Nanotechnology is used for reproducing and repairing damaged tissue by means of various nanomaterial-based scaffolds, proliferation of stimulated cells, growth factors, artificial implants and organ transplant therapy which help with life extension.

Antibiotic resistance: Combinational therapy decreases antibiotic resistance. Nanoparticles of zinc oxide reduce antibiotic resistance and boost the ciprofloxacin antibacterial activity against various microorganisms, thereby adhering to various proteins which are responsible for antibiotic resistance [33].

Nanopharmaceuticals: Nanoparticles can be used for early stage detection of diseases and for diagnostic applications. The major concern in the pharmaceutical industry is to deliver the required dose of the active ingredient to only a particular site of disease. This concern of site-specific delivery can be overcome by nanopharmaceuticals, including the reduction of toxic systemic side effects.

Pharmaceutical companies, in order to maintain profit, need to deliver high-quality products for which they utilize the nanospray-drying process for encapsulation of pharmaceuticals. To reduce drug discovery and diagnostic time, nanopharmaceuticals can be used in developing a cost-effective drug discovery process [34].

Modified medicated textiles: Antibacterial textiles have been developed with newer antibacterial cotton using nanotechnology. The good antibacterial activity of inorganic nanostructured materials has been of main focus in the application of these materials to textiles [35].

4.8 TOXICOLOGICAL AND ETHICAL ISSUES IN NANOMEDICINE

Significant advances have been achieved in the development of nanomedicine. However, many hurdles exist regarding ethical, social and toxicological aspects of nanomaterials in the treatment and diagnosis of disease. The toxicity of nanomaterials and accompanying ethical issues are described briefly for those nanostructures used in biomedical applications [36].

Toxicity issues: Immunotoxicity, inflammation, oxidative stress, carcinogenicity and genotoxicity occurs when nanomedicines are exposed to the physiological system and these ill effects are a major concern in the development of nanotechnology. Such potential toxicological effects are generally not evaluated by the researchers during nanomedicine product development. The toxic properties of such particles are influenced by size, shape, chemistry, charge of particles, solubility and degree of agglomeration. Due to a nanoparticle's exceptionally small size and increased specific surface area, its toxicity profiles are different from bigger particles which have a similar chemical composition. In addition, the route of administration can also affect the toxicity of particles. The undesirable side effects of several nanomaterials have been reported, but the complete evaluation of acute and chronic toxicity of nanoparticles has not yet been undertaken. Investigations have reported that crystalline silica oxides nanoparticles may cause size-dependent irritation, inflammation, fibrosis and carcinogenesis. Furthermore, gold nanoparticles have induced apoptosis and necrosis, as observed in some in vitro studies. It was noted that anionic-functionalized gold nanoparticles exert higher cytotoxicity compared to cationic-functionalized gold nanoparticles. Size-dependent toxic effects have been revealed by injecting 13 nm-sized polyethylene glycol gold nanoparticles in mice which induced acute inflammation and apoptosis in the liver, in addition to fatigue, severe distress, loss of appetite and weight and change of fur color induced by 8 to 37 nm-sized gold nanoparticles. The route of administration is an important key factor in determining the toxicity of the nanoparticles as the fullerene toxicity in rats was observed in intraperitoneal injection, whereas the oral administration did not produce any toxicity. The single- and multi-walled carbon nanotubes provoke oxidative stress, cytotoxicity, necrosis and apoptosis. The nanoparticles are reported to possess carcinogens and genotoxins, as the nanoparticles destroy the genetic material by DNA interaction through diffusion by a nuclear pore complex or producing a reactive oxygen species. Interactions with DNA have been observed with nanoparticles sized between 1 nm and 10 nm whereas the large-size nanoparticles interact with DNA during the mitosis process, either by chemically or mechanically binding. In silico analysis has shown that nanoparticles may alter repair proteins like RFC3, PMS2 and PCNA in DNA, which can further lead to mutation. Many factors are responsible for the generation of free radicals i.e. nanoparticles surface in aqueous suspensions, by metal ion release, nanoparticle-induced stress and the activation of inflammatory cells. Reactive oxygen species can oxidize the base pairs of DNAs, giving rise to DNA base lesions that result in the breakdown of the DNA strand. The formation of base lesions can be responsible for cross-linking during replication, resulting in mutation and carcinogenicity. Research on nanoparticles' toxicity is yet to establish whether nanoparticles produce cancer and other developmental and adverse effects in humans. As the use of nanomaterials is increasing in the pharmaceutical industry, there is a current need to clarify the potential adverse health effects [37, 38, 39].

4.9   NANODIAGNOSIS

Nanoparticles are widely used in the diagnosis of disease by conventional imaging using ultrasound, computed tomography, radiography and magnetic resonance imaging. However, all these techniques can detect cancer once the tumor cell occurs as a visible physical entity. Over the past decade, there has been a shift from anatomical imaging that identifies gross pathology to molecular imaging that is used to identify cancerous cells at an early stage. This is at molecular level, before the change in the cell phenotype. However, molecular imaging helps in the identification of changes during oncogenesis, thereby helping in predicting the molecular therapy for treatment of cancer in the patient. It also provides repeated noninvasive monitoring of the progression of disease and transformation. Molecular imaging is a current need in the medical diagnostic field, as it can recognize the disease stage, providing early diagnosis; the use of pathological processes gives fundamental information, and thus determines the efficacy of therapy [40]. Nanoparticles can move through the blood and lymphatic vessels of the human body due to their potential for molecular imaging. Antigen antibodies, nucleic acid hybridization and gene expression serve as targets for biological interaction to discover the particular site in the body. Fluorescent, isotope-tagged, magnetic and magneto-fluorescent nanoparticles are of prime interest for molecular imaging. Ultrasound, optical, CT, MRI and nuclear imaging have been used for nanotechnology-enabled molecular imaging. There are some merits and demerits of each imaging modality. The selection of imaging technique depends upon the specific molecular process that is to be targeted. The use of magnetic nanoparticles in bioassays for separation and preconcentration has a great impact. Various biomolecules and the location of cancerous cells can be detected by magnetic micro/nanoparticles. Magnetic field-assisted targeting of nanoparticles and the treatment of hyperthermia can be achieved using these nanoparticles. It is their use as contrast-enhancing agents in MRI that helps in the diagnosis of cancer, hypoperfusion region visualization, cell labeling, targeted molecular imaging and in the detection of apoptosis, angiogenesis and gene expression [41]. The fluorescent, magnetic and inorganic nanoparticles combinations have wide applications, in which one nanocomposite is expected to allow the engineering of nanoscale devices. Some of the areas where these nanocomposites can be used are multi-modal magnetic-fluorescent assays for in vitro- and in vivo- bioimaging, as agents in nanomedicine and to visualize and simultaneously treat various diseases. The sensitivity, adaptability of fluorescence-based detection methods can be enhanced by nanotechnology. However, there are various defined structural traits of nanoparticles which assist in the improvement of cancer detection assays. The dynamic and sensitivity range of a specific assay, in particular, can be determined by using the optical property which is relevant in the aim of fluorescence-based biosensors for the diagnosis of cancer, the force and steadiness of fluorescence emission along, with the efficiency of fluorescence quenching in "off-on" probes. Quantum dots are accurate in the absorption ability thereby producing high quantum yields. Dots with a high fluorescence yield of quantum did not show any signs of cellular toxicity. Gold nanoparticles are considered to be outstanding fluorescence quenchers, while mesoporous silica nanoparticles or fluorophore-encapsulating polymerics are helpful in the delivery of photosensitizing agents and chemotherapeutics [42].

Dots are used in the management and detection of cancer biomarkers, cells and tissues through fluorescence. The role in diagnosis and therapy is due to their huge flexibility and versatility. Identification of the various cellular and molecular functions in living organisms can be achieved using radio-labeled imaging agents. Diagnosis, therapeutics and investigation can be undertaken by loading or labeling different kinds of particles with various radio nuclides. The diagnosis and treatment of various disorders can be achieved with the use of radioisotopes. Radiological imaging techniques – in contrast to nuclear medicine – use radiotracers which are given to the patient. The images obtained by emitted radiation help in the detection of the metabolic pathway or functions of an organ system. Dynamic or static imaging techniques are generally used in the analysis of in vivo function of the organs in the body. Functional abnormality or tumors can be identified early, and therapy planning can be achieved as per the gathered information. A major role of medicine and biology is to choose the radionuclide compounds. The surface labeling of the nanoparticles after encapsulation of radio-labeled nanoparticles serves as an approach for the labeling of radionuclides [43].

4.10 CLINICAL APPLICATIONS OF NANOPARTICLES

Nanoparticles are widely used in clinical practice that includes inorganic and organic materials. Liposomal systems, polymer-drug conjugates and micelles systems are various nanocarriers used in clinical practice. Nanoparticles play a vital role in clinical use as they target drug delivery, tissue engineering, drug release at a controlled rate and increase the bioavailability of several therapeutic agents. These are widely used in pharmaceutical product development.

4.10.1 NANOTECHNOLOGY TOOLS AND NANO MACHINES IN MEDICAL RESEARCH

Nanopore sequencing: Pore nanoengineering and assembly form the basis of this rapid technique. The DNA-charged strands are drawn by a 1–2 nm-sized pore through the complex of hemolysin proteins by applying a small electric potential, that is then incorporated into a bilayer of lipid that separates the two conductive compartments. Each base can be identified by recording current time profiles followed by its translation into electronic signals. Over 1,000 bases can be sequenced per second. Gene diagnosis of pathogens and detection of single nucleotide polymorphisms are some other potential applications of this technology [44].

Microneedles: A combination of fusion bonding, photolithography and anisotropic plasma etching can be used to produce micro-machined lancets and needles with adjustable angles of bevel, dimensions of channel and wall thickness which can be attained from single crystal silicon. Cellular and painless drug injection and diagnostic procedures like glucose monitoring can be performed with this method [45].

Microchips: Micro-fabricated devices consist of micrometer-scale pumps, flow channels and valves. After the implantation of the device in the patient, it can be used for a long time and can help in the controlled release of drugs. The mechanism of drug release is based on electrochemical dissolution through membranes of anode covering micro-reservoirs that contain the drugs. The rate of the drug release depends upon the material used for the membrane. The microchip devices comprise of a 1.2 cm diameter, 500 m thickness with 36 drug reservoirs; to achieve a controlled drug release, the membrane was fabricated with poly (L-lactic acid) [46].

Carbon nanotubes: Carbon nanotubes consist of graphite sheets in a tubular rolled form and are obtained as single- or multi-walled nanotubes. The length and diameter of single-walled nanotubes normally differ in the range of 0.5–3.0 nm and 20–1000 nm. However, multi-walled nanotubes dimensions are 1.5–100 nm and 1–50 m. Carbon nanotubes' hydrophilicity may be increased by surface fictionalization. The ionic and molecular movement achieved by carbon nanotubes makes them capable of sequencing electronic nucleic acid and molecular sensor fabrication. Carbon nanotubes possess the ability to pass through the cell membrane as "nanoneedles" without unsettling the cell membrane and entering the mitochondria and cytosol. Such mechanisms are not well understood. Superoxide dismutase mimetic properties are displayed by various carbon nanotubes' tris-malonic acid, a derivative of the fullerene C60, that is used in the protection of cell culture from injury, neurons' dopaminergic degeneration in Parkinson's disease and in ischemia of the nervous system. The fullerene C60 mainly acts by its action on catalytic superoxide dismutase. Furthermore, single-walled carbon nanotubes have been reported to have potassium channel subunits blocking action in a dose-dependent manner [47].

4.11   THE FUTURE OF NANOMEDICINE

Change in the scale and methodology of drug delivery and vascular imaging is possible with nanotechnology. In the following ten years, the NIH Roadmap's "Nanomedicine Initiatives" predicts medical benefits from nanoscale technologies. It includes the improvement of nanoscale laboratory-based drug discovery and diagnostic platform procedures such as nanoscale cantilevers for nanopore sequencing, microchip devices, chemical force microscopes, etc. The purpose of developing a multifunctional nanometer scale is to help in diagnosis, the distribution of therapeutic agents, and observation of the cancer treatment progress by the National Cancer Institute. It includes a plan to engineer the development of contrast agents in order to visualize cancer cells at a cellular level. Nanodevices have the inherent ability to analyze the evolutionary diversity and biological changes that occur in multiple cancer cells, thus leading to the formation of tumors in the body. In order to design effective nanoparticles, there is a need to understand the physiological and physicochemical processes, as well as the interactions with the microenvironment. The selection of nanoparticles depends upon the target site and disease state. The toxicity profile should be carried out before the selection of nanoparticles as a carrier, so that the effective drug delivery can be achieved. Nanomedicine's future relies on a rational plan of nanotechnology material development and the effective use of tools through a complete understanding of the biological processes prior to use as a carrier for drug delivery.

4.12   DISADVANTAGES OF NANOMATERIALS

   (i) *Particle instability*: Nanoparticles are normally encapsulated in other matrices, and are thermodynamically metastable, thus prone to transformation. This leads to phase change, poor corrosive resistance and the deterioration of nanomaterials.
  (ii) *Explosive nature*: Due to a high surface area, metal particles act as strong explosives when meeting oxygen as a part of exothermic combustion. This is a major factor that explains their explosive property.
 (iii) *Impurity*: Due to their highly reactive properties. they interact with impurities. So, nanoparticles' encapsulation is essential when there is any development with such chemical methods. The nanoparticles' stabilization is a major concern as the presence of non-reactive species engulfs the reactive nano-entities. Due to this engulfing phenomenon, the secondary impurities become a part of the synthesized nanoparticles. This in turn leads to the synthesis process of achieving pure nanoparticles, which may prove difficult. Environmental changes can lead to the formation of nitrides,

oxides, etc. during the synthesis of nanoparticles which can further lead to the deterioration of the nanomaterials. Hence maintaining the purity of the nanoparticles is a challenge. This needs to be considered before using as a carrier for drug delivery.

(iv) *Biologically unsafe*: Nanomaterials are normally considered unsafe. But their toxicity is well reported due to their increased surface activity. Nanomaterials are also responsible for producing irritation, and can act as a carcinogenic. If inhaled, they may be trapped inside the lungs and cannot be expelled out from the body. This is due to their low molecular weight.

(v) *Complexity in isolation and synthesis*: It is difficult to maintain the particle size of nanoparticles after their synthesis. Due to their instability, they are encapsulated by using a stable material/molecule. However, nanoparticles in free form are difficult to isolate. The chances of impurity incorporation are high during their processing.

(vi) *Recycling and disposal*: Safe and secure nanoparticle disposal policies are in the early stages of development. The absence of a disposal procedure can lead to the generation of biowaste, which is harmful for the environment due to its toxicity and various interactions with other metals [48].

REFERENCES

1. Rajneesh, Pathak, J., Singh, V.; Kumar, D.; Singh, S. P.; Sinha, R. P. DNA in Nanotechnology. In: *Nanomaterials in Plants, Algae, and Microorganisms.* 2018. 1:79–89. https://doi.org/10.1016/b978-0-12-811487-2.00004-9.

2. Wadhwa, A.; Mathura, V.; Lewis, S. A. Emerging Novel Nanopharmaceuticals for Drug Delivery. *Asian Journal of Pharmaceutical and Clinical Research.* 2018. 11(7): 35–42. https://doi.org/10.22159/ajpcr.2018.v11i7.25149.

3. Patil, A.; Mishra, V.; Thakur, S.; Riyaz, B.; Kaur, A.; Khursheed, R.; Patil, K.; Sathe, B. Nanotechnology Derived Nanotools in Biomedical Perspectives: An Update. *Current Nanoscience.* 2018. https://doi.org/10.2174/1573413714666180426112851.

4. Ko, F. K.; Wan, L. Y. Nanofiber Technology: Bridging the Gap Between Nano and Macro World. In: *Nanomaterials Handbook, 2nd Edition.* 2017. https://doi.org/10.1201/9781315371795.

5. Murty, B. S.; Shankar, P.; Raj, B.; Rath, B. B.; Murday, J.; Murty, B. S.; Shankar, P.; Raj, B.; Rath, B. B.; Murday, J. Applications of Nanomaterials. In: *Textbook of Nanoscience and Nanotechnology.* 2013. https://doi.org/10.1007/978-3-642-28030-6_4.

6. Han, J.; Chen, R.; Wang, M.; Lu, S. S.; Guo, R.; Atta, S.; Tsoulos, T. V.; Fabris, L.; Nelson, A.; Ha, D. H.; et al. Fundamentals of Nanomaterials. *Nano Letters.* 2012. https://doi.org/10.1039/c1cc13658e.

7. Kesharwani, P.; Jain, K.; Jain, N. K. Dendrimer as Nanocarrier for Drug Delivery. *Progress in Polymer Science.* 2014. https://doi.org/10.1016/j.progpolymsci.2013.07.005.

8. Sharma, M.; Gao, S.; Mäder, E.; Sharma, H.; Wei, L. Y.; Bijwe, J. Carbon Fiber Surfaces and Composite Interphases. *Composites Science and Technology.* 2014. https://doi.org/10.1016/j.compscitech.2014.07.005.

9. Semonin, O. E.; Luther, J. M.; Beard, M. C. Quantum Dots for Next-Generation Photovoltaics. *Materials Today.* 2012. https://doi.org/10.1016/S1369-7021(12)70220-1.

10. Qu, X.; Alvarez, P. J. J.; Li, Q. Applications of Nanotechnology in Water and Wastewater Treatment. *Water Research.* 2013. https://doi.org/10.1016/j.watres.2012.09.058.

11. Wang, R.; Zhan, H. L.; Li, D. Z.; Li, H. T.; Yu, L.; Wang, W. Application of Endoscopic Tattooing with Carbon Nanoparticles in the Treatment for Advanced Colorectal Cancer. *Zhonghua Wei Chang Wai Ke Za Zhi = Chinese Journal of Gastrointestinal Surgery.* 2020. https://doi.org/10.3760/cma.j.issn.1671-0274.2020.01.010.

12. Gelderblom, H.; Verweij, J.; Nooter, K.; Sparreboom, A. Cremophor EL: The Drawbacks and Advantages of Vehicle Selection for Drug Formulation. *European Journal of Cancer.* 2001. https://doi.org/10.1016/S0959-8049(01)00171-X.

13. Peiris, P. M.; Bauer, L.; Toy, R.; Tran, E.; Pansky, J.; Doolittle, E.; Schmidt, E.; Hayden, E.; Mayer, A.; Keri, R. A.; et al. Enhanced Delivery of Chemotherapy to Tumors Using a Multicomponent Nanochain with Radio-Frequency-Tunable Drug Release. *ACS Nano.* 2012. https://doi.org/10.1021/nn300652p.

14. Radovic-Moreno, A. F.; Lu, T. K.; Puscasu, V. A.; Yoon, C. J.; Langer, R.; Farokhzad, O. C. Surface Charge-Switching Polymeric Nanoparticles for Bacterial Cell Wall-Targeted Delivery of Antibiotics. *ACS Nano.* 2012. https://doi.org/10.1021/nn3008383.

15. Draz, M. S.; Fang, B. A.; Zhang, P.; Hu, Z.; Gu, S.; Weng, K. C.; Gray, J. W.; Chen, F. F. Nanoparticle-Mediated Systemic Delivery of SiRNA for Treatment of Cancers and Viral Infections. *Theranostics.* 2014. https://doi.org/10.7150/thno.9404.

16. MacDiarmid, J. A.; Brahmbhatt, H. Minicells: Versatile Vectors for Targeted Drug or Si/ShRNA Cancer Therapy. *Current Opinion in Biotechnology.* 2011. https://doi.org/10.1016/j.copbio.2011.04.008.

17. Wójcik, P.; Berlicki, Ł. Peptide-Based Inhibitors of Protein-Protein Interactions. *Bioorganic and Medicinal Chemistry Letters.* 2016. https://doi.org/10.1016/j.bmcl.2015.12.084.

18. Nagda, D.; Rathore, K. S.; Bharkatiya, M.; Sisodia, S. S.; Nema R. K. Bucky Balls: A Novel Drug Delivery System. *Journal of Chemical and Pharmaceutical Research.* 2010.

19. Ul Islam, Salman; Ahmed, M. B.; Ul-Islam, Mazhar; Shehzad, A.; Lee, Y. S. Switching from Conventional to Nano-Natural Phytochemicals to Prevent and Treat Cancers: Special Emphasis on Resveratrol. *Current Pharmaceutical Design.* 2019. https://doi.org/10.2174/1381612825666191009161018.

20. Nie, S.; Xing, Y.; Kim, G. J.; Simons, J. W. Nanotechnology Applications in Cancer. *Annual Review of Biomedical Engineering.* 2007. https://doi.org/10.1146/annurev.bioeng.9.060906.152025.

21. Song, J.; Yang, X.; Yang, Z.; Lin, L.; Liu, Y.; Zhou, Z.; Shen, Z.; Yu, G.; Dai, Y.; Jacobson, O.; et al. Rational Design of Branched Nanoporous Gold Nanoshells with Enhanced Physico-Optical Properties for Optical Imaging and Cancer Therapy. *ACS Nano.* 2017. https://doi.org/10.1021/acsnano.7b02048.

22. Banik, B. L.; Fattahi, P.; Brown, J. L. Polymeric Nanoparticles: The Future of Nanomedicine. *Wiley Interdisciplinary Reviews: Nanomedicine and Nanobiotechnology.* 2016. https://doi.org/10.1002/wnan.1364.

23. Wong, H. L.; Wu, X. Y.; Bendayan, R. Nanotechnological Advances for the Delivery of CNS Therapeutics. *Advanced Drug Delivery Reviews.* 2012. https://doi.org/10.1016/j.addr.2011.10.007.

24. Mandal, A.; Bisht, R.; Pal, D.; Mitra, A. K. Diagnosis and Drug Delivery to the Brain: Novel Strategies. In: *Emerging Nanotechnologies for Diagnostics, Drug Delivery and Medical Devices.* 2017. https://doi.org/10.1016/B978-0-323-42978-8.00004-8.

25. Di Stefano, A.; Iannitelli, A.; Laserra, S.; Sozio, P. Drug Delivery Strategies for Alzheimer's Disease Treatment. *Expert Opinion on Drug Delivery.* 2011. https://doi.org/10.1517/17425247.2011.561311.

26. De Maio, F.; Palmieri, V.; De Spirito, M.; Delogu, G.; Papi, M. Carbon Nanomaterials: A New Way Against Tuberculosis. *Expert Review of Medical Devices.* 2019. https://doi.org/10.1080/17434440.2019.1671820.

27. Priyadarsini, S.; Mukherjee, S.; Mishra, M. Nanoparticles Used in Dentistry: A Review. *Journal of Oral Biology and Craniofacial Research*. 2018. https://doi.org/10.1016/j.jobcr.2017.12.004.

28. Ozak, S. T.; Ozkan, P. Nanotechnology and Dentistry. *European Journal of Dentistry*. 2013. https://doi.org/10.1055/s-0039-1699010.

29. Zarbin, M. A.; Montemagno, C.; Leary, J. F.; Ritch, R. Nanomedicine for the Treatment of Retinal and Optic Nerve Diseases. *Current Opinion in Pharmacology*. 2013. https://doi.org/10.1016/j.coph.2012.10.003.

30. Zhang, W.; Wang, Y.; Lee, B. T. K.; Liu, C.; Wei, G.; Lu, W. A Novel Nanoscale-Dispersed Eye Ointment for the Treatment of Dry Eye Disease. *Nanotechnology*. 2014. https://doi.org/10.1088/0957-4484/25/12/125101.

31. Xu, Q.; Kambhampati, S. P.; Kannan, R. M. Nanotechnology Approaches for Ocular Drug Delivery. In: *Middle East African Journal of Ophthalmology*. 2013. https://doi.org/10.4103/0974-9233.106384.

32. Leung, C.; Hodel, A. W.; Brennan, A. J.; Lukoyanova, N.; Tran, S.; House, C. M.; Kondos, S. C.; Whisstock, J. C.; Dunstone, M. A.; Trapani, J. A.; et al. Real-Time Visualization of Perforin Nanopore Assembly. *Nature Nanotechnology*. 2017. https://doi.org/10.1038/nnano.2016.303.

33. Sirelkhatim, A.; Mahmud, S.; Seeni, A.; Kaus, N. H. M.; Ann, L. C.; Bakhori, S. K. M.; Hasan, H.; Mohamad, D. Review on Zinc Oxide Nanoparticles: Antibacterial Activity and Toxicity Mechanism. *Nano-Micro Letters*. 2015. https://doi.org/10.1007/s40820-015-0040-x.

34. Arpagaus, C.; Collenberg, A.; Rütti, D.; Assadpour, E.; Jafari, S. M. Nano Spray Drying for Encapsulation of Pharmaceuticals. *International Journal of Pharmaceutics*. 2018. https://doi.org/10.1016/j.ijpharm.2018.05.037.

35. Fouda, M. M. G.; Abdel-Halim, E. S.; Al-Deyab, S. S. Antibacterial Modification of Cotton Using Nanotechnology. *Carbohydrate Polymers*. 2013. https://doi.org/10.1016/j.carbpol.2012.09.074.

36. Farjadian, F.; Ghasemi, A.; Gohari, O.; Roointan, A.; Karimi, M.; Hamblin, M. R. Nanopharmaceuticals and Nanomedicines Currently on the Market: Challenges and Opportunities. *Nanomedicine*. 2019. https://doi.org/10.2217/nnm-2018-0120.

37. Zhang, X. Q.; Xu, X.; Bertrand, N.; Pridgen, E.; Swami, A.; Farokhzad, O. C. Interactions of Nanomaterials and Biological Systems: Implications to Personalized Nanomedicine. *Advanced Drug Delivery Reviews*. 2012. https://doi.org/10.1016/j.addr.2012.08.005.

38. Caster, J. M.; Patel, A. N.; Zhang, T.; Wang, A. Investigational Nanomedicines in 2016: A Review of Nanotherapeutics Currently Undergoing Clinical Trials. *Wiley Interdisciplinary Reviews: Nanomedicine and Nanobiotechnology*. 2017. https://doi.org/10.1002/wnan.1416.

39. Adiseshaiah, P. P.; Hall, J. B.; McNeil, S. E. Nanomaterial Standards for Efficacy and Toxicity Assessment. *Wiley Interdisciplinary Reviews: Nanomedicine and Nanobiotechnology*. 2010. https://doi.org/10.1002/wnan.66.

40. Thakor, A. S.; Gambhir, S. S. Nanooncology: The Future of Cancer Diagnosis and Therapy. *CA: A Cancer Journal for Clinicians*. 2013. https://doi.org/10.3322/caac.21199.

41. Stephen, Z. R.; Kievit, F. M.; Zhang, M. Magnetite Nanoparticles for Medical MR Imaging. *Materials Today*. 2011. https://doi.org/10.1016/S1369-7021(11)70163-8.

42. Chinen, A. B.; Guan, C. M.; Ferrer, J. R.; Barnaby, S. N.; Merkel, T. J.; Mirkin, C. A. Nanoparticle Probes for the Detection of Cancer Biomarkers, Cells, and Tissues by Fluorescence. *Chemical Reviews*. 2015. https://doi.org/10.1021/acs.chemrev.5b00321.

43. Kim, M.; Lee, J. H.; Kim, S. E.; Kang, S. S.; Tae, G. Nanosized Ultrasound Enhanced-Contrast Agent for In Vivo Tumor Imaging via Intravenous Injection. *ACS Applied Materials and Interfaces*. 2016. https://doi.org/10.1021/acsami.6b02115.

44. Jain, M.; Koren, S.; Miga, K. H.; Quick, J.; Rand, A. C.; Sasani, T. A.; Tyson, J. R.; Beggs, A. D.; Dilthey, A. T.; Fiddes, I. T.; et al. Nanopore Sequencing and Assembly of a Human Genome with Ultra-Long Reads. *Nature Biotechnology.* 2018. https://doi.org /10.1038/nbt.4060.

45. Ma, G.; Wu, C. Microneedle, Bio-Microneedle and Bio-Inspired Microneedle: A Review. *Journal of Controlled Release.* 2017. https://doi.org/10.1016/j.jconrel.2017.02 .011.

46. Sutradhar, K. B.; Sumi, C. D. Implantable Microchip: The Futuristic Controlled Drug Delivery System. *Drug Delivery.* 2016. https://doi.org/10.3109/10717544.2014.903579.

47. Hasnain, M. S.; Nayak, A. K. Background: Carbon Nanotubes for Targeted Drug Delivery. In: *Springer Briefs in Applied Sciences and Technology.* 2019. https://doi.org /10.1007/978-981-15-0910-0_1.

48. Alagarasi, A. Application of Quantum Mechanics to Nanomaterial Structures. In: *Introduction to Nanomaterials. National Centre for Catalysis Research.* 2011. John Wiley & Sons, Inc. 68–134. https://doi.org/10.1002/9781118148419.

5 Utility of Nanotechnology in Various Disciplines

Shipra, Priyanka and Vibha Aggarwal
Punjabi University Neighborhood
Campus, Rampura Phul, India

CONTENTS

5.1   INTRODUCTION

Nanotechnology is the science of the nanoscale: objects around a nanometer in size. Nanotechnology is commonly considered to deal with particles in the size range <100 nm, and with the nonmaterial manufactured using nanoparticles. Nanotechnology was being used a very long time before the field was officially defined. James Clark Maxwell (Scottish physicist and mathematician, 1831–1879) and Richard Adolf Zsigmondy (Austrian-German scientific expert, 1865–1929) were early supporters of the field. Other significant patrons in the primary portion of the twentieth century include Irvin Langmuir (American scientific expert and physicist, 1881–1957) and Katherine B. Blodgett (American physicist, 1898–1910). A speech titled "There's Plenty of Room at the Bottom", given by Richard Feynman (American physicist, 1918–1988) in 1959, is considered to be the earliest discussion on this topic. Feynman talked about the significance "of controlling things from a more minor perspective" and how they could "disclose to us quite a bit of incredible enthusiasm about the abnormal wonders that happen in complex circumstances". The term "nanotechnology" was utilized first by the Japanese researcher Norio Taniguchi (1912–1999) in a 1974 paper on generation innovation that highlights the request for a nanometer. The American specialist K. Eric Drexler (b. 1955) is credited with the advancement of atomic nanotechnology, prompting nanosystem machinery manufacturing. The creation of a filtering burrowing magnifying lens during the 1980s by IBM Zurich researchers, and afterward the nuclear power magnifying instrument enabled researchers to see materials at a remarkable nuclear level. The accessibility of an ever-increasing number of ground-breaking PCs around this time empowered enormous scale reenactments of material frameworks utilizing supercomputers. These examinations gave insight into nanoscale material structures and their properties. In the late 1990s and mid-2000s, practically all industrialized countries undertook nanotechnology activity, prompting an overall expansion of nanotechnology exercises. Over the last few decades, nanotechnology has made tremendous progress. There is a developing acknowledgment of the significance of instructing future researchers and scientists in this rising field (Aithal and Aithal, 2015). Globally, nanotechnology research and development are highly active, and hundreds of products are already using nanotechnology devices.

5.2   APPLICATIONS OF NANOTECHNOLOGY

Materials that display diverse physical properties have come about because changes at the nanoscale have opened the entryway to numerous new applications. Nanotechnology has developed from the research laboratory and is now used in the design and production of many industrial products and systems. New nanotechnologies will tackle both human fundamental needs and comfort needs. Food, water, energy, clothing, shelter, health, and the environment are the basic needs of the human being and automation in all fields, space travel and increased lifespan, etc. are comfort needs. A significant number of these applications are still in different phases of research, yet some are as of now accessible industrially. Nanotechnology has the potential to bring about significant benefits, such as improved health, better use of natural resources, and reduced environmental pollution (Abhilash, 2010; Chaturvedi and Dave, 2014).

5.2.1   ENVIRONMENT

The high price of oil, concerns about the environment as a consequence of increased greenhouse pollution, the will to protect the planet against environmental disasters, and the demand to develop the power effectiveness of the systems that are being used now have paid close attention to the alternative energy sources. Soil and water contamination, including the effects of toxic waste in waste disposal areas, is a major concern worldwide. Scientists are developing new catalysts with the use of nanotechnology for waste management and the conversion to healthy components of hazardous gaseous material. These designs benefit from the large surface area of nanomaterials, and the new characteristics and reactivity of the nanoscale. Carbon nanotubes and other nanomaterials are being used to develop new effective filters to capture arsenic, heavy metals, and other hazardous substances. The use of nanotechnology can reduce air emissions is two important ways: catalysts that are actually in use and are continually developed, and nanostructured membranes that are under development. Catalysts used to transform vapors that escape from cars or industrial plants into harmless gasses may improve performance and costs with the use of nanotechnology. This is due to the fact that nanoparticle catalysts have a bigger surface than catalysts from larger particles, so they interact with the reacting chemicals. The wider area allows more chemicals to potentially interact with the catalyst, which makes the catalyst more productive (Diallo et al., 2011). Nanotechnology is being used to develop solutions to three very different water quality issues. One challenge is the elimination of industrial water pollution, such as TCE, a cleaning solvent, from groundwater. To make the contaminating chemical harmless, nanoparticles can be used to convert it through a chemical reaction. Studies have proven that this method can be used to reach contaminants scattered through underground pools at a far lower cost than processes that require water to be pumped from the ground in order to be treated. The separation of salt or metals from water is another problem. A method of deionization using electrodes made up of nano-sized fiber allows the conversion of salt water into drinking water at a reduced cost and with lower energy requirements. The third issue is that standard filters are not working on virus cells. A filter with a diameter of a few nanometers that should be able to remove virus cells from the water is currently being created (Zhang and Elliot, 2006; Brame et al., 2011; Rabban et al., 2016).

5.2.2 MEDICAL

Nanotechnology should be able to positively enhance progress made in medical services in terms of early detection, precise malignancy diagnostics, viable immunizations, enhanced medication conveyance, propelled inserts, and different applications. Biosensors have emerged which allow for the early detection of several life-threatening diseases using a mixture of nanomaterials, new hardware manufacturing techniques, and advances in signal processing. Such instruments use carbon nanotubes or silicone nanowires that can be used to recognize the signature of a specific disease. It is foreseen that nano-biosensors utilizing this methodology will be mass-delivered with PC section systems. The role of nanotechnology in therapeutics will also be important. The development of advanced drugs using nanotechnology concepts and selective drug delivery are two fields where nanotechnology is supposed to have an impact. People in need of artificial components of their bodies – legs, limbs, ligaments, or organs – can expect to use nanomaterials, better composite nanoparticles, and other mechanically appropriate nanomaterials to be more reliable and rejection-proof. Some of the required features provide improved electrical (and other force) reactions. This would help develop durable artificial materials which are long-lasting. The development of contrasting agents for cell imaging is part of medical nanotechnology applications. This agent allows cell imaging and facilitates biomedical research (detection and diagnosis of diseases), and medical diagnostics. There are also preparations for the use of a nanorobot for assisting with the repair of defective/damaged parts within the human body that may be detected by nanotechnology (Salata, 2004; Gupta et al., 2012; George, 2015).

5.2.3 Consumer Goods Applications

Nutritious foods are one of the fundamental needs of humanity in this world. People around the world are looking for nutritious foods because of various reasons. Agricultural nanotechnology innovation is expected to solve food-sector problems and to maximize agricultural productivity. The need for clean, affordable drinking water can be helped by nanotechnologies, by rapid low-cost impurity detection and water purification. Nanopores (small pores in an electrically insulating membrane which can be used as a single molecular detector) have improved freshwater filtration through the application of nanotechnology. Nanopores have been produced so small that the smallest contaminant can be removed. These water purification devices use UV chemicals to prevent toxins like pesticides, solvents, and bacteria from being radiated in the water flowing through them. Nanopores can be produced precisely so that the system is filtered effectively with minimal influence on the flow rate.

In uses such as bullet-proof shoes, tennis rackets and golfing, cricket and bowling, nanotechnology is used for splitting tougher and lighter fabrics. Nanoscale titanium dioxide and zinc oxide have been used in sunscreen for years, to protect the skin against the sun.

5.2.4 Communication and Information System

Nanotechnology contributes to information and communication by introducing new data storage methods which enable the design of new semiconductors and optoelectronic devices, along with integrated circuits. Space is utilized by PCs and processors to store data and execute the necessary functions. Every memory bit contains a parallel worth and a few arrangements of bits are deciphered as a guidance or snippet of data. Digital devices are increasingly simpler and smaller, requiring more lightweight parts. The creation of complex devices of extremely small size is facilitated by different types of memory devices developed by nanotechnology. Nanotechnology has made for much advancement in computer memory, increased capacity, lower power consumption, and higher speed (Hullmann, 2007). Nano Random Access Memory (NRAM) utilizes carbon nanotubes to assess the condition of the memory component, including data bit. NRAM – a patented computer memory technology – is predicted to be used in a wide variety of applications at a low cost. Another type of non-volatile storage which benefits from nanotech properties is Ferroelectric-RAM and FRAM. FRAM is like ordinary incorporated circuit memory; nevertheless, the gadget is worked from a dielectric, instead of a ferroelectric polymer layer. A ferroelectric material consists of molecules which have an inborn electrical polarization. The natural polarization of the rail-electric matter allows FRAM memory cells to absorb fewer powers and can be designed for smaller sizes and so replace traditional dielectrics with ferro-electric material. Millipede memory is a third type of memory improvement using nanotechnology. It is designed to replace memory magnets, such as hard drives. In order to store details, millipede memory uses several small impressions in a polymer band. Millipede memory utilizes nuclear power sensors to recuperate memory content, which identify the nanospaces caught in film.

5.2.5 Military

In the defense sector, there are clear expectations with regard to nanotechnology. Indeed, the next major field that will soon find multiple applications in the military field is nanotechnology. The entirety of the world's significant military forces is vigorously engaged in innovative work with the use of nanotechnologically implanted materials and frameworks. Currently, nanotechnological research centers around improving medicinal offices and delivering lightweight, solid, and multi-utilitarian materials that encourage both assurance and upgraded availability in a network-driven combat area. A significant strategic benefit over the enemy would be provided by all miniaturized military systems. Soldiers can use nanotechnology in two main ways. The first is the miniaturization of existing devices to make them less energy-efficient and harder to hide, not only smaller but also lighter. Furthermore, new materials should be created and modified for military purposes. A lot of heavy equipment needs to be carried by soldiers, and so their uniforms are not bullet-proof. In the production of nano-battle suits, several nanotechnology R&D teams are exclusively concerned. The battle suits can be as perfect as a stretchy polyurethane layer which includes medical instruments and communication devices. Normal body motions can generate energy for communication. This material would also provide much better strength than current materials and provide effective bullet protection. Nanotechnology allows smaller sensors to be used in different segments. Including these nanosensors with neural networks, for example, can help detect and track extremely small concentrations of airborne chemicals. A selection of these sensors can help to determine the nature and extent of the possible risk when explosive agents are identified at the frontier forces on the front line. Nano-drones, like any mobile device, have cameras, sensors, and facial recognition. Military nano-drones might also carry tiny enough grams of explosive to penetrate the brain and kill the skull. These nano-drones would make surgical accuracy easier. If equipped as a squad, these nano-drones can infiltrate houses, vehicles, ships, evade humans, bullets, and almost any countermeasure, which is deadly enough to destroy. The medication, tension, and release of medicinal medications and hormones when required would also be a form of application. This application includes connecting these devices to the brain cortex or nervous organs, auditory receptors, motor nerves, or muscles to decrease the soldiers' reaction time.

5.2.6 Business and Industry

Due to high priority investment in research in many countries, nanotechnological products are diversifying and growing exponentially. Nanotechnology influences companies, which aims to deliver new and improved technologies and processes and helps enter the new markets with creativity. Nanoparticles, including agriculture, foodstuffs, energy, electronics, pharmaceuticals, and chemical and biomedical materials, are used in several industries. Many industries produce a higher income for nanoparticles; for example, chemical-mechanical polishing, electrically conductive coatings and optical fibers, sun panel coats, magnetic tapes, screens, car catalyst supports, biolabeling, etc. Nanotechnological products and services are expected to have a huge effect on most sectors in the coming years and reach large quantities of the consumer market (Maine et al., 2014).

5.2.7 TEXTILES

For the textile industry, nanotechnology also has real business opportunities. This is primarily because conventional methods used to deliver different textile properties to consumers often have no lasting effect, and are losing their function after washing or wearing. Nanotechnology can provide high resistance to fabrics because of the wide area-to-volume ratio of nanoparticles and high surface energy, which can thus improve fabric affinity and increase the lifespan of the function. Furthermore, a nanoparticle coating does not affect your respiratory ability or hand sensation. Nanotechnology is used to manufacture materials with improved proprieties such as resistance to rust, soil, and wind. The material used for these applications includes small, fiber-like structures or nano-whiskers. The droplets are bigger than the spacing between whisks, which is why they can be brushed or washed on top of the fabric. In addition, nanotechnology for improved textiles is used in anti-static fabrics, the controlled release of perfumes, anti-microbials and antifungals, fabrics that increase skin moisture, and transparent textiles. The creation of smart and interactive textiles (SMIT) that detect electrical, thermal, chemical, magnetic, or other stimulation is to be used by nanotechnology in the future.

5.2.8 SPACE

Nanotechnology can be useful in the construction of innovatory light-weighted radiation-resistant sensors, devices, and integrated circuits. In future space missions, nanotechnology will be significant. A few examples are nanosensors, significantly improved high-performance structures, or highly effective propulsion systems. Nanotechnology can be the key to enhancing spaceflight. The progress made in materials for lightweight solar sails and the space-lift cable will greatly reduce the cost of entering and flying in orbit. New materials as well as nanorobots and nanosensors can also enhance the performance of astronauts, spacecraft, and tools for exploring planets and moons. The field of radiation protection can make a most important input to human space flights with the use of nanotechnology. According to NASA, the dangers of regional radiation exposure are an important factor affecting people's capacity to engage in long-term space missions. Therefore, much research is focused on developing anti-astronaut procedures against these dangers. Spacecraft manufacturers are seeking materials to facilitate the construction of multifunctional spacecraft hulls. Advanced nanomaterials like the newly developed and isotopically enriched boron nanotubes could open up the path to future spacecraft with integrated nanosensor hulls which provide effective radiation protection and energy storage. Many rocket motors today focus on chemical propulsion. Every existing spacecraft uses some type of chemical rocket and most use it to control its behavior. Nevertheless, rocket engineers are actively exploring new forms of space propulsion systems. Electric propulsion (EP), which includes field emission electrical propulsion (FEEP), colloid thrusters, and other field emission thruster (FET) versions, is a heavily researched area. In contrast with conventional chemical rockets,

EP systems significantly reduce the required propellant weight, which can improve payload capacity or minimize the starting weight. A principle of nanotechnology, along with the employment of EP, suggests the use of propellant nanoparticles with a static charge and acceleration. The development of extremely scalable thruster array will work in millions of micron-size nanoparticular thrusters on one square centimeter. Carbon nanotubes are used to make lightweight solar sails using sunlight strain on the spacecraft's mirror-like solar cell, solving the problem of providing enough fuel to launch into orbit and driving spaceships during interplanetary missions. Nanotechnology can help to install a network of nanosensors to find traces of water or other contaminants in large areas of planets such as Mars.

5.2.9 Fuel and Energy

Energy is a basic tool for human survival, along with food and water. Energy demand in the world is expected to grow by 50 percent by 2025 with fossil fuels producing the majority of energy. A major challenge is matching human energy consumption with environmental cost. Innovations in nanotechnology of renewable energy meet human beings' entire energy required for their basic needs and comfort. Fuel scarcity such as diesel and gasoline can be resolved by nanotechnology by making low-garden raw material fuel production economical, improving engine miles, and making standard raw-material fuel production more efficient (Knell, 2011). Incandescent light bulbs are one remarkable result of this. Such lamps, manufactured at the end of the 19th century, are slowly being replaced with devices with more perceptible illumination at the same electrical input level. Fluorescent lamps, high-intensity discharge lamps, and LEDs are alternative sources of light. Innovations in nanotechnology are intensively implemented to reduce the costs of manufacturing such alternative light bulb designs. While developing other alternative energy systems, nanomaterials are starting to play a prominent role. Many of today's worldwide solar cells rely on bulk crystalline technology that competes with the silicon raw material computer industry. This is not a desirable situation. Alternative work recommendations cover emerging nanomaterials such as performance improvements with the use of quantum dots and transparent substrate processing of solar cells (e.g. rubber, thin sheets of material). Several different types of batteries already use nanotechnology that renders them less flammable, faster charging, more powerful, lighter, and with a higher power density, as well as having more electric loads (Sabet et al., 2016). The use of nanocomposites to use wind power is built with a lightweight and high-strength composite for wind turbine blades. Nano-based science options are being sought in order to turn dissipate heat into usable electrical power in automobiles, homes, computers, and power plants. Scientists are designing thin-film solar panels with machine cases and lightweight piezoelectric nanofilms woven to clothes to generate energy from friction, wind and body temperature to power mobile electronic devices (Rabbani et al., 2016; Pratsinis, 2006).

5.2.10 Transportation

Personal mobility is a fundamental necessity for people and a fundamental precondition of contemporary society. Concerns of passenger safety, adaptive traffic control systems, pollutant mitigation, and productive recycling at the end of the value-added chain are becoming highly important to ensure the conservation of scarce resources. Nanotechnology contributes significantly to the production of the novel automobile, aerospace, and water transport materials and processes. In the automotive industry, we rely on most research and development based on nanotechnology more frequently than air or water. Body parts, emissions, frames, and pneumatic structures as well as car interiors, electrical sensors, engines, and drive trains are subject to nanotechnology. Many companies rely on air transport because they are capable of transporting large amounts of goods and people all over the world in quick times that can only be imagined for other transport modes. Aerospace needs high levels of safety and perfection as small production or operational defects put many lives at risk and therefore heavy investment and care are taken in these industries. Aircraft materials are supposed to have high output power, high tensile resistance, and low-density corrosion resistance. Nanotechnology uses in the aerospace industry include: low weight nanocomposites, nano high-strength and enhanced electronics and displays with lower power consumption, multifunctional materials with sensors, sophisticated air purification filters and membranes, and various other applications. The function of marine shipping includes passenger journeys, platforms for arms transport, rail carrier, and many more. The biggest problem in marine transport is sea water and air degradation. The maritime microbial fouling and corrosion of the board and waterline due to long periods spent in the water are also of concern in the field of water transport. There are many innovations which aim to avoid corrosion and fouling, but they do not seem to be very successful. Exploration into the use of nanotechnology still offers us positive results in this area.

5.3 SUMMARY

Nanotechnology has significantly improved many industries and technology sectors, including IT, environmental science, transportation, medicine, and home and food security. Nanotechnology has caused enormous changes to its outlook in the transport industry. By using nanotechnology, transports are made more efficient, intelligent, stronger, and more sustainable. It can be the gateway to a new world in terms of agriculture, food, building materials, electrical engineering, medical engineering, and mechanical engineering. While the natural system's replication is one of the technology's most exciting areas of expertise, scientists remain committed to understanding their enormous complexities. Through solving any problem in society, such as food, electricity, drinking water, health, and the environment, nanotechnology is going to be a disruptive form of innovation that will ultimately contribute to the longevity of human life.

REFERENCES

Abhilash, M. 2010. Potential applications of nanoparticles. *International Journal of Pharmacy and Biological Sciences* 1:1–12.
Aithal, P.S. and S. Aithal. 2015. Ideal technology concept and its realization opportunity using nanotechnology. *International Journal of Application or Innovation in Engineering & Management* 4(2):153–164.

Brame, Jonathon, Qilin Li and Pedro J.J. Alvarez. 2011. Nanotechnology-enabled water treatment and reuse: Emerging opportunities and challenges for developing countries. *Trends in Food Science and Technology* 22(11):618–624.

Chaturvedi, S. and P.N. Dave. 2014. Emerging applications of nanoscience. *Materials Science Forum*, 781:25–32.

Diallo, Mamadou and C. Jeffrey Brinker. 2011. Nanotechnology for sustainability: Environment, water, food, minerals, and climate. In: Roco, M.C., Mirkin, C.A. and Hersam, M.C. (ed.) *Nanotechnology Research Directions for Societal Needs in 2020*, Springer, Dordrecht, Netherlands, 221–259.

George, S. 2015. Nanomaterial properties: Implications for safe medical applications of nanotechnology. In: Kishen A (ed.) *Nanotechnology in Endodontics*, Springer, Switzerland, 45–69.

Gupta, A., A. Arora, A. Menakshi, A. Sehgal and R. Sehgal. 2012. Nanotechnology and its applications in drug delivery: A review. *Webmedcentral: International Journal of Medicine and Molecular Medicine* 3(1):2867.

https://www.nanotechmag.com/nanotech-in-the-army/.

https://www.understandingnano.com/nanotech-applications.html .

Hullmann, A. 2007. Measuring and assessing the development of nanotechnology. *Scientometrics* 70(3):739–758.

Knell, M. 2011. Nanotechnology and the sixth technological revolution. In: Cozzens, S., Wetmore, J. (ed.) *Nanotechnology and the Challenges of Equity, Equality and Development*, Springer, Netherlands, 127–143.

Maine, E., V. Thomas, M. Bliemel, A. Murira and J. Utterback. 2014. The emergence of the nanobiotechnology industry. *Nature Nanotechnology* 9(1):12–15.

Pratsinis, S.E. 2006. Overview-nanoparticulate dry (flame) synthesis and applications. *TechConnect Briefs*, 1:301–307.

Rabbani, M.M., I. Ahmed and S.J. Park. 2016. Application of Nanotechnology to Remediate Contaminated Soils. In: Hasegawa, H., Rahman, I.S.M., Rahman, M.A. (ed.) *Environmental Remediation Technologies for Metal-Contaminated Soils*, Springer, Japan, 219–229.

Sabet, M., S. Hosseini, A. Zamani, Z. Hosseini and H. Soleimani. 2016. Application of nanotechnology for enhanced oil recovery: A review. *Defect & Diffusion Forum 367*, 149–156.

Salata, O.V. 2004. Applications of nanoparticles in biology and medicine. *Journal of NANO Biotechnology* 2(3):16–22.

Zhang, W.X. and D.W. Elliot. 2006. Applications of iron nanoparticles for groundwater remediation. *Remediation Journal* 16(2):402–411.

6 Renewable Energy through Nanotechnology

Nada W, Dania S, Sharon Santhosh,
and Asha Anish Madhavan
Amity University, Dubai, United Arab Emirates

CONTENTS

6.1 INTRODUCTION

Renewable energy technologies are future contenders for a sustainable generation of energy. The reason for current constraints in renewable energy is the intermittent nature of these resources and their availability [1]. The use of conventional energy sources has led to its rapid depletion despite paving the way for economic progress. Environmental issues corresponding to the use of conventional resources have led researchers to investigate and develop new and efficient technologies that are renewable and environmentally friendly [2]. This has seen a reduction in the cost of fuels and a shift to alternative sources of energy that will enable us to not only reduce greenhouse gas emissions but also ensure a cost-efficient, reliable and well-timed energy supply.

6.2 RENEWABLE ENERGY

Renewable energy sources have the ability to be replenished continuously – this is considered its major advantage. These sources of energy are either directly derived from the sun, such as photoelectric, photochemical and thermal sources or indirectly received from the sun through means of hydropower, wind and biomass. The other mechanism through which energy is derived is through movements in the environment such as tidal and geothermal energy. Renewable energy does not include organic and inorganic waste, fossil fuels or fossil waste [3]. The various renewable energy sources are illustrated in Figure 6.1 [4, 5].

6.3 TYPES OF RENEWABLE ENERGY SOURCES

6.3.1 BIOMASS

Biomass functions similarly to fossil fuels but they are involved in the conversion of organic materials originating from trees, plants and crops into useful energy such as heat, biofuels and electricity. This type of energy utilizes the waste-to-energy conversion, which deals with the elimination of a large quantity of solid waste to a useful product. It is widely available and is a carbon-neutral energy source. Another advantage of biomass is that it can be directly burnt to harness energy or it can be converted to different types of biofuels by serving them as feedstock. It is easy to transport and store biofuels, which makes them an essential

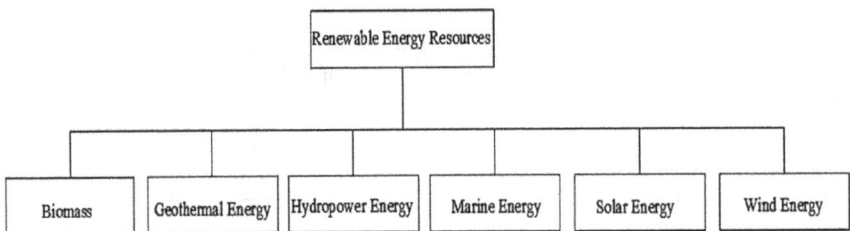

FIGURE 6.1 An outline of the various renewable energy sources.

material that is mixed with other energy production, from sources such as wind. They can generate power and heat on demand through the gasification or combustion of biomass under controlled circumstances, which ensures the reduction in the emission of greenhouse gases whilst maintaining a cost-effective production of energy.

6.3.2 Geothermal Energy

These power plants capture the heat stored in the earth to generate steam and produce electricity. This thermal energy, which is utilized, is stored in the interiors of the earth in the form of steam or liquid water in the rocks. Tectonically active areas have the greatest and more powerful geothermal sources, which are associated with high temperature systems, i.e. more than 180°C. It is also viable in intermediate temperature systems in continental areas with a temperature range of 80 180°C. It is cheaper than other renewables and has a continuous generation of energy at a relatively lower fuel cost [6].

6.3.3 Hydropower Energy

Hydropower is the energy that is derived from flowing water, making good use of the elevation difference along rivers and other water bodies. It also benefits from the gravitational potential energy that comes from the natural gradient of river slopes. This type of energy is prevalent in Canada, Brazil, China and the U.S. due to the higher number of rivers in these areas, thereby making it very site-specific [7]. The most dominant type of hydro energy is dams, which are sourced by solar radiation.

6.3.4 Marine Energy

Marine energy deals with hydrokinetic energy, which includes three types of energy: wave, tidal and current energy [8]. Wave energy is harnessed from shore-based areas whereas waves are created from wind blowing over the waters. Tidal energy is obtained from the flow of tides due to the gravity of the moon and current energy is obtained from moving ocean/river currents.

6.3.5 Solar Energy

Solar energy is harvested directly from the heat emissions from the sun. This type of energy source can be active or passive. Active solar entails utilization for cooking, water heating and generating electricity. Passive solar entails design of buildings to maximize solar insolation, planning agricultural plots for optimum energy captivation. Solar sources are dependent on the location and are variable and unpredictable depending on the weather and cloud formation. Efficiency has been slowly but steadily improving with improving technology. This source requires coupling with storage since peak demand of energy occurs in the afternoon. Nevertheless, it requires low maintenance and costs. In addition, it is important since there is no greenhouse gas emission after production of the solar panels, and it is highly accessible since the commercial solar photovoltaic (PV) facilities tend to be modular. Globally, this source is massive and inexhaustible as long as the sun lasts.

6.3.6 WIND ENERGY

This is one of the most prominent types of energy source that uses wind to spin a turbine and generate electricity. The rotational motion from the turbine generates AC current. The wind power needed through extraction of kinetic energy exponentially increases with an increase in wind speed.

$$P_{wind} = 0.5 \times \rho \times A \times V^3 \tag{6.1}$$

Where ρ is the density of moving air, **A** is area covered by turbine rotor and **V** is the velocity of wind in meters/second. Wind energy is highly dependent able on the location and hence output is unpredictable. It is more efficient at lower average wind speeds as it needs lower initiation energy to get started. Nevertheless, it has low maintenance with very limited safety risks. Its distribution of energy makes this an easily scalable source. Wind is a significant contributor to the national electricity in many countries like Germany, Denmark, the U.S and China. It is prevalent in China, Germany, India, France and Italy [9]. A comparison of different renewable energy resources is summarized in Table 6.1.

6.4 RENEWABLE ENERGY CURRENT SCENARIO

With the advent and growth of photovoltaics and wind in recent years, electricity is the most significant part of the renewable energy sector. Nevertheless, electricity contributes only about a fifth of the energy consumption worldwide and energy transition is heavily affected by the role of renewables in the heating and transport sectors. Bioenergy, being the largest contributor to renewable energy, is analyzed through market assessments globally. Consumption of renewable energy technologies is shown in Figure 6.2.

The renewable energy sector is expected to grow by 1/5th and reach 12.4% in 2023. The fastest growing sector in renewables is electricity, which is expected to provide 30% of power by 2023 as compared to 24% from 2017. Solar PV, wind, hydropower and bioenergy are the fastest growing renewable technologies in order of succession and are predicted to fulfill 70% of the global energy growth. Hydropower contributes to 16% of electricity demand globally and is the largest source of renewable energy followed by wind, which is 6%, solar PV, which is 4% and bioenergy, which is 3% [10].

6.5 ROLE OF NANOTECHNOLOGY IN RENEWABLE ENERGY

In order to control and expand the properties of materials, scientists have been developing nanoscale materials, which have a physical size ranging between 1 nm and 100 nm. Nanomaterials may exist in different forms like nanoparticles, nanofilms, nanotubes, nanofibers, nanorods and nanoflowers. As the material size decreases,

TABLE 6.1

Comparison Chart of Different Renewable Energy Resources [2]

Energy Source	Advantages	Disadvantages	Negative Impact on the Environment
Biomass Energy	Used to burn waste products, abundant.	Air pollution, therefore not cost-effective.	Emission of greenhouse gases such as methane during biofuel production. Soil productivity deterioration may occur, and hazardous waste is generated.
Geothermal Energy	No air pollution and unlimited energy supply.	Expensive and location-dependent. Constant maintenance required.	Change in landscape, pollution of waterways.
Hydropower Energy	Abundant, easy storage, inexpensive and clean.	May lead to flooding and can be used only where water bodies are present	Results in alteration in the ecosystems, cultural impacts and affects weather.
Marine Energy	Energy is collected that would otherwise be overlooked.	Occupies large regions and is expensive.	Decrease in motion of water. Marine life is affected, and sea patterns may change.
Solar Energy	Renewable and never-ending energy supply. No pollution or harmful gas emissions.	Requires good storage and backup. It is not always cost- effective or reliable, as it depends on sunshine availability.	Harmful resources
Wind Energy	Free energy source. No harmful emission of gases. Inexpensive building of farms.	Requires constant wind loads. Occupies a large portion of land; is very site-specific.	Harming of birds and other wildlife by rotor blades.

Reprinted from Omar Ellabban, Haitham Abu-Rub, Frede Blaabjerg, "Renewable energy resources: Current status, future prospects and their enabling technology", Renewable and Sustainable Energy Reviews, 39, pp. 748–764, 2020; with permission from Elsevier.

the surface-to-volume ratio increases [11, 12]. The atoms on the surface may contain one or more dangling bonds, which form bonds with neighboring molecules to reduce the total energy and stabilize the material. This leads to enhanced chemical stability, high solubility and a low melting point when compared to their bulk counterparts [12]. Intrinsic properties like electronic band structures such as the Bohr radius in semiconductors also are altered. However, due to confinement, there may occur a quantum tunneling effect, which causes the semiconductor bands to be discrete or quantized. This will inherently lead to changes in the optical properties of

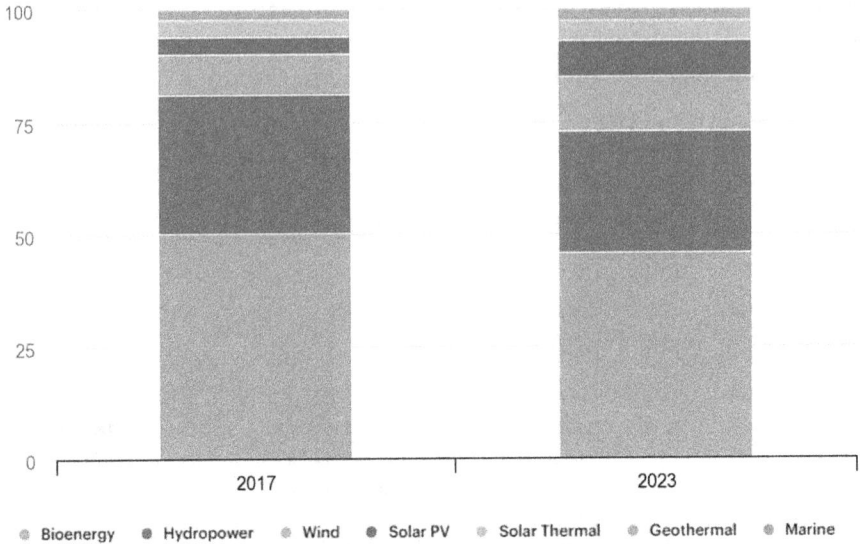

FIGURE 6.2   Consumption of renewable energy technologies, 2017 compared to 2023 [9].

the semiconductors, which can be modified as per the applications. The fundamental properties of nanomaterials, such as large surface-to-volume ratio, high surface area, stability and short migration distance make it suitable for versatile applications. However, significant side reactions and agglomeration of the material can occur in the material due to the large surface area and large surface energy of nanomaterials. There are also issues with homogeneity control but recent advances in technology such as inkjet printing and roll-to-roll manufacturing have helped combat the limitations significantly.

6.6   HARVESTING SOLAR ENERGY WITH THE APPLICATION OF NANOTECHNOLOGY

As we know, the sun is continuously irradiating the earth with an intensity of 1.2×10^5 TW, whereas the current worldwide energy consumption is just 12 TW which is only 0.001% of the energy that we receive from the sun [13, 14]. But at the same time, in the renewable energy sector, solar PV contributes only 0.5% of total volume. Silicon solar panels have shown a power conversion efficiency of ~25% whereas the maximum efficiency calculated for a silicon solar cell is 33.7% under AM 1.57. Though the efficiency of silicon solar cells is good, the manufacturing cost is very high. Further, there are environmental hazards for the processing of silicon [15]. Nanotechnology has played a major role in improving cell efficiency, energy conversion and conservation. This can be obtained via modifying light interacting semiconductors, tweaking photo catalyst and many other methods. A comprehensive discussion of nanotechnology-modified solar cells is represented in the following subsections.

6.6.1 Dye-Sensitized Solar Cell (DSSC)

Dye-sensitized solar cells, which belong to the third-generation solar cells, are completely powered by the nanomaterials. DSSC is a promising recently developing technology for low-cost, high-efficiency solar cells [16]. It is highly flexible and environmentally friendly. Although the photovoltaic energy conversion efficiencies recorded for DSCs are lower than those measured for silicon-based solar cells, their efficiency can be improved with the use of nanotechnology. The significant impression of this technology is mirroring natural photosynthesis in a low-cost manner and by environmentally friendly technology.

6.6.2 Main Components of DSSC

Substrate is made up of transparent conducting oxide (TCO) that is coated on a glass substrate. Its main purpose is to collect a current, support the structure and act as a sealing layer to the outside environment. The most commonly used TCOs are FTO or ITO. A wide variety of large band gap semiconductor oxides, such as ZnO, Nb_2O, TiO_2 and SnO_2, are used for the fabrication of working electrodes in DSSC [17]. The most commonly used metal oxide is TiO_2 because of its advantages like thermal and chemical stability, low cost, abundance, non-toxicity and many more. The TiO_2 electrode is used mainly for the key processes of light absorption, light scattering and charge transport and for the suppression of charge recombination. Dyes are used as sensitizers, which help to provide the required energy to excite electrons from the valence band to the conduction band. TiO_2 ensures that it falls in the UV spectrum since it has a bandgap value of 3.2 eV. UV region also only comprises 5% of the entire solar spectrum. The main properties of an efficient sensitizer or dye are chemically inert and stably irreversible and strong chemisorption with TiO_2, higher reduction potential, higher molar extinction coefficient, higher absorption and lower oxidation potential. Electrolytes are mainly composed of redox couples, blocking agents, organic solvents and ionic liquids. They play an important role in determining the effective cell efficiency by functioning as regeneration of oxidized dye. Acetonitrile, propylene and ethylene carbonate are some of the non-protonic solvents used. The purpose of counter electrodes is to collect electrons from external circuit and to aid the reduction of the redox electrolyte. Counter electrodes display efficient electrochemical activity, which improves cell performance. Carbon-based counter electrodes have been developed, such as CNTs and conducting polymers like PEDOT but they are still not comparable with platinum. An illustration of DSSC is given in Figure 6.3.

6.6.3 Architecture and Working Principle

Three processes characterize DSSCs: photon absorption, charge separation and electron transport. DSSCs comprise a working electrode, which is made of conducting substrate material with a thin nanostructural mesoporous semiconductor layer like ZnO or TiO_2; and a counter electrode usually made of platinized FTO. Dyes, which are monolayer in nature, are adsorbed on the surface of the semiconductor to

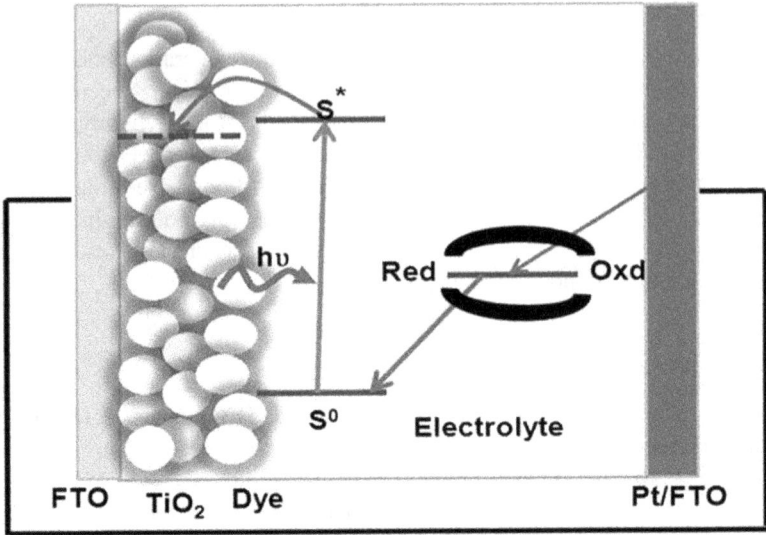

FIGURE 6.3 Illustration of DSSC.

improve sensitization. Electrolytes are usually redox couples such as I⁻/I³⁻. Electrons are injected into the conduction band (CB) of semiconductor due to the excitation of dye on illumination. The oxidized dye is regenerated by I⁻ in the electrolyte. Likewise, I_3^- is reduced which leads to the regeneration of I⁻ is regenerated at the counter electrode with electrons from the external circuit [18–20]. The key reactions are:

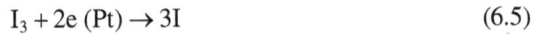

$$S + h\upsilon \rightarrow S \tag{6.2}$$

$$S^* \rightarrow S^+ + e\,(TiO_2) \tag{6.3}$$

$$2S^+ + 3I \rightarrow 2S + I_3 \tag{6.4}$$

$$I_3 + 2e\,(Pt) \rightarrow 3I \tag{6.5}$$

Since there is no permanent chemical transformation, DSSC is an example of a regenerative type photo electrochemical cell [21]. The solar cell IV curve is shown in Figure 6.4.

Quantitative valuations of the performance of solar cells are based on the overall light-to-electricity conversion efficiency (η). The product of photocurrent density (Jsc), the open-circuit photo voltage (Voc), the fill factor (FF) and the intensity of the incident light (Inc) provide the effective efficiency of the DSSCs.

$$\eta = \frac{Voc \times Jsc \times FF}{Inc} \tag{6.6}$$

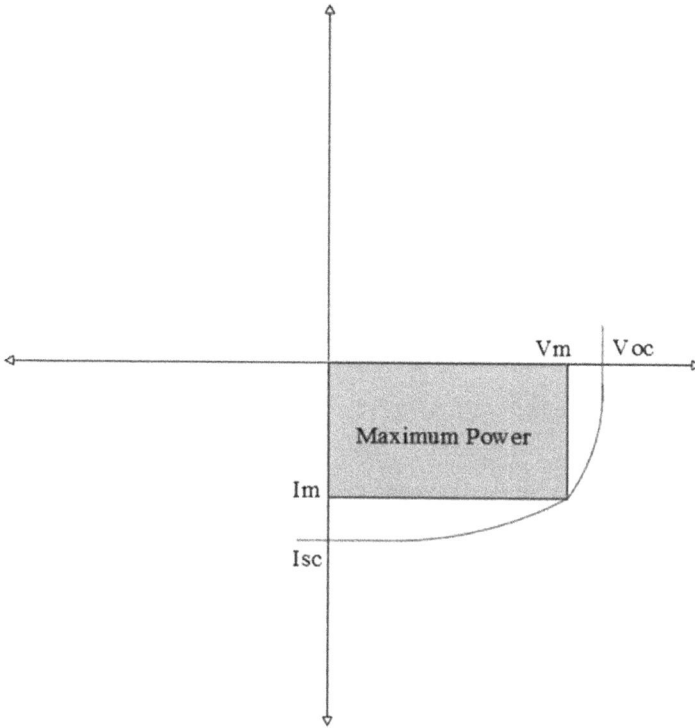

FIGURE 6.4 The solar cell IV curve as the incident light increases.

The fill factor is derived from the IV curve as shown in Figure 6.3.

$$FF = \frac{I_m \times V_m}{I_{sc} \times V_{oc}} \qquad (6.7)$$

where *Vm* is the maximum output voltage and *Im* is the maximum output current.

6.6.4 MAIN PROCESSES IN DSSC

The science involved in the process of energy generation in DSSC is multi-disciplinary, i.e. it includes chemistry, physics, biology, material science and many more. A detailed description of the process is explained below.

6.6.5 CHARGE GENERATION AND CHARGE SEPARATION

The bandgap of the semiconductor and the energy difference between the HOMO and the LUMO of the dye is the range of 0.5eV to 3.5eV in which DSSC can absorb photons. The resultant excitation is based on the excited state lifetime of Ru-based dye in the order of nanoseconds. The excited electron present in the LUMO of the dye is transferred to the CB of TiO_2 and the dye is oxidized. The electron transfer occurs through the ester link between the dye and TiO_2. The rate of electron injection depends on the absorption coefficient of the dye, the light intensity and the injection efficiency of the dye.

6.6.6 ELECTRON TRANSPORT AND RECOMBINATION

The electrons from TiO_2 are transferred to the conducting substrate and the resultant photocurrent is detected in the external circuit. The electrons then move through a trapping/de-trapping process in the TiO_2 layer and Coulombic trapping occurs due to the interaction between the electrons and the cations of the electrolyte. Figure 6.5 depicts the DSSC charge transport and recombination processes.

6.7 QUANTUM DOT-SENSITIZED SOLAR CELL (QDSSC)

Quantum dot-sensitized solar cells (QDSSCs) constitute a form of technology under DSSC. The primary difference between DSSCs and QDSSCs is the difference in the surface conditions of the nanocrystal sensitizer that exist in DSSC and QDSSC. This small distinction affects the charge transfer properties at the QD/MO/Electrolyte interface.

6.7.1 COMPONENTS OF QDSSC

The efficiency of light collection and electron injection in QDSSCs are dependent on the size, structure and morphology of the semiconductor. Hence, photoanodes were used which can be used in different morphologies such as nanotubes, mesoporous structures, spheres and nanoflowers. These nano-sized structures have excellent properties, such as good corrosion resistance, good chemical stability, good

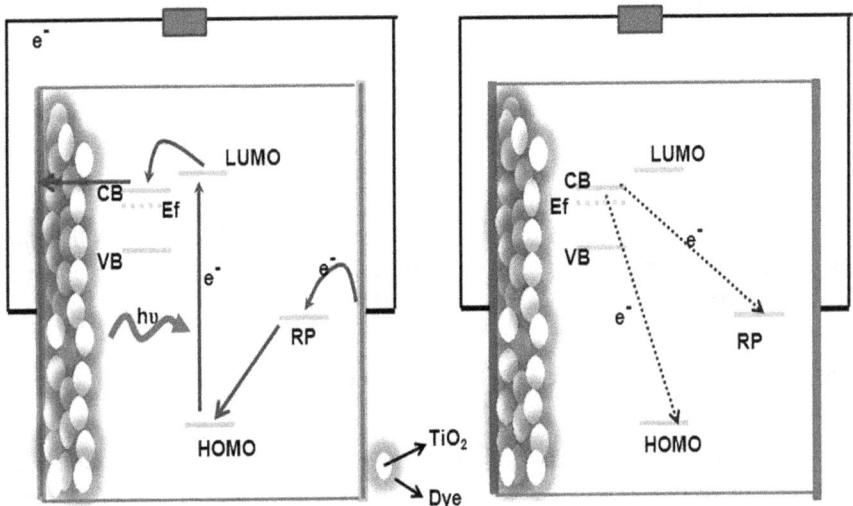

FIGURE 6.5   Schematics of DSSC charge transport and recombination processes.

hardness, low cost and good charge transport characteristics. ZnO and TiO_2 are commonly used in the semiconductors of QDSSCs. Quantum dots (QDs) are used as sensitizers in QDSSCs. The different QDs used are CdTe, CDS and ZnSe. They are very versatile and cheap in their synthesis and have the ability to change their bandgaps over a large range of spectrums. QDs have significantly high extinguishing coefficients relative to dyes used in DSSCs. Sulfide/polysulfide aqueous solution is the most commonly used electrolyte system in QDSSCs. Even QDSSCs use platinized counter electrodes made of noble metals such as Au, Al or Pt. The functions of counter electrodes are the same as those of DSSCs, with little to no difference.

6.7.2 Working Principle of QDSSC

When light energy falls on the metal oxide layer, photons are absorbed and produce excitons. These excitons dissociate at the MO/QD interface whose band-offset must be greater than the exciton-binding energy [22]. The electron then transports to the CB of the MO, which then gets transferred to the TCO (transparent conducting oxide). In parallel, the hole generated in the QD is reduced by the reductant species (Re) in the electrolyte. The Re then turns into oxidant species (Ox) and diffuses towards the counter electrode due to the loss of an electron to a corresponding hole. Since there is a reduction in the holes in the QD, it gets ready to absorb other photons falling on the surface for more exciton formation. This cyclic photo-conversion cycle continues as long as the solar cell is illuminated and there is no net change in the free energy of the electron as long as the chemical stability of the different materials is maintained and controlled, consisting of electrode/electrolyte interface. [22] Figure 6.6 shows the quantum dot-sensitized solar cells.

The following steps summarize different processes leading to the QDSSC photovoltaic effect:

1. $QD + h\nu \rightarrow QD^*$ (formation of exciton due to absorption of photon)
2. $QD^* + MO \rightarrow QD^+ + MO^-$ (Electron transfer from QD to MO)
3. $QD^+ + Re \rightarrow QD + Ox$ (charged QD being reduced to neutral QD)
4. $Ox + e^- \rightarrow Re$ (Reduction of Oxidant from e - supplied by CE) [22]

Another significant difference between QDSSC and DSSC is the electron injection to MO, which is carried out by an entropic driving force. Electron transfer usually occurs to increase the entropy. From the timespan of electron transfer processes mentioned below, we can understand that the recombination lifetimes in QDSSC are faster than DSSC for the different charging processes. Therefore, DSSCs have an advantage over QDSSCs in terms of recombination.

The electron transfer processes' timespan of DSSC and QDSSC are detailed below [22]:

- The timespan for electron injection to metal oxide from sensitizer for DSSC was calculated to be 10^{-13} whereas for QDSSC was found to be less than 10^{-12}.
- The electron-hole recombination timespan for DSSC was 10^{-7} whereas for QDSSC was found to be between 10^{-11} and 10^{-6}.

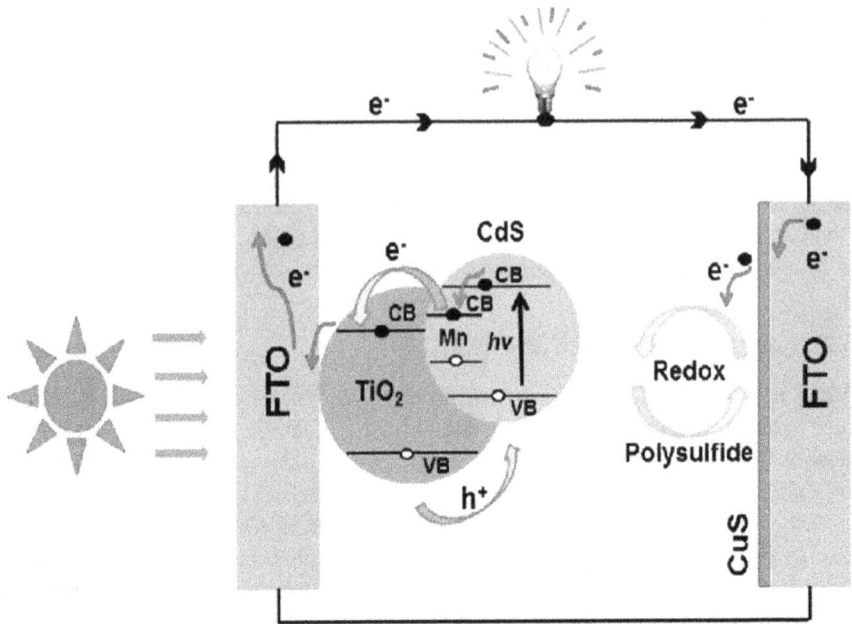

FIGURE 6.6 Schematic of quantum dot-sensitized solar cells [19]. [Republished with permission of Royal Society of Chemistry, from Chandu V. V. M. Gopi, M. Venkata-Haritha, Soo-Kyoung Kim and Hee-Je Kim, "A strategy to improve the energy conversion efficiency and stability of quantum dot-sensitized solar cells using manganese-doped cadmium sulfide quantum dots", *Dalton Transactions*, 44(2), pp. 630–638, 2020; permission conveyed through Copyright Clearance Center, Inc.]

- Electron transfer towards the electrolyte from metal oxide for DSSC was calculated to be 10^{-4} and for QDSSC it was calculated to be between 10^{-6} and 10^4.
- The time taken for evacuation of holes in DSSC was found to be 10^{-8} and for QDSSC, it was calculated to be 10^{-11}.
- The timespan for the electron trapping of conduction band electrons in sensitizer for QDSSC was calculated to be between 10^{-12} and 10^{-10}.
- The timespan for electron transfer to electrolyte from the sensitizer of QDSSC was calculated to be between 10^{-6} and 10^{-4}.

6.7.3 PEROVSKITE SENSITIZED SOLAR CELL (PSC)

Studies on perovskite solar cells have attracted a great deal of research because of their exceptional lifetimes, mobility of charge carriers and light absorption properties, which enable the production of solar cell devices that have low cost, industrially scalable technology and most importantly, have high efficiencies. Perovskite has a molecular structure of ABX_3 and is a derivative of the $CaTiO_3$ compound (calcium titanate) [23]. Since the cubic grid-screen layered octahedral structures and special optical, thermal and electromagnetic properties, perovskite materials have attracted considerable interest. The main characteristics of these solar cells are:

- Cell thickness is negligible relative to its length and width, thereby taking into consideration the heat losses only from the top and bottom surfaces and not the sides of the cell.
- Thermal capacity is neglected since the cell thickness is significantly small.
- The small thickness of the cell enables the touch resistances to not affect the average cell temperature [23, 24].

6.7.3.1 Principle of the Cell

The following features of PSC make it important in the determination of its efficiency. The materials firstly have a strong pH, lower exciton binding energy and high coefficients of optical absorption. The second is that perovskite can absorb solar energy efficiently as a light-absorbing sheet. Thirdly, a large dielectric constant is used to transfer and capture electrons and holes efficiently. Finally, the transmission distance can be as much as 100 nm or more and even more than 1 µm at the same time, leading to the transmission of electrons and holes [25]. Such characteristics result in high open-circuit voltage and short-circuit current density when used in solar power plants. The layer of perovskite absorbs first photons during the exposure to sunlight and generates excitons. The excitons in perovskite solar cells can produce current or recombine into excitons, which are free carriers due to the difference in the exciton binding energy [25]. The diffusion distance and lifetime of a carrier are long, due to the low carriers' recombination probabilities of $CH_3NH_3PbI_3$ ($MAPbI_3$) and other masked materials and greater carrier mobility. The longer distance of diffusion and life cycle of carriers provide superior solar cell efficiency. The electron transport material (ETM) and a hole transport material (HTM) then absorb these electrons and holes. Electrons from perovskite material are transmitted to TiO_2, which is used for ETM layers and collected finally by FTO. The holes are simultaneously transferred to the HTM layer and the metal electrode is collected. The FTO and metal electrodes are connected to the outer circuit, and the photocurrent is generated. In summary, the operating principle of PSCs is: the perovskite layer absorbs the light of the occurrence, producing the electron and the hole extracted and carried by ETMs and HTMs respectively [25]. Finally, electrodes that form PSCs collect these load carriers. For optimum performance of PSCs, uniform film deposition is essential to enhance the optoelectronic properties. Triple cation perovskite films are thermally more stable as they are less influenced by fluctuating variables like temperature and solvents. Flexible PSCs have also been constructed by using flexible substrate, transparent electrode, a perovskite and a charge carrier (electron or hole). These layers are essential for achieving higher efficiencies and flexibility [25]. A schematic diagram of general PSC is given in Figure 6.7.

6.8 ORGANIC SOLAR CELL (OSC)

Organic solar cells have drawn a great deal of interest recently due to their flexibility, low cost, versatility and improving power conversion efficiencies; however, they still have poorer efficiencies as compared to crystal silicon solar cells due to higher

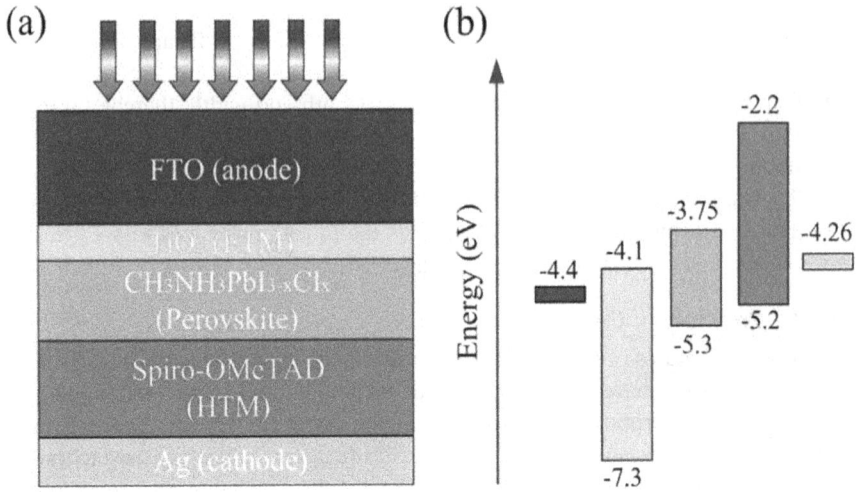

FIGURE 6.7 Schematic diagram of general PSC [26]. [Reprinted from Yun Da, Yimin Xuan, Qiang Lia, "Quantifying energy losses in planar perovskite solar cells", *Solar Energy Materials and Solar Cells*, 174, pp. 208., 2020; with permission from Elsevier].

bandgaps in organic solar cells. Organic solar cells are based on small molecules or conjugated polymers, which are fabricated by economically viable techniques such as spin coating, printing and deposition techniques. Kalowekamo and Baker reported that 15% is the highest photo-conversion efficiency (PCE) recorded for OSCs, which make them uncompetitive with inorganic solar cells, but they are competitive in terms of cost, fabrication and novel applications which is otherwise impossible with inorganic solar cells [27].

6.8.1 ARCHITECTURE AND WORKING PRINCIPLE OF OSC

Organic cells are primarily made up of carbon chains with heteroatoms such as oxygen or hydrogen and they have different intermolecular interactions with respect to each other. They are comprised of excitons, which have strong binding energy, which makes the mobility of charge carriers smaller in OSCs as compared to inorganic solar cells. Organic materials are usually applied directly from solution to the glass substrate. It is coated with a transparent conducting oxide such as ITO upon which the active layer is deposited. The active layer consists of either donor and acceptor materials applied as a single or bilayer material or by mixing the donor and acceptor together to create a blend for the bulk heterojunction layer. The last layer is made up of silver, aluminum or other metal that has an appropriate work function for efficient electron transport.

The four stages involved in the working of an OSC are:

(i) Light absorption and exciton formation. This depends on the absorption coefficient of the material.

(ii) Exciton diffusion occurs due to differences in the concentration of charges present in the donor and acceptor layer of the OSC. The diffusion length of excitons also plays a key role in this process.

(iii) Exciton dissociation occurs when the work function produced by the electrodes is sufficiently large enough to overcome the exciton binding energy and separate the electron-hole pair.

(iv) Charge transport occurs to the respective electrodes if recombination is reduced to a minimum.

The working of the organic solar cell is described in Figure 6.8. The deficiencies showcased by OPVs are usually related to low absorption in the active layer and short diffusion length. Nevertheless, their simple manufacturing processes, high efficiency per unit cost ratio and mechanical stability make them a target for extensive research, which is why new technologies such as tandem cells and multi-junction OSCs have been fabricated to minimize the limitations to a great extent.

6.9 FUEL CELLS (FC)

Fuel cells use sources such as hydrogen gas, methanol, ethanol, natural gas and hydrocarbons to create electrical energy through electrochemical reactions between a fuel and an oxidizing agent. FC technologies are robust, versatile and can be classified based on temperature as molten carbonate fuel cells (MCFCs) or solid oxide fuel cells (SOFCs). Fuel cells are also classified based on the electrolytes used in the particular type of technology [28].

6.9.1 COMPONENTS OF FUEL CELLS

A fuel cell is a system that transforms chemical energy of fuel into electricity (usually natural or biogas hydrogen) and oxidants (air or oxygen). A fuel cell

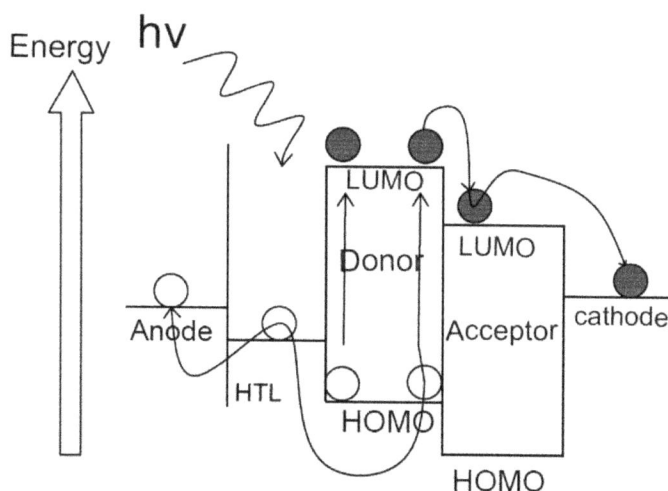

FIGURE 6.8 Working of organic solar cell.

functions like a battery in principle. Nevertheless, compared to a battery, a fuel cell does not run down or recharge (although cell stacks may need to be regularly replaced). It produces power and heat when providing fuel and an oxidizer. In fuel cells does not contain any movable parts, which makes it a silent and reliable power source. A cell stack is a "sandwich" chemical power generator consisting of three parts: anode, cathode and an electrolyte. The construction of a fuel cell is given in Figure 6.9.

6.9.2 Working Principle

A fuel cell works through the fuel cell anode by passing hydrogen by the cathode and oxygen. The hydrogen molecules at the anode site are divided into electrons and protons. The protons pass through the membrane of the electrolyte, whereas the electrons pass a circuit to produce electric current and excessive heat. The cathode creates water molecules together with protons, electrons and oxygen. Because of its high efficiency, cells are extremely clean, with power, excess heat and water as their only by-products.

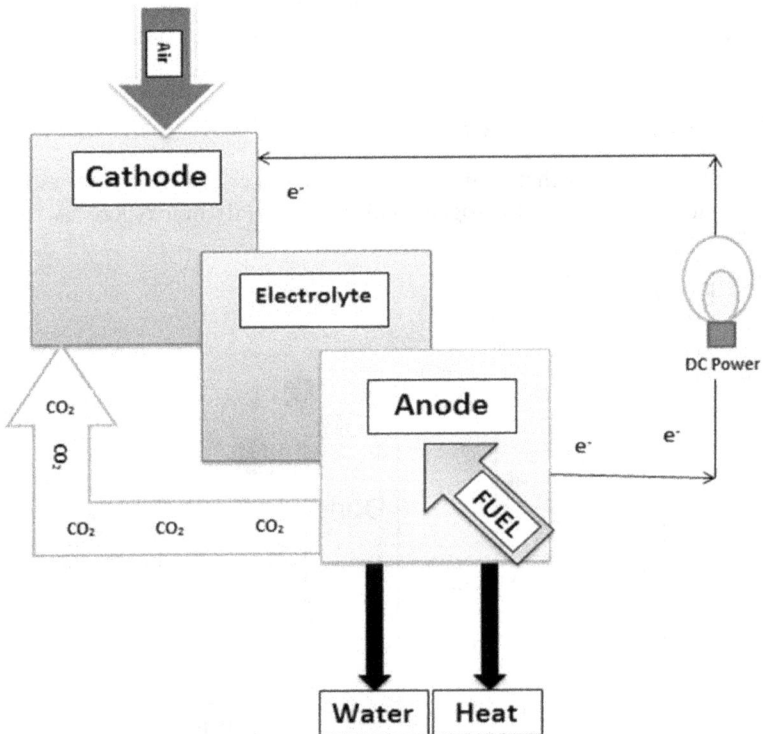

FIGURE 6.9   Construction of fuel cells.

There are slight changes in the working of different fuel cells. For a carbonate molten fuel cell, the chemical reactions are as follows [29]:

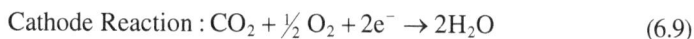

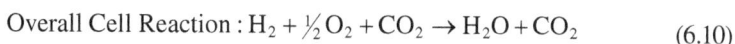

$$\text{Anode Reaction}: CO_3^2 + H_2O + CO_2 + 2e^- \qquad (6.8)$$

$$\text{Cathode Reaction}: CO_2 + \tfrac{1}{2} O_2 + 2e^- \rightarrow 2H_2O \qquad (6.9)$$

$$\text{Overall Cell Reaction}: H_2 + \tfrac{1}{2}O_2 + CO_2 \rightarrow H_2O + CO_2 \qquad (6.10)$$

Fuel cell systems are widely ranging in size and power, from compact mobile recharge systems, combustion engine and multi-megawatt power supply systems to the utility grid [30]. Platinum-based electrocatalysts have primarily been used for Hydrogen Evolvement Reactions (HER) and Oxygen Reduction Reactions (ORR), respectively. The cell performance and its longevity are primarily affected due to the corrosion that is supported by carbon, which leads to the degradation of platinum (Pt) catalyst. Limiting factors of Pt-based catalysts led to the incorporation of TiO_2 nanoparticles, which improve stability and porosity of the catalyst. TiO_2 nanoparticles also ensure that Pt particles do not agglomerate into clusters and efficiently distribute the particles. It also enables thermal and oxidative stability and improves the Pt oxygen reducing activity by enhancing oxygen spill and surface diffusion [31]. Titania nanowires were more effective than spherical TiO_2 particles [32]. CNTs are used due to their excellent biocompatibility, electrical conductivity, thermal conductivity and mechanical and catalyst strength which enables an increased surface area [33].

6.10   WIND ENERGY

Wind energy is another type of environmentally friendly energy, which uses wind as an energy source. This system converts the kinetic energy of wind into mechanical energy, which is further utilized for practical use. Recently, more than 23 billion kWh of clean and cheap electricity has been produced annually around the world. Performance is based on the utilization of wind power i.e. with an average wind speed of 13–15 m/s. The energy provided by a wind turbine is proportional to the square of its blade length. The blade size must increase in order to increase the power output produced by the wind turbine. This is because wind power output is directly proportional to the square of blade length. Nanocomposites are now being used to improve the characteristics of high-performance blades as nanomaterials contribute excellent stiffness-to-weight ratios and strength-to-weight ratios. They also contributed good barrier diffusion properties and excellent water-repellent properties [34]. One example is the synthesis of polymers such as diacetylenes which have a low molecular weight. These polymers have exceptional process ability and stability. Losses due to tribological issues such as scuffing, wear, spalling, etc. that occur in gearboxes can greatly be reduced by using nano-enhanced coatings and lubricants. Greco et al. experimentally assessed the behavior of boron nitride-based nano-enhanced lubricants to mitigate any losses and failures that affect gearbox applications. Superhydrophobic and ice-phobic materials were developed to improve the wind turbine operations during rain and snow [35]. Carbon nanotubes can be used as a blade surface coating wherein the nanotubes heat up on application of a small electric current, which prevents the formation of ice [35, 36]. Wind farms that are offshore based have to endure irregular weather patterns. Novel nanocomposite elastomers were developed as sealants which can be integrated with offshore wind farms [33–36]. Nanotechnology can contribute towards the optimization of wind power utilization using high-strength lighter materials for rotor blades, composed of nanomaterials, tribological coatings and wear protection layers of bearings and gearboxes.

6.11 CONCLUSIONS

This chapter thereby provides an overview of the applications of nanotechnology in the renewable energy sector. From the applications mentioned above, it can be comprehended that nanotechnology plays a significant role in improving the efficiency of renewable energy technologies, ranging from improving their surface areas to effectively optimizing storage properties. Major technologies to benefit from nanomaterial applications are solar cells, wind turbines and fuel cells, which lead to decreasing the environmental impact posed by the fabrication and operation of these technologies. The current global scenario is shifting towards renewable energy as a source of sustenance for electricity and other needs. It also provides an insight into the importance of using renewable technologies while highlighting the issues or risks associated with the ongoing use of fossil fuels and their massive impact on the environment. Therefore, efforts to steadily shift towards renewable energy production backed by the application of nanotechnology can thereby be concluded to be a boon for mankind and for the environment respectively.

REFERENCES

1. A. Evans, V. Strezov and T. Evans, "Assessment of Utility Energy Storage Options for Increased Renewable Energy Penetration", *Renewable and Sustainable Energy Reviews*, 16(6), pp. 4141–4147, 2012.
2. O. Ellabban, H. Abu-Rub and F. Blaabjerg, "Renewable Energy Resources: Current Status, Future Prospects and Their Enabling Technology", *Renewable and Sustainable Energy Reviews*, 39, pp. 748–764, 2014.
3. "TREIA-Texas Renewable Energy Industries Alliance", *TREIA-Texas Renewable Energy Industries Alliance*, 2020.
4. S. Müller, A. Brown and S. Ölz, Is.muni.cz, 2011.
5. S.R. Bull, "Renewable Energy Today and Tomorrow", *Proceedings of the IEEE*, 89(8), pp. 1216–1226, 2001.
6. T. Hammons, "Geothermal Power Generation Worldwide: Global Perspective, Technology, Field Experience, and Research and Development", *Electric Power Components and Systems*, 32(5), pp. 529–553, 2004.
7. Q. Wang and T. Yang, "Sustainable Hydro Power Development: International Perspective and Challenges for China", *International Conference on Multimedia Technology(ICMT)*, Hangzhou, pp. 5564–5567, 28, 2011.

8. "Marine Energy - National Hydropower Association", National Hydropower Association, 2020.
9. "Renewable Energy Consumption by Technology, 2017 Compared to 2023 – Charts – Data & Statistics - IEA", IEA, 2020.
10. H. Ritchie and M. Roser, "Renewable Energy", *Our World in Data*, 2017.
11. C. Burda, X. Chen, R. Narayanan and M. El-Sayed, "Chemistry and Properties of Nanocrystals of Different Shapes", *Chemical Reviews*, 105(4), pp. 1025–1102, 2005.
12. A. Alivisatos, "Perspectives on the Physical Chemistry of Semiconductor Nanocrystals", *The Journal of Physical Chemistry*, 100(31), pp. 13226–13239, 1996.
13. J. Turner, "A Realizable Renewable Energy Future", 1999.
14. "International Energy Outlook 2019", *EIA.Gov*, 2020.
15. K. Tupper and J. Kreider, "A Life Cycle Analysis of Hydrogen Production for Buildings and Vehicles", *MRS Proceedings*, 895, pp. 117–154, 2005.
16. D. Cahen, G. Hodes, M. Grätzel, J. Guillemoles and I. Riess, "Nature of Photovoltaic Action in Dye-Sensitized Solar Cells", *The Journal of Physical Chemistry. Part B*, 104(9), pp. 2053–2059, 2000.
17. A. Parvizi-Majidi, "Whiskers and Particulates", *Comprehensive Composite Materials*, 1, pp. 175–198, 2000.
18. M. Grätzel, "Highly Efficient Nanocrystalline Photovoltaic Devices", *Platinum Metals, Review*, 38, 151, p. 4, pp, 1994.
19. C.V.M. Gopi, M.V. Haritha, K. Soo-Kyoung and K. Hee-Je, "A Strategy to Improve the Energy Conversion Efficiency and Stability of Quantum Dot-Sensitized Solar Cells Using Manganese-Doped Cadmium Sulfide Quantum Dots", *Dalton Transactions*, 44, pp. 630–638, 2014.
20. B. Gregg, S. Chen and S. Ferrere, "Enhanced Dye-Sensitized Photoconversion Efficiency via Reversible Production of UV-Induced Surface States in Nanoporous TiO2", *The Journal of Physical Chemistry. Part B*, 107(13), pp. 3019–3029, 2003.
21. D. Matthews, P. Infelta and M. Grätzel, "Calculation of the Photocurrent-Potential Characteristic for Regenerative, Sensitized Semiconductor Electrodes", *Solar Energy Materials and Solar Cells*, 44(2), pp. 119–155, 1996.
22. R. Prasad, "An Introduction to Quantum Dot Sensitized Solar Cells (QDSSC)".
23. A. Mutalikdesai and S. Ramasesha, "Emerging Solar Technologies: Perovskite Solar Cell", *Resonance*, 22(11), pp. 1061–1083, 2017.
24. S. Sharma, K. Jain and A. Sharma, "Solar Cells: In Research and Applications—A Review", *Materials Sciences and Applications*, 06(12), pp. 1145–1155, 2015.
25. D. Zhou, T. Zhou, Y. Tian, X. Zhu and Y. Tu, "Perovskite-Based Solar Cells: Materials, Methods, and Future Perspectives", *Journal of Nanomaterials*, 2018, pp. 1–15, 2018.
26. Y. Da, Y. Xuan and Q. Li, "Quantifying Energy Losses in Planar Perovskite Solar Cells", *Solar Energy Materials and Solar Cells*, 174, pp. 1–8, 2018.
27. Y. Huang and C. Luscombe, "Towards Green Synthesis and Processing of Organic Solar Cells", *The Chemical Record*, 19(6), pp. 1039–1049, 2019.
28. W. Wang, J. Qu, P. Julião and Z. Shao, "Recent Advances in the Development of Anode Materials for Solid Oxide Fuel Cells Utilizing Liquid Oxygenated Hydrocarbon Fuels: A Mini Review", *Energy Technology*, 7(1), pp. 33–44, 2018.
29. "Fuel Cell Basics — Fuel Cell & Hydrogen Energy Association", *Fuel Cell & Hydrogen Energy Association*.
30. Z. Wen, S. Ci, S. Mao, S. Cui, G. Lu, K. Yu, S. Luo, Z. He and J. Chen, "TiO2 Nanoparticles-Decorated Carbon Nanotubes for Significantly Improved Bioelectricity Generation in Microbial Fuel Cells", *Journal of Power Sources*, 234, pp. 100–106, 2013.
31. C. Montero-Ocampo, J. Garcia and E. Estrada, "Comparison of TiO_2 and TiO_2-CNT as Cathode Catalyst Supports for ORR", *International Journal of Electrochemical Science*, 8(12), pp. 12780–12800, 2013.

32. Z. Wen, S. Ci, S. Mao, S. Cui, G. Lu, K. Yu, S. Luo, Z. He and J. Chen, "TiO2 Nanoparticles-Decorated Carbon Nanotubes for Significantly Improved Bioelectricity Generation in Microbial Fuel Cells", *Journal of Power Sources*, 234, pp. 100–106, 2013.
33. S. Thomas, "Nanotechnology in Wind Energy Engineering", *Wind Engineering*, 2013.
34. A. Greco, K. Mistry, V. Sista, O. Eryilmaz and A. Erdemir, "Friction and Wear Behaviour of Boron Based Surface Treatment and Nano-Particle Lubricant Additives for Wind Turbine Gearbox Applications", *Wear*, 271(9–10), pp. 1754–1760, 2011.
35. K. Knausgard, "Superhydrophobic Anti-Ice Nanocoatings", Norwegian University of Science and Technology, pp. 67–124 2012.
36. A. Dantas de Oliveira and C. Augusto Gonçalves Beatrice, "Polymer Nanocomposites with Different Types of Nanofiller", *Nanocomposites – Recent Evolutions*, IntechOpen, pp. 104–119, 2018.

7 Applications of Nanostructured Materials for High-Temperature Wear and Corrosion Resistance in Power Plants

Hitesh Vasudev, Lalit Thakur and Harmeet Singh
Lovely Professional University, Phagwara, India; National Institute of Technology, Kurukshetra, India; and Guru Nanak Dev Engineering College, Ludhiana, India

CONTENTS

7.1 INTRODUCTION

7.1.1 NICKEL-BASED MATERIALS

In the modern era, advanced engineering applications have emerged, such as machinery operated at a high temperature including rocketry, aerospace engine parts, dies and so on [1–3]. The material required for these applications has to withstand elevated temperature ranges from 200°C to 1200°C. The properties that can be retained at these temperature ranges are a prior requirement of the material to be used in these applications. Superalloys are heat-resistant nickel-based alloys and iron-nickel alloys that can withstand a temperature of up to 1000°C [4]. These materials exhibit a combination of properties like high strength, corrosion resistance and creep resistance, and can bear thermal cycling due to their metallurgical stability [5–6]. The combination of such properties is unmatched to that of any other metallic materials.

The advent of gas turbines highlighted the need for special materials to be used in high-temperature conditions. The materials have to withstand hostile environments at cyclic oxidation at elevated temperatures [7]. This problem has been solved to some extent with the introduction of superalloys like Inconel-718, Inconel-625 and Inconel-800 [8]. The main reason for the use of superalloys is its structure- retaining property at a high-temperature range of up to 1100°C [9]. In the earlier stages, researchers focused on the Ni and Cr alone for corrosive resistant applications. These materials have been alloyed with some other elements such as Nb, Al, Mo and Fe, etc., to form superalloys. The addition of such elements responded well against corrosion-related problems [10]. Furthermore, the ceramics were blended with elements like Ni, Cr and Co for obtaining a multiple set of properties like toughness and hardness. This combination of properties could only be possible through the use of composite materials, but in optimized proportions [11]. The failure of numerous engineering applications initiates from its surface. The thermal degradation at the surface of the material also affects the subsurface. It involves the formation of pores and cracks on the surface, and oxygen enters these paths to cause oxidation in the sub-surface [12]. There is a need to develop such materials and methods for protection against thermal degradation of the surface at elevated temperature [13]. The generation of high power in power plants, high speed in cutting and forming processes with machine tools, furnace parts, turbocharger housings, boilers and steam turbines are some high temperature applications [14].

7.1.2 Superalloys

Superalloys are mainly used in engines of jet turbines, combustion chambers, vanes and blades and it constitutes 50% weight of components used in engines. These are also used in other engineering applications where high temperature strength is required. These applications are rocket engines, liquefaction systems, steam turbine power plants, metal processing equipment, reciprocating engines, chemical and petrochemical plants and pollution-control equipment [15].

In general, the nickel-based alloys are used in high temperature engineering applications and then iron-based alloys are also used. The extensive use of nickel-based alloy is due to its metallurgical properties. The nickel-based alloy has a face-centred cubic (FCC) structure at room temperature and this structure does not change even at high temperature, while in the case of iron and cobalt, the structure changes after transformation temperature. Therefore, due to the stability in the structure of nickel-based alloys, these are preferred for high temperature applications. The superalloys have a surface stability up to 85% of their melting points temperature.

Literature on the subject reveals that the high velocity oxy-fuel (HVOF) process is capable of depositing better nickel-based coating than other thermal spray processes in terms of lower porosity levels, higher hardness, homogenous structure and excellent bond strength [16].

7.1.3 COMPOSITE MATERIALS

The desired properties required for high temperature applications can be achieved with the use of composite-coating combinations. In composite coatings, usually a hard nano-reinforcement like Al_2O_3, WC, SiC, TiO_2 and Cr_3C_2 is embedded in a softer matrix such as Ni, Cr and Co [17–18]. Initially, Ni-20Cr coating was deposited by HVOF for oxidation resistance and it showed good performance up to 650°C. However, the diffusion of Fe from the work piece to the coating took place [19]. Furthermore, the Ni-50Cr coating reduced the problem of Fe diffusion to the substrate [20]. Both these coatings belong to the category of mnemonics and are widely used for high-temperature oxidation resistance. In some applications like rocketry parts, turbochargers, gas turbine blades and machine tools, protection against erosive wear is also required along with oxidation resistance. Hence, a requirement of a hard phase becomes a secondary requirement. In high temperature applications, the optimum combinations of properties are required, and this may be possible with composite coatings. The use of composite coatings with Ni as a matrix and alumina as a reinforcement showed better resistance against slurry erosion of up to 2.5 times [21]. The SEM micrographs of nano- and micron-sized Al_2O_3 and composite powders are shown in Figure 7.1. According to the Hall-Petch relation, the yield strength of the material increases with a decrease in the particle size. Therefore, nanoparticles provide good metallurgical and mechanical properties and multiple sets of properties are obtained by tailoring the materials as in the case of composite materials.

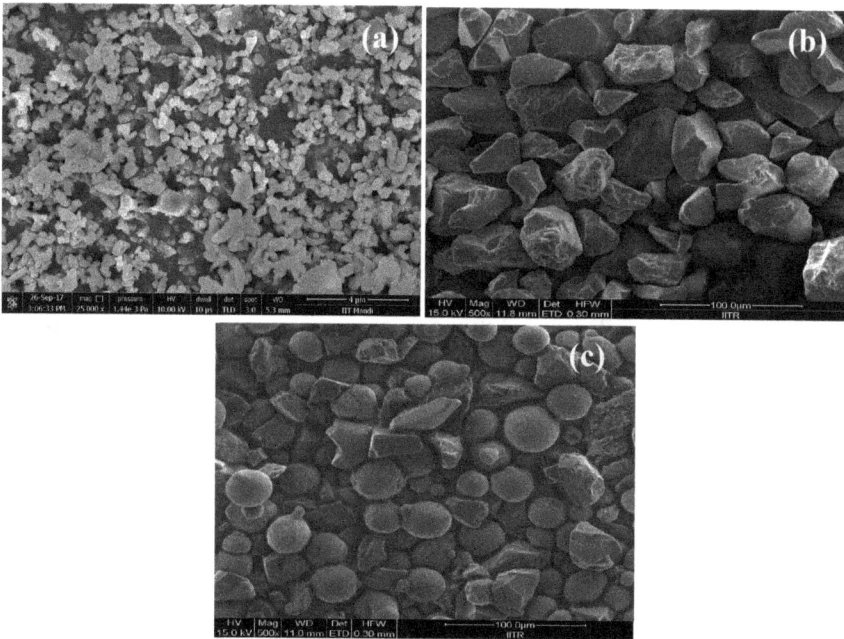

FIGURE 7.1 (a) SEM micrograph of nano-Al_2O_3 powder, (b) SEM micrograph of micron-sized Al_2O_3 powder, and (c) SEM micrograph of composite Ni-based Alloy+ Al_2O_3 powder.

7.2 MATERIALS FOR HIGH TEMPERATURE APPLICATIONS

7.2.1 CHALLENGES AT ELEVATED TEMPERATURES

New challenges in engineering have been introduced with the invention of gas turbine engines [22]. High-temperature capabilities were the requirements for this engine. The use of iron-based alloys has been replaced by nickel-based alloys. These nickel-based alloys have special properties to withstand high temperature conditions. Further advances in high temperature-resisting materials have allowed the aircraft industry to produce engines of small and intermediate size for commercial aircrafts [23]. The temperatures in the rim sections of the gas turbine vary in a high range, from 760°C to 815°C [24]. Therefore, such materials are required for these ranges and at the same time, more economical costs are required. The new nickel-based materials are much efficient and are a compromise between economics and performance.

To meet the requirements of high temperature oxidation-like problems, nickel-based alloys can be used and they provide good corrosion resistance. The need for good mechanical performance at 650°C temperature is accomplished with the use of nickel-based alloys and particularly Inconel series of Ni-Cr-Fe alloys. The Inconel series alloys belong to the class of Ni-Cr based alloys, and they cover a range of mechanical properties and compositions. At high temperatures, Ni and Cr provide resistance against oxidation, corrosion, carburizing and other damage mechanisms. These materials have good creep resistance and mechanical strength at moderate temperatures. To increase the corrosion resistance and mechanical strength, there are some elements like Nb, Al, Ti, Co and W which are added to Inconel alloy [25–26].

Inconel-718 is a relatively modern alloy, being adopted by industries in 1965. It is a hardenable alloy and contains amounts of Nb, Fe and Mo. Al and Ti are present in small proportions. It is particularly used in rocket engines, gas turbines and extrusion dies and containers.

7.2.2 PROTECTIVE COATINGS

A material has to withstand aggressive conditions when working at high temperatures and when subjected to oxidation, erosion and corrosion. Material may stop working due to such failures, and therefore requires some surface modification in order to prevent the material from surface degradation.

A number of properties are required at the same time to avoid failure with the material. With the use of a combination of elements or a different class of coatings like composite coatings, this requirement can be fulfilled. A coating enhances the service life of the component. The coatings can be deposited and sometimes used for decorative purposes, thus benefiting from its aesthetic elements. The main advantages of coatings have been summarized by Heath et al. (1997) and are given below [27]:

- Any material from polymer to metallic can be coated. Material selection can be optimised according to the particular condition under which a particular component has to work, such as corrosion and erosion. Surface properties can be different from the bulk material property.

- Functionally graded materials can be by applied by combining the materials required for combating corrosion and erosion resistance. Multi-layered coatings can also prove helpful in some applications. For instance, a bond coat is used between the top coat and substrate to avoid the mismatch of a thermal coefficient of expansion for two different materials. Therefore, coating materials can be tailored according to the particular requirement for any material.
- Unique microstructures and alloys can be achieved with thermal spray techniques which are not possible with the bulk materials. These include composites and corrosive resistant phases in the coatings.
- The coating cost is lower than the bulk material, and a coating improves the service life of a component. Therefore, the cost/performance ratio of the coating in the case of thermal spray coatings is worth considering.
- Thermal spray coatings can even be used on-site for repair and maintenance purposes with equipment. Hence, this technique has gained popularity for surface protection against corrosion or wear rates.

7.3   PROBLEMATIC AREA OF SURFACE FAILURE

In actual practice, 90% of engineering problems are derived from the working surfaces of engineering components from corrosion, erosion and wear. To mitigate these problems, the surface properties of the engineering components can be changed by using suitable techniques such as coating, cladding and glazing. The developed coatings must exhibit properties such as high wear resistance, corrosion resistance, anti-sticking behavior at high temperatures and higher thermal conductivity. Protective coatings provide up to 10 to 50 times' increase in a tool's life.

7.3.1   HIGH TEMPERATURE OXIDATION

This is essentially the formation of oxides at high temperature mainly due to the presence of oxygen. The surface of the material gets damaged due to oxidation. The fast production rates of the mold require immediate cooling of less than 20 seconds, and this results in the thermal cycling operation in cast iron [28]. Corrosion removes the large particles from the grain boundaries and these particles take place at the surface of the component, hence decreasing the surface quality of the component. Grain boundaries act as an obstacle to the dislocation which prevents the material from being plastically deformed. It is reported that nickel-based alloys have such properties that can mitigate the problem of oxidation at high temperature by forming such phases, thus providing a protective coating on the surface of the material. The structure of these alloys remains dense even at higher temperatures. The stability of these elements depends on the free surface energy of the element and the shape of the particles of the material to be used for the coating. Studies of coatings by thermal spraying process and microwave cladding have shown that the bonding of the material in the form of powder particles is dependent upon the size and shape of particles. The spherical particles of the powders have the lowest surface energy due to a low volume-to-area ratio, whereas the use of nanoparticles for the coating of a surface provides more strength to the coating. This is due to the fact that as the size of the grains decreases, the volume of the grain boundaries increases. This acts as a barrier for dislocations.

7.3.2 BEHAVIOR OF COATINGS IN A HIGH TEMPERATURE OXIDATION CONDITION

At high temperature, the oxidation of metals is the addition of oxygen to the metal surface and that causes the formation of oxides. The rate of oxidation of metals at a high temperature range depends on the nature of the oxide layer that is formed on the surface of the metal. Oxidation is a very significant high temperature reaction; the oxidation of metals and alloys occurs when they are heated in a very high-oxidizing atmosphere, like in air or oxygen. Simply put, an oxidation reaction is given by the interaction of metal with oxygen, thus forming an oxide. The rate of oxidation of the metallic materials and alloys increases with an increase in temperature [29].

The thermal barrier coating (TBC) that is used to reduce the temperature imposed on the hot section of the component is subjected to high temperatures, such as gas turbines and diesel engines [30]. The TBC consists of a metal bond coat (MNiCrAlY, M=Ni Co) and a ceramic top layer (ZrO_2/Y_2O_3). The bond coat is used to mediate the contact between the top coat and the metal alloy substrates, while the top coat of zirconia coating offers an excellent thermal shock resistance as a thermal barrier coating. In this study, the bond coat of Ni22Cr10Al1.0Y powders was deposited on M2 steel substrates by atmospheric plasma spray technique repeated at three different feed rates, i.e. 0.5, 1.0, 1.5 rpm, while other parameters were kept constant. The surface morphology demonstrates an overlapping splat and otherwise appears to be poorly consolidated by fine particles, with no definite splat structure. Results show that an increase in the feed rate resulted in an increase of the thickness of the bond coat and surface roughness, but a decrease in the hardness of the coatings. Studies also show that the deposition of ZrO_2-$8Y_2O_3$ has a higher resistance towards hot corrosion when compared to the application of bond coat only. Nanostructured coatings exhibit more mechanical strength as compared to conventional coatings. It also allows the uniform distribution elements like Al to form Al_2O_3 and thereby provides oxidation resistance. The uniform oxide layer of Al_2O_3 is formed and acts as a barrier against the diffusion of oxygen in the surface and sub-surface of the substrate.

A comparison of slurry and the air jet erosion behavior of HVOF sprayed WC-CoCr coatings was undertaken [31]. They compared conventional powders and nanopowders for the performance evaluation of coatings. WC nanoparticles were mixed and the air jet erosion testing was conducted. The nanocoatings showed an improved erosion behavior as compared to conventional coatings, due to its higher hardness and fracture toughness. The morphology of the eroded surfaces was mainly due to the formation of grooves, the pull out and the lip formation.

The three combinations of composite coatings with the use of nickel as matrix and alumina as reinforcement have been mixed [32]. The 20%, 40% and 60% alumina by weight percentage was added in a nickel matrix. The powders were sprayed with high-velocity flame spray (HVFS) process and the erosion resistance of the coatings was evaluated. It was found that the increase in the alumina content increased the highness and porosity levels in the coatings. In the erosion test it was found that the erosion rate was minimum in the case of 40% alumina coatings and higher in the case of 60% alumina coating. This was most significant due to the high hardness of alumina that leads to the brittleness in the coatings. Due to high brittleness in the case of 605 alumina coatings, the coating detached from the surface.

The process of oxidation of metals at high temperature refers to the addition of oxygen to the metal surface and that causes the formation of oxides. The oxidation rate of metals at high temperature range varies according to the oxide layer which is formed on the metal surface. Oxidation is a very important reaction carried out at high temperature and the oxidation of alloys or metals occurs when they are exposed to a very high temperature heat under oxidizing temperatures, like in the presence of oxygen or air. Basically, an oxidation reaction is given by the interaction of metal with oxygen to become an oxide. The rate of oxidation of the metallic materials and alloys increases with an increase in temperature. The behavior of coatings and bulk materials has been reviewed and is presented as follows:

The coating of intermetallic-ceramic NiAl-40Al$_2$O$_3$ was performed in the laboratory using a tester for elevated temperature erosion [33]. Fly and bottom ash collected from the boilers under operation were utilized in the form of erodent substances. Effects of high temperature, impact angle, as well as velocity, were investigated on the behavior of erosion–corrosion effects. This impact of erosion and corrosion on the coatings was evaluated with respect to other thermal-sprayed coatings such as AISI 1018 low carbon steel. The thermal shock resistance and the hardness of this coating were measured additionally. The experiments estimated that the coating of NiAl-40 Al$_2$O$_3$ HVOF revealed higher resistance to erosion and had excellent resistance to thermal shock, specifically at high temperature and higher impact angles. When compared with the coatings of chromium carbide cermet, the erosion characteristics of NiAl-40 Al$_2$O$_3$ HVOF coating were less sensitive to the exposure of varying impact angles as well as test temperatures. The wasting erosion due to elevated temperature on these coating was escalated by particle velocity, as per the power law which means having a lower exponent velocity than the chromium carbide.

The behavior of erosion on the coatings sprayed with a plasma on Ni-based superalloy [34]. To bond the used coat, all coating was NiCrAlY. Experiments on erosion were performed on both specimens, uncoated as well as coated with plasma sprayed superalloy under room temperature. The characterization of the coating was achieved using an X-ray Diffractometer Machine (XRD), micro-hardness optical microscope and Scanning Electron Microscope (SEM). Another conclusion was made regarding the micro-hardness and average porosity of coating of Ni-20Cr: Comparatively, of all coatings it was highest; and specifically for Ni$_3$Al the values were lowest among every coated or uncoated material taken into consideration. The rate of erosion was higher at a 30° impact angle, and showed ductile behavior on a 90° or higher impact angle. Other rankings related to the coatings, however, were the same: For any of the set impact angles where the erosion resistance occurred in the sequence as Ni$_3$Al, NiCrAlY, then Ni-20Cr.

The erosion and corrosion aspects of sprayed plasma coating as well as laser re-melted Stellite-6 coatings in the boiler fired with coal [35]. When experimenting regarding the combined mechanism of erosion-corrosion in some boilers, the degradation in heat-exchanger tubing as well as metal containment walls were noticed.

For enhancing the physical properties, a promising technique was adopted and suggested where the surface coating was re-melted using laser. Their study estimated the behavior of erosion-corrosion (E-C) on the laser re-melted Stellite-6 (St-6) as well as plasma-sprayed coating on the steel tubes of the boiler under the coal-fired boiler environment. The experimental research was carried out in a superheater zone of boiler heater with coal raising the temperature to 755°C. This experiment was repeated for ten cycles of 100 hr each, and added one hour of cooling at an ambient temperature. The results showed higher resistance in the case of E-C in coated steels as compared to the uncoated steels. A higher resistance to degradation was tested in the T11 coated steel and then the laser-re-melted coating.

The coating made from Ni–Al$_2$O$_3$ composite was created with a technique called the Sediment Co-Deposition (SCD) using a solution of Watt's type having particles of nano Al$_2$O$_3$. The oxidation and corrosion resistance at high temperature were investigated on the coatings of the resulting composites. The authors concluded that the combination of Ni medium with the particles of nano-Al$_2$O$_3$ helped in refining the crystals of Ni, and also reformed the preferred compounds used for coatings. Also, their study provided enhanced oxidation and corrosion resistance after mixing the Ni matrix particles, with the nano-Al$_2$O$_3$. The deposited particles of nano-Al$_2$O$_3$ have a crucial part in increasing protection against oxidation and corrosion. The resistance from oxidation and corrosion of coatings made from Ni-Al$_2$O$_3$nano-composites made using the SCD method was better than those made with the CEP method. Comparatively, the coating made from Ni-Al$_2$O$_3$ and pure Ni composites made with the CEP technique, and the coatings of Ni–7.58 wt.% Al$_2$O$_3$ composites fabricated using the SCD method, showed enhanced resistance to corrosion and increase in oxidation resistance. Along with this, the authors also discussed the corrosion mechanism as well as oxidation resistance at high temperatures of coatings made from Ni-Al$_2$O$_3$nano-composites.

The Inconel-718 powder thermally sprayed on the mild steel substrate through the "Air Plasma Spray" technique [37]. The thickness of coatings of size 200 μm and 250 μm were fabricated using this technique with the powder particles of 50 μm size. A microstructure study, a slurry erosive wear test and a microhardness test were conducted on the developed coatings, in 3.5% of the NaCl medium. The experiments of microstructural tests showed that the Inconel-718 coating was uniform, consisting of minimum porosities, and had a good bond. When undergoing the slurry erosive wear test on resistance, the developed coatings proved to be better than the uncoated mild steel material. Also, the surface hardness was higher than the substrate and was observed to be increasing with the thickness of the coating. The authors experimented on SEM tests on slurry eroded surfaces of both the Inconel-718 coating and uncoated mild steel.

The NiCrAlY coatings to be used on steel substrate underwent the process of high velocity oxygen fuel spraying [38]. The parameters of the spraying process were modified for studying different effects on the properties of coatings. Coatings and powder microstructures were experimented on using combinations of X-ray diffraction, energy dispersive spectroscopy, optical and scanning electron microscopy analysis. A micro-hardness as well as bond strength test was carried out on the developed coatings. It was discovered that powder feed rate, spray distance, oxygen flow rate, as

well as the propane flow rate were all significantly effective on the coating properties. Conclusively, with the aim of reducing the oxygen content and increasing the
bond strength, the optimal parameters were chosen to be the preferred parameters of
the spraying process.

The study of varying the behavior of high temperature oxidation on the cast
iron has been analyzed [12]. Spheroidal and flake graphite cast irons having close
composition were tested under high temperature oxidation in order to assess the
morphology of the graphite and the effects of distribution on oxidation behavior
under high temperature. The temperature chosen was 400°C–750°C to perform
the cyclic oxidation tests in air. Low carbon steel was also tested for comparison
purposes. It was discovered that the oxidation resistance under high temperature
affected the morphology of the graphite. The oxidation resistance under high temperature was found to be higher in the case of spheroidal graphite cast iron as
compared to the flake graphite cast iron. Graphite flakes present the ideal locations
for iron oxide development and are nearly interconnected; the iron oxides develop
extremely fast which permeates on the borders of graphite flakes, thus resulting in
the subsurface oxidation. Because of this subsurface oxidation, the parabolic rate
of flake graphite cast iron is nearly 5 times higher than spheroidal graphite cast
iron. However, the oxide layer thickness and the parabolic rate constants of spheroidal graphite cast iron are equal to that of low carbon steel. Therefore, in terms
of oxidation resistance under high temperature, the graphite flakes have a negative
impact on cast iron.

The authors used an ultrasonic gas atomization process, depositing the rare earth-
modified NiCrAlY powders on stainless steel through the high-velocity oxygen fuel
method of spraying [38]. The experiments were carried out with respect to NiCrAlY
coatings and its impacts of RE on thermal shock, resistance and other such properties of microstructure. It was concluded that NiCrAlY powders are uniformly distributed and are refined after using RE, while there is a reduction in the amount of
un-melted particles. Furthermore, the coatings modified with RE depicted enhanced
distribution uniformity as well as microhardness. The value of microhardness in the
proposed coating was observed to be higher when 0.9 wt.% was added, making it
34.4% higher as compared to that without RC coating. There was also an increase in
adhesive strength observed which reached a maximum with the addition of 0.6wt.%
and RE, the result being 18.8% higher when compared to uncoated RE substrates.
The adhesive strength is decreased with higher RE. On the other hand, the thermal
life cycle of coatings with NiCrAlY is raised significantly by adding RE. Also, an
optimal resistance for thermal shock was observed with 0.9 wt.% coating, ultimately
being 21.2% higher when compared to the substrate without the coating of RE.

The 60wt.%NiCrSiB-40 wt.% Al_2O_3 coating was applied to the base material
of AISI-304 through APS ("Atmospheric Plasma Spraying technique"). The coated
surface was then tested using an X-ray diffractometer (XRD), optical microscope
and a scanning electron microscope (SEM). In addition, the coating surface roughness, density, porosity and micro-hardness of the coating were estimated. A pull-off
adhesion test was used to gauge the adhesion strength of the coating sprayed. Using
an apparatus for hot air jet erosion, the behavior of erosion on sprayed coating was
tested at temperatures of up to 450°C. The rates of erosion of uncoated and coated

material were calculated at both 30° and 90° impact angles. For all the eroded samples the SEM images were collected to test the mechanism of erosion. From this study it was concluded that such a coating provides protection to the substrate at the given angles.

The "Carbon Nano Tube" (CNT)-supported Al_2O_3 coatings were successfully deposited on an ASME-SA213-T11 steel tube boiler with plasma spray coatings using Ni-Cr as a bond coat. After investigation, it was shown that the porosity of CNT- Al_2O_3-combined coatings decreases with an increase in the CNT substance. They also found that CNTs were uniformly distributed in the Al_2O_3 matrix and are stable chemically during spray formation. Also, there was no reaction to aluminum carbides or oxides, even at extremely high temperatures. The mechanical properties and slurry erosion behavior of high velocity oxy-fuel and plasma sprayed Cr_2O_3-50 % Al_2O_3 coatings on CA6NM turbine steel under hydro-accelerated conditions were studied for the performance of composite coatings [39].

The mixtures of clads of nickel-alumina powdered material are fabricated from SS-304, an austenitic substrate of stainless steel through a process of microwave heating at the frequency of 2.45 GHz. The hybrid heating process was implemented for the processing of metal matrix composite clads at the power value of 900 W. For formation of metal matrix composite clads, the exposure time varied between 60–300 s. It was then observed at an exposure time of up to 300 s. The coatings of a 0.6 mm average thickness were developed successfully. The microstructural evaluation of microwave-prepared clads uncovered the consistent dispersion of alumina-powdered contaminants within the nickel medium. The evolved clads were totally free of any sort of interfacial splits and had metallurgically bonded with the SS-304 substrate. Analysis of x-ray diffraction established the existence of alumina powdered FeNi3 as well as chromium carbide phases, which in turn contributed to the increased micro-hardness of evolved clad. The micro-hardness of the clad was discovered to be 4 times that of the austenitic value of stainless steel that makes it ideal for anti-wear purposes.

The authors reported the use of the Air Plasma Spray (APS) technique to analyze the ceramic oxide formed by the SSA: "Samarium Strontium Aluminate" process which is coated on the Inconel (NiCrAlY) 718 superalloy bond. Thickness formed with TGO (thermally grown oxide) process was regulated at varying pre-oxidation times at 10 h, 20 h and 30 h at a temperature of 1050°C. The maximum thickness (TGO) was obtained after 30 hours on a per-oxidized sample. The % elastic recovery of samples rose from 45% to 50% at the rising time of pre-oxidation at a temperature of 1050°C. The experiments of isothermal oxidation were implemented on peroxidized samples for fifteen hours in air under 1100°C. The rate constant of parabolic oxidation reduced to 3.90×10-5 from 6.08×10^{-5} mg^2 cm^{-4} s^{-1} in 10-20 h and again increased till 4.55×10^{-5} mg^2 cm^{-4} s^{-1} after thirty hours of pre-oxidized SSA samples under 1100°C. There was nearly a 65% decrease in weight gain for SSA-based TBCs as opposed to the standard YSZ TBCs prior to the oxidation process under 1100°C in air. The SEM cross-sectional outcomes showed that an SSA TBCs threshold thickness lay between 5.3–5.8 μm. On the interface of ceramic top coat and TGO, the failure of SSA TBCs was discovered because of the $SmAlO_3$ formation using spinel oxides.

7.4   CONCLUSION

Thermal cycling is a serious concern in components operating at elevated temperatures. Various forms of materials are used to protect the surface of components from high-temperature oxidation. The role of nanostructured coatings is directly related to the particle size, where the yield strength of a material increases, along with a decrease in particle size. Moreover, the nanostructured coatings help to decrease the diffusion of oxidation in the surface and sub-surface of the components. It also allows the uniform distribution elements like Al to form Al_2O_3, thereby providing oxidation resistance. The uniform oxide layer of Al_2O_3 is formed and acts as a barrier against the diffusion of oxygen in the surface and sub-surface of the substrate. Hence, the use of material to a nanoscale enhances the mechanical and microstructural properties of a material.

REFERENCES

1. Vasudev, H., Thakur, L., Singh,H. 2017 A review on tribo-corrosion of coatings in glass manufacturing industry and performance of coating techniques against high temperature corrosion and wear. *i-Manager's Journal on Material Science* 5:38–48.
2. Praveen, A.S., Sarangan, J., Suresh, S., Subramanian, J.S. 2015 Erosion wear behaviour of plasma sprayed $NiCrSiB/Al_2O_3$ composite coating. *International Journal of Refractory Metals and Hard Materials* 52:209–218.
3. Kaur, M., Singh, H., Prakash, S. 2011 Surface engineering analysis of detonation-gun sprayed Cr3C2-NiCr coating under high-temperature oxidation and oxidation-erosion environments. *Surface and Coatings and Technology* 206(2–3):530–541.
4. Vasudev, H., Thakur, L., Bansal, A., Singh, H., Zafar, S. 2019 High temperature oxidation and erosion behaviour of HVOF sprayed bilayer Alloy-718/NiCrAlY coating. *Surface and Coatings and Technology* 362:366–380.
5. Sidhu, B.S., Prakash, S. 2003 Evaluation of the corrosion behaviour of plasma-sprayed Ni_3al coatings on steel in oxidation and molten salt environment at 9000C. *Surface and Coating and Technology* 166(1):89–100.
6. Sidhu, B.S., Prakash, S. 2006 Evaluation of the behavior of shrouded plasma spray coatings in the platen superheater of coal-fired boilers. *Metals and Material Transactions A* 27:37a–19.
7. Fecht, H., Furrer, D. 2000 Processing of nickel-base superalloys for turbine engine disc applications. *Advanced Engineering Materials* 2(12):777–787.
8. Hardwicke, C.U., Lau, Y.C. 2013 Advances in thermal spray coatings for gas turbines and energy generation: A review. *Journal of Thermal Spray Technology* 22(5):564–576.
9. Tellkamp, V.L., Lau, M.L., Fabela, A., Laverniaa E.J. 1997 Thermal spraying of nano-crystalline Inconel 718. *Nanostructured Materials* 9(1–8):489–492.
10. Vasudev, H., Singh, P., Thakur, L., Bansal, A. A. 2020 Mechanical and microstructural characterization of microwave post processed Alloy-718 coating. *Materials Research Express* 6:1265f5.
11. Bansal, A., Vasudev, H., Sharma, A.K., Kumar, P. 2019 Investigation on the effect of post weld heat treatment on microwave joining of the Alloy-718 weldment. *Materials Research Express* 6(8):086554.
12. Lin, M.B., Wang, C.J., Volinsky, A.A. 2011 High temperature oxidation behavior of flake and spheroidal graphite cast irons. *Oxidation of Metals* 76(3–4):161–168.
13. Pawlowski, L. 2008 *The Science Engineering of Thermal Spray Coatings*, 2nd ed., John Wiley & Sons Ltd., London.

14. Khanna, A.S. 2002 *High Temperature Oxidation and Corrosion*, 1st ed., ASM International, Metals Park, OH.

15. Davis, J.R. 2004 *Handbook of Thermal Spray Technology*, 2nd ed., ASM International, Metals Park, OH.

16. Kaur, M., Singh, H., Prakash, S. 2009 High temperature corrosion studies of HVOF sprayed Cr_3C_2-NiCr coating on SAE-347H boiler steel. *Journal of Thermal Spray Technology* 18(4):619–632.

17. Chen, M., Zhu, S., Wang, F. 2015 High temperature oxidation of NiCrAlY, nanocrystalline and enamel-metal nano-composite coatings under thermal shock. *Corrosion Science* 100:556–565.

18. Naeimi, M.H.F., Tahari, M.S.M. 2018 High-temperature oxidation behavior of nanostructured CoNiCrAlY – YSZ coatings produced by HVOF thermal spray technique. *Oxidation of Metals* 90(1–2):153–167.

19. Sundararajan, T., Kuroda, S., Nishida, K., Itagaki, T., Abe, F. 2004 Behaviour of Mn and Si in the spray powders during steam oxidation of Ni-Cr thermal spray coatings. *ISIJ International* 44(1):139–144.

20. Sundararajan, T., Kuroda, S., Abe, F. 2005 Steam oxidation of 80Ni-20Cr high-velocity oxyfuel coatings on 9Cr-1Mo steel : Diffusion-induced phase transformations in the substrate adjacent to the coating. *Metallurgical and Materials Transactions: Part A* 36(8):2165–2174.

21. Grewal, H.S., Singh, H., Agrawal, A. 2013 Understanding Liquid Impingement erosion behaviour of nickel-alumina based thermal spray coatings. *Wear* 301(1–2):424–433.

22. Krzyzanowski, M., Beynon, J.H., Farrugia, D.C.J. 2010 *Oxide Scale Behavior in High Temperature Metal Processing*, John Wiley & Sons.

23. Campbell, F.C. 2006 Manufacturing technology for aerospace structural materials. *Superalloys* 211–272.

24. Delaunay, F., Berthier, C., Lenglet, M., Lameille, J. 2000 SEM-EDS and XPS studies of the high temperature oxidation behaviour of inconel 718. *Mikrochimica Acta* 132:337–343.

25. Ramesh, C.S., Devaraj, D.S., Keshavamurthy, R., Sridhar, B.R. 2011 Slurry erosive wear behaviour of thermally sprayed Inconel-718 coatings by APS process. *Wear* 271(9–10):1365–1371.

26. Singh, H. 2009 *Plasma Spray Coatings for Superalloys, Characterization and High Temperature Oxidation Behavior*, VDM Verlag.

27. Bala, N. 2010 Investigations on the hot corrosion behaviour of cold spray and HVOF spray coatings on T22 and SA 516 steels. Ph.D. Thesis, Mechanical Engineering Department, Punjab Technical University, Jalandhar, India.

28. Akdogan, A.N., Durakbasa, M.N. 2008 Thermal cycling experiments for glass moulds surface texture lifetime prediction-evaluation with the help of statistical techniques. *Measurement* 41(6):697–703.

29. Davis, J.R. 2004 *Handbook of Thermal Spray Technology*, ASM International, Metals Park, OH.

30. Istikamah, S., Fazira, M.F., Talib, R.J. 2011 Atmospheric plasma spray of NiCrAlY bond coat with different feed rates. *Solid State Science and Technology* 19(1):32–39.

31. Thakur, L., Arora, N. 2013 A comparative study on slurry and dry erosion behaviour of HVOF sprayed WC-CoCr coatings. *Wear* 303(1–2):405–411.

32. Grewal, H.S., Singh, H., Agrawal, A. 2013 Microstructural and mechanical characterization of thermal sprayed nickel–alumina composite coatings. *Surface and Coatings and Technology* 216:78 –92.

33. Wang, B., Lee, S.W. 2000 Erosion-corrosion behaviour of HVOF NiAl– Al_2O_3 intermetallic ceramic coating. *Wear* 239(1):83–90.

34. Mishra, S.B., Prakash, S., Chandra, K. 2006 Studies on erosion behaviour of plasma sprayed coatings on a Ni-based superalloy. *Wear* 260(4–5):422–432.

35. Chatha, S.S., Sidhu, H.S., Sidhu, B.S. 2012 Characterisation and corrosion-erosion behaviour of carbide based thermal spray coatings. *The Journal of Minerals and Materials Characterization and Engineering* 11(6):569–586.

36. Feng, J., Ferreira, M.G.S., Vilar, R. 1997 Investigation on the corrosion and oxidation resistance of Ni–Al_2O_3 nano-composite coatings prepared by sediment co-deposition. *Surface and Coatings and Technology* 88(1–3):212.

37. Mathapati, M., Ramesh, M.R., Doddamani, M. 2017 High temperature erosion behavior of plasma sprayed NiCrAlY/WC-Co/cenosphere coating. *Surface and Coatings Technology* 325:98–106.

38. Chen, S.F., Liu, S.Y., Wang, Y., Sun, X.G., Zou, Z.W., Li, X.W., Wang, C.H. 2014 Microstructure and properties of HVOF-sprayed NiCrAlY coatings modified by rare earth. *Journal of Thermal Spray Technology* 23(5):809–817.

39. Goyal, K. 2018 Experimental investigations of mechanical properties and slurry erosion behaviour of high velocity oxy fuel and plasma sprayed Cr_2O_3-50 % Al_2O_3 coatings on CA6NM turbine steel under hydro accelerated conditions. *Tribology - Materials, Surfaces and Interfaces* 5831:1–10.

8 Sensitization of Organic Dyes to Be Used in the Fabrication of DSSC

Shivani Arora Abrol and Cherry Bhargava
Lovely Professional University, Phagwara, India

CONTENTS

8.1   INTRODUCTION OF AN ORGANIC DYE

A photon absorber layer is formed by dye molecules forming a single layer in between the glass substrates. This layer is spread between the TiO_2 layer which is coated on the FTO glass substrates for enhanced conductivity. A layer of polymer electrolytes is positioned in between the glass film to penetrate the pores of the TiO_2, thus completing the circuit of the cell. On top of the cell is positioned a conductive electrode which is coated with a thin catalytic coating of graphite to act as a counter electrode, and light photons are made incidental for illumination from the TiO_2 side of the electrode. In contrast, a denser coating of natural dyes was considered in several studies; but electric charges move rapidly and uneasily within many organic natural materials, and it was witnessed that for charge injection, an active charge is effective only with tremendously thin coatings. Thus, for absorbing all of the light, thick organic dye films do not transfer photoexcited charges as well as thin films. Consequently, a solar cell made from an interconnected sequence of layers of thin film can be considered to be more effective than a solo dense layer of dye.

In this postulation, dyes were extracted from natural substances like red cabbage and onion peels and used as sensitizers for the dye-sensitized solar cells. The properties of the dyes as a part of the fabricated cells are studied here.

8.2 STEPS FOR FORMATION OF AN ORGANIC DYE

Step 1. Measurement of the raw material (natural organic substances) to be used for the dye (for example, in this case onion peels and red cabbage) using a scientific scale for accurate measurement was done. 6 gm of small pieces of onion peels and 147 gm of chopped cabbage (red) were measured.

Step 2. The measured onion peels and red cabbage were immersed in 250 ml and 400 ml respectively of distilled water.

Step 3. The samples which were made ready were placed in BOD incubator (in the dark) at 45° C for 48 hours so that the dye sensitized with this procedure contains extracts from the organic materials, which can act as conducting molecules when light falls on them.

Step 4. After cooling to room temperature, the dispersions were filtered through filter papers to extract the anthocyanin (dye) for use as sensitizers.

Step 5. All dye solutions were stored in incubation in the dark.

Figure 8.1 shows the various steps for formation of an organic dye.

8.3 RESULTS AND DISCUSSIONS

The sensitized anthocyanin was extracted from the preparation of dye using natural and organic substrates like red cabbage and onion peels. It was seen that the

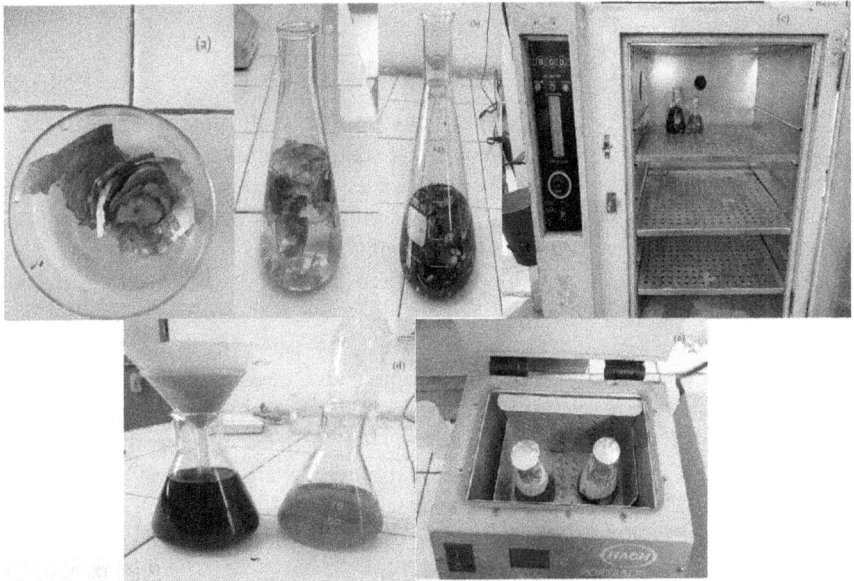

FIGURE 8.1 Steps for formation of an organic dye. (a) Collection of raw material, (b) preparation of solution in distilled water, (c) incubation for 48 hrs in dark, (d) filtration of leeched dye, (e) the filtered dye stored in the dark.

absorption level of the photons of light rose. There are specific functional groups found present in natural dyes which are effectively adsorbed into the conducting catalytic layer of TiO_2.

8.3.1 ABSORPTION PROPERTIES

It was further understood that the functional groups, i.e. the OH group found in molecules of anthocyanin sensitized dye extracted from these natural organic materials, lead to decent absorption properties. Further, C=O group presence also leads to improvement in absorption. All the extracted dyes also exhibited a spectral shift from the UV to the visible region.

8.3.2 FILL FACTOR

An emission peak at 565 nm was observed for anthocyanin dye and in the light spectra it was observed that it was red-shifted by 15 nm as compared to the sensitized dye from red cabbage.

For analysis, fill factor was calculated as:

$$FF = \left(V_m J_m\right) / V_{oc} . J_{sc} \qquad (8.1)$$

And further efficiency was estimated by:

$$\eta = [(V_{oc.} J_{sc.} FF)/P_{in}] \times 100 \qquad (8.2)$$

where P_{in} = radiation power incident on cell, J_{sc} = short circuit current density,
V_{oc} = open circuit voltage, J_m = maximum current density, V_m = maximum voltage

8.3.3 OPEN CIRCUIT VOLTAGE

The DSSC's open-circuit voltage (V_{oc}), when observed by a multimeter, and making use of the onion and red cabbage fabricated dyes, was observed to be 0.48 V and 0.51 V respectively. This gave a moderate efficiency and fill factor. However, the abstraction process of anthocyanin compound from natural resources yielded by-products such as sugars, sugar alcohols, organic acids, amino acids, and proteins. These impurities caused the acceleration of anthocyanin dilapidation during stowage. Research is being carried out to find a solution to this problem.

8.4 CONCLUSION

Natural organic dyes can be effortlessly and safely sensitized by simple techniques. The UV-visible absorption and photoluminescence properties of the extracted dyes considered here gave long-lifetime stability and the highest possible efficiency in addition to a low-cost attribute.

BIBLIOGRAPHY

1. B. O'Regan and M. Grätzel, "A low-cost, high-efficiency solar cell based on dye-sensitized colloidal TiO2 films", *Nature*, 353(6346), pp. 737–740, 1991.
2. M. Ammar Ahmed, S. H. Mohamed, M. M. Yousef et al. "Dye sensitized solar cells based on extracted natural dyes", *Journal of Nanomaterials*, pp. 1–10, 2019. Article ID 1867271n, Hindawi.
3. S. Hao, J. Wu, Y. Huang and J. Lin, "Natural dyes as photosensitizers for dye-sensitized solar cell", *Solar Energy*, 80(2), pp. 209–214, 2006.
4. J. Zhang, H. S. Tan, X. Guo, A. Facchetti and H. Yan, "Material insights and challenges for non-fullerene organic solar cells based on small molecular acceptors", *Nature Energy*, 3(9), pp. 720–731, 2018.
5. M. K. Kashif, M. Nippe, N. W. Duffy et al. "Stable dye sensitized solar cell electrolytes based on cobalt(ii)/(iii) complexes of a hexadentate pyridyl ligand", *Angewandte Chemie International Edition*, 52(21), pp. 5527–5531, 2013.
6. G. Calogero, J.Barichello and I. Citroetal, "Photoelectrochemical and spectrophotometric studies on dye-sensitized solar cells (DSCs)and stable modules (DSCMs) based on natural apocarotenoids pigments", *Dyes and Pigments*, 155, pp. 75–83, 2018.
7. Jihuai Wu, Sanchun Hao, Zhang Lan et al. "An all-solid-state dye-sensitized solar cell-based poly(N-alkyl-4-vinyl-pyridine iodide) electrolyte with efficiency of 5.64%", *Journal of the American Chemical Society*, 130(35), pp. 11568–11569, 2008.
8. K. F. Chen, C. H. Liou, C. H. Lee and F. R. Chen, "Development of solid polymeric electrolyte for dssc device", *2010 IEEE*, pp. 003288–003290.
9. M. Shaheer Akhtar, Zhen Yu Li, Woojin Lee and O-Bong Yang, "Effective inorganic-organic composite electrolytes for efficient solid-state dye sensitized solar cells", *2013 IEEE* pp. 2414–2416.
10. R. A. Senthil, J. Theerthagiri and J. Madhavan, "Organic dopant added polyvinylidene fluoride based solid polymer electrolytes for dye-sensitized solar cells", *Journal of Physics and Chemistry of Solids*, 89, pp 78–83, 2016. Elsevier.
11. Ramanpreet Kaur Aulakh, Sana Sandhu, Sandeep Kumar Tanvi et al. "Designing and synthesis of imidazole based hole transporting material for solid state dye sensitized solar cells", *Synthetic Metals*, 205, pp. 92–97, 2015. Elsevier.
12. Masanobu Chiku, Shoji Tomita, Eiji Higuchi and Hiroshi Inoue, "Preparation and characterization of organic-inorganic hybrid hydrogel electrolyte using alkaline solution", *Polymers*, 3(4), pp. 1600–1606, 2011.
13. Aswani Poosapati, Eunhwa Jang, Deepa Madan, Nathaniel Jang, Liangbing Hu and Yucheng Lan, "Cellulose hydrogel as a flexible gel electrolyte layer", *MRS Communications*, Materials Research Society, 9, pp. 122–128, 2019.
14. Chun-Chen Yang and Lin Sheng-Jen, "Alkaline composite PEO–PVA–glass-fibre-mat polymer electrolyte for Zn–air battery", *Journal of Power Sources*, 112, pp. 497–503, 2002. Elsevier.
15. Zahra Seidalilir, Rasoul Malekfar, Hui-Ping Wu, Jia-Wei Shiu, Eric Wei-Guang Diau, "High-performance and stable gel-state dye-sensitized solar cells using anodic TiO2 nanotube arrays and polymer-based gel electrolytes", *ACS Applied Materials and Interfaces*, 7, pp. 12731–12739, 2015.

9 Conduction Mechanism and Performance Evaluation of Advance Nanoscale Semiconductor Devices

Jeetendra Singh, Balwant Raj, and Balwinder Raj
NIT Sikkim, India; Punjab University SS Giri Regional
Centre, Hoshiarpur, India; and NITTTR, Chandigarh, India

CONTENTS

9.1   INTRODUCTION

MOSFET is the basic unit component of ICs, so there have been revolutionary changes that occur in the MOSFET structure at each generation node to make them faster in performance, smaller in size, and attain a higher degree of reliability in ICs for analog, digital, and biomedical circuit applications. As the physical size of the CMOS technology has continuously scaled down, the drain and source start coupling to each other, which results in the gate losing control over the channel. The low control of the gate electrode over the channel in short-channel devices produces many undesirable effects; these are known as Short Channel Effects (SCEs) [1]. These SCEs are responsible for the degradation of classical MOS device performance in the nanometer regime. Therefore, the classical MOS device scaling reaches its fundamental limits and further scaling below 22 nm is impossible.

The certain downsides of MOSFETs at or below 22 nm node are: high off current (I_{off}) due to low electrostatic gate control over the channel, I_{off} increment over the fundamental limits 60 mV/decade sub-threshold swing, parasitic capacitance and resistance effects at room temperature [2–5]. So, we have to think about new advanced MOS structures so as to maintain its performance at an optimized level. The ITRS (International Technology Road Map for Semiconductors) [6] is focused on continued scaling of CMOS technology which has become the primary bottleneck for the performance advancement in semiconductor devices in the IC industries. With these scaling efforts, it is also difficult to imagine and afford the continuously increasing process equipment and factory costs for the next 15 years. So, ITRS addresses the post-CMOS technologies that must reduce the cost per function and simultaneously increase the performance of ICs. Although the new post-CMOS approach does not only not scale the number of semiconductor devices in ICs, it is also limited by a complex set of process parameters and design paradigms.

However, there are various investigations and developments which have evolved from generation to generation in semiconductor devices. The five devices, TFETs, FinFETs, CNTFETs, NWFETs, and memristors, impact the semiconductor technology at the next level and perform a crucial role in the progress of recent semiconductor technology. Therefore, there is a need to understand and investigate these devices from a primitive level to more advanced stages through their structure, conduction mechanism, and performance evaluations. This chapter discusses the device structure, their conduction mechanism, and also analyzes the performance of TFETs, FinFETs, CNTFETs, NWFETs, and memristors in one platform. These devices will then be able to be widely investigated and understood by various researchers, and will be frequently adopted by different industries in the advancement of novel circuitry applications.

9.2   TUNNEL FIELD EFFECT TRANSISTORS (TFETS)

9.2.1   DEVICE GEOMETRY TUNNEL FETS

Tunnel Field Effect Transistors' (TFETs') device geometry primitive consists of a PIN configuration as shown in Figure 9.1, it has an intrinsic channel 'I' sandwiched between two highly doped P+ and N+ regions, which work as a source and drain for the tunnel FET [7]. Here, the electron tunneling occurs through the very thin intrinsic channel between source and drain and this tunneling phenomenon is approximated by WKB. So, the thermionic emission, unlike traditional MOS, is omitted in the conduction of the electron in the tunnel FETs and band-to-band tunneling is incorporated. The tunneling mechanism is accomplished in the presence of the high field. A low voltage can establish a high field in a very narrow dimension and renders a large current [8]. The band-to-band tunneling phenomenon offers several advantages for TFETs over the traditional thermionic emission-based MOSFETs, like steep and low leakage current, and enables the TFETs to be used for low-power systems [9].

9.2.2   CONDUCTION MECHANISM OF TUNNEL FETS

The conduction of electrons in the TFETs can be easily understandable from the energy band diagram shown in Figure 9.2. The off-state condition is represented by Figure 9.2a, in which the gate-to-source voltage is kept at zero (VGS=0) and drain-to-source voltage is kept at 1V (VDS=1V). It can be seen that in the case of the off state from Figure 9.2a the potential barrier between the conduction band and the valance band is wide enough, and it prevents the tunneling of the electrons. The on state of the transistor is shown in Figure 9.2b in which both voltages (gate to source and drain to the source) are kept at 1V (VGS=1V, VDS=1V). In this case, the gate

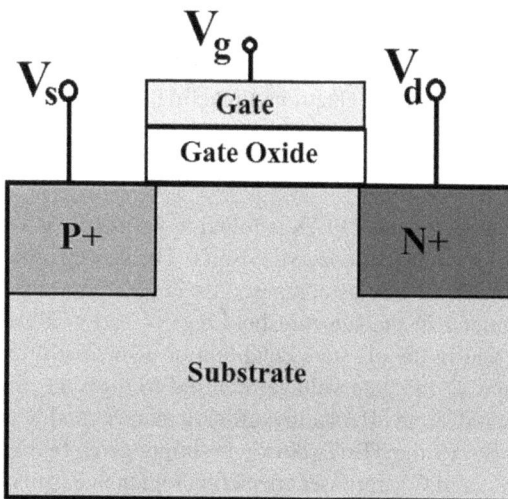

FIGURE 9.1   Tunnel FETs' (TFETs') device structure.

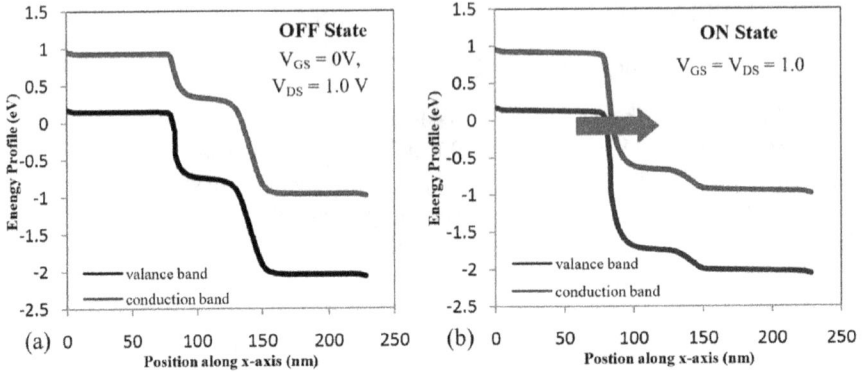

FIGURE 9.2   Energy band diagram TFETs (a) in off-state condition; (b) in on-state condition.

voltage builds a narrower potential barrier between the conduction and valance band and provides a direct tunneling path. Here, the applied voltage at the gate should be greater than the threshold voltage; as the gate voltage increases, the probability of the electron tunneling and thus the current will increase [10, 11].

9.2.3 PERFORMANCE MERITS OF TUNNEL FETs

As TFETs show, the non-linear transfer characteristics due to the band-to-band tunneling of the electrons, the device possesses some advantages such as high switching speed, low leakage, temperature independence, immunity to SCEs [12], etc.

9.2.3.1 Low OFF-State Current

Since Tunnel FETs avoid thermionic emission, these FETs have a very low off-state current, which is not possible for the traditional MOS structure because these MOS devices are never fully OFF due to the presence of the thermionic emission mechanism. But in the case of the TFETs, the potential barrier in the off-state condition fare well enough to prevent the electrons tunneling, and renders a much lower off-state current or one sub-threshold.

9.2.3.2 Sub-Threshold Slope

The sub-threshold slope of the TFETs is found to be as low as 60mv/dec, which is the limited value of the conventional MOSFETs. The sub-threshold slope has a different definition for MOSFETs as compared to TFETs. The sub-threshold slope is an important parameter in the sub-threshold region, and it is used to measure the amount of the current in the off-state condition or below the threshold region. It is given as the amount of the gate voltage required to drop the current one decade. The less sub-threshold slope offers a low off-state current and hence a better device performance. The SS for the TFETs device is simply given by Equation 9.1 [13]. In this equation, the C_D and C_{OX} represent depletion region and gate oxide capacitances respectively. k, q, and T represent Boltzmann constant, electric charge, and temperature in kelvin respectively.

$$ss = \frac{kT}{q} \ln(10)\left(1 + \frac{C_D}{C_{OX}}\right) \qquad (9.1)$$

9.2.3.3 Drain-Induced Barrier Lowering (DIBL)

The large applied voltage at the drain end lowers the source channel barrier and affects the drain current, and is known as drain-induced barrier lowering. This phenomenon is significant at the short channel device, and due to this, the drain current is controlled by the drain voltage along with the gate voltage [14]. Mathematically, the DIBL relation is given by Equation 9.2. This represents the ratio of the difference of the threshold voltages to the drain voltages at two different drain voltages in the transfer characteristics.

$$DIBL = \frac{V_{T(VD=1V)} - V_{T(VD=0.1V)}}{V_{D=1V} - V_{D=0.1V}} \qquad (9.2)$$

9.3 FINFETS

TFETs have a good handling capability of SCEs because of their low sub-threshold slope, which lies below the ideal value of 60mV/Dec; due to this property, the leakage current is significantly reduced in the tunnel FETs [15]. However, it is seen that the TFETs offer less current as compared to the conventional MOSFETs; therefore, it is necessary to make some modifications in the conventional MOS devices, which can provide a small leakage current like TFETs. The leakage can be reduced in the conventional MOS devices by providing more gate control over the channel. Multi-gate technology or developing Fins below the surrounded gate of the MOS devices is one of the solutions to achieve more over the channel by the gate and this will result in low off-state current.

9.3.1 DEVICE STRUCTURE OF FINFETS

Further advances in the bulk MOS structure have led to multi-gate technology, the so-called "FinFet". The multi-gate structure has the advantage of improved SCEs and the channel is controlled by multiple sides of the gate as compared to conventional MOSFET. It has better electrostatic gate control and reduces the significant amount of current leakage components. Such enhanced gate control also results in a reduction of output conductance in the saturation region. This is beneficial for analog circuits for voltage gain improvements and also improves the noise margin for the digital circuits. Moreover, this technique also renders a high drive current, which is beneficial for the high-speed circuit's applications.

Figure 9.3a shows the schematic journey of a single gate (SG) to double-gate (DG) to tri-gate FinFET. Whereas, Figure 9.3b shows the final finFET structure in which the fin is surrounded by a gate and gives large electrical control to the channel. The DG-MOSFET has two symmetrical gates and can be illustrated by different gate orientations like vertical, planar, and mixed-mode [16]. Initially, the use of a

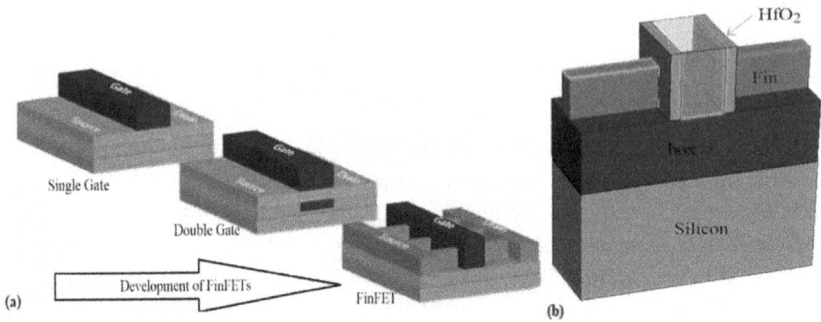

FIGURE 9.3 (a) Evolution of FinFET structure; (b) FinFET device structure.

double gate was given by Balestra et al. (1987). They used a double gate with silicon on insulator (SOI) MOSFET that connects the front and back gate in full depletion mode [17, 18].

The Si thickness of this device is thick when compared to the depletion region generated by two gates and there is no interaction between the two inversion regions generated by both gates. So, the double gate works like two bulk MOSFETs and are connected in parallel. In this structure, if the thickness of the Si is taken to be less than the generated depletion layers by two gates, then the whole silicon layer depleted and MOSFET works in fully depleted mode.

9.3.2 CONDUCTION MECHANISM OF FINFETS

The Si thickness of this device is taken to be thick when compared to the depletion region generated by two gates and there is no interaction between the two inversion regions generated by both gates. So, a double gate working like two bulk MOSFETs is connected in parallel. In this structure, if the thickness of the Si is taken to be less than the generated depletion layers by two gates, then the whole silicon layer is depleted and MOSFET works in a fully depleted mode. In addition, carriers are not confined near to the silicon-oxide interface, but are distributed across whole silicon volume. The performance parameters like drive current, transconductance, and switching speed of the device are improved because the minority carriers available in the middle region of the channel experience less scattering. In addition, DG MOSFETs are claimed to have improved SCEs, as compared to conventional MOSFET or even a fully depleted single gate SOI MOSFET [19]. Wind et al. demonstrate that two gates in a DG MOSFET jointly have better control of the charge carries, and screen the fields away from the channel, which is generated by the drain [20]. Wong et al. further claimed that DG MOSFET shows better scaling characteristics as compared to bulk MOSFET. Instead of having the above advantages of DG MOSFETs, it still suffers from a considerable number of SCEs in the deep nanometer regimes. So, the tri-gate FinFET structure is developed against SCEs for further miniaturization of a device to enhance the circuit reliability. Andrade et al. [21] studied the SCEs parameters like sub-threshold swing, DIBL, and threshold voltage variation study for tri-gate bulk FinFETs. Sun et al. also compared the tri-gate MOSFET, SOI FinFET, and ground-plane bulk MOSFET designs and give notable results for further research on the 3-D platform [22].

9.3.3 Performance Metrics of FinFETs

The threshold voltage adjustment becomes a challenging obstacle to the device in high-performance (HP) applications. Adjustment of the gate work function is one of the solutions to achieve the appropriate threshold voltage. At appropriate off-current (Ioff), the highest overdrive and the reduction of the overall parasitic capacitance present another big challenge for devices in various applications [23]. The non-planar device like 3-D FinFET with various engineering techniques is needed to optimize the performance metrics through the 3-D simulation process [24]. For static performance analysis, it is important to study the DIBL and SS parameters to suppress SCEs in the nano-scale regime. For short-channel devices, the drain to source (VDS) has a strong impact on the inversion layer or on the band bending of the device. Therefore, the threshold voltage is potentially affected by the drain bias along with the gate bias. This threshold voltage variation effect on the device is known by DIBL [25]. DIBL effects are generated due to the interaction of drain and source depletion regions near the channel surface and it reduces the barrier potential of source end. It is evaluated by Equation 9.2, sub-threshold current gives information about the standby power dissipation of CMOS circuits. The MOSFETs which are designed for high-speed digital applications and low power consumptions purpose are generally operated in the sub-threshold region [26, 27]. So, a steeper sub-threshold slope (SS) becomes the primary requirement for switching applications. The SS parameter can be expressed by Equation 9.3.

$$ss = \left[\frac{\partial \log_{10}(I_D)}{\partial V_{GS}} \right]^{-1} \tag{9.3}$$

The DIBL representation from ID-VD characteristics of the device is shown in Figure 9.4a. The gate voltage difference between saturated drain current and the linear drain current is depicted by DIBL [28]. Figure 9.4b shows the sub-threshold slope evaluation and representation from I_D-V_D graph.

9.4 CARBON NANOTUBE FETS (CNTFETS)

Carbon Nanotubes (CNT) are essentially rolled graphene sheets with 1 to a few nm diameters. These sheets can be single-walled or may be multi-walled depending upon the applications [29]. CNTFETs use CNT as the channel material. In single wall CNTFET, the channel contains the only cylindrical tube of the CNT [30]. CNTs support 1-D transport of the carriers on the surface and thus provide excellent gate control over the channel. The CNTFETs exhibit linear dependence of the gate-source voltage and drain current and thus provide high linearity above the threshold voltage [31, 32]. CNTFETs are found to be one of the best alternatives to

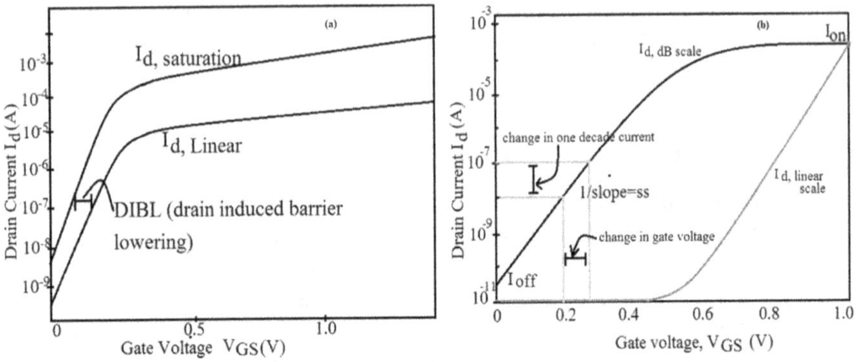

FIGURE 9.4 Representation of (a) DIBL and (b) subthreshold (SS), from ID-VD characteristics.

conventional MOSFETs. A large free path in CNTFETs enables them to be used in high-frequency applications.

9.4.1 DEVICE GEOMETRY OF CNTFETS

Figure 9.5 shows two types of CNTFETs; Figure 9.5a shows the planer structure and Figure 9.5b shows the cylindrical structure. In the planer structure of CNTFET, the drain and the source are connected through a carbon nanotube, which works as the channel material. This CNT is gated from the backside to control the channel; the CNT and the gate are separated by gate oxide similar to the MOSFET's structure [33]. The diameter of the CNT is taken as the 'd_{nt}' and the gate oxide thickness is shown by dox. In the cylindrical structure, the CNT is surrounded by the gate all around and provides strong control of the gate over the channel. The CNT under the gate electrode is lightly doped whereas the source and drain side are highly doped with the same type of doping. The planer structure is found to be simpler and more popular because of its easy construction, while the cylindrical structure is complex to form with available fabrication techniques [34].

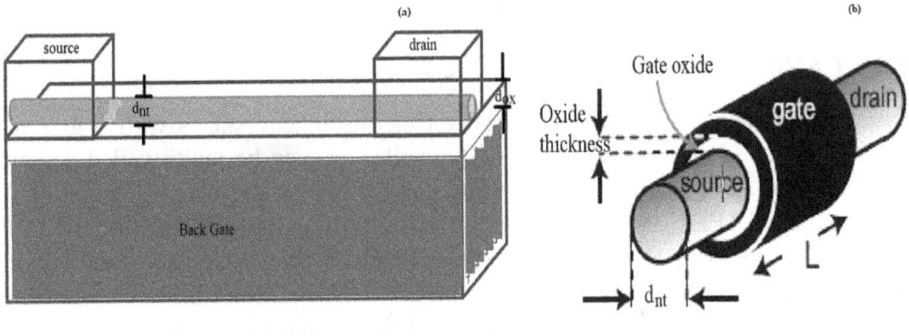

FIGURE 9.5 Device structure of (a) planer and (b) coaxial CNTFETs [33].

9.4.2 CNTFET CONDUCTION OPERATION

In the simple CNTFETs, the thermionic emission enables the charge carrier to cross the barrier developed between the source and the channel and eases the electron to move from the source to the channel, similar to the MOSFETs. However, in the tunnel CNTFETs, the inter-band tunneling is incorporated [35]. Here, the conduction mechanisms of tunnel CNTFETs are discussed. Figure 9.6a shows the cross-sectional view of the tunnel CNTFET in which the CNT under the gate electrode is taken as intrinsic and the source region is highly doped with p-type material and the drain side is highly doped with n-type material [36]. This allows for the formation of the PIN structure. The energy band diagram of the p-type CNTFETs is depicted in Figure 9.6b shows both the on state and the off-state conditions. Zener phenomenon enables the electron to tunnel from the conduction band to the valance band [37]. In the case of the off-state condition, when both V_{GS} and V_{DS} are zero, the electrons of the source available in the conduction band do not find any direct path to the tunnel to the valance band of the channel, since there is a large gap or high potential barrier between them. Therefore, in this condition, the electron-tunneling probability is very low, and results in a very low off-state current. While in the case of the on state, when both the voltages V_{GS} and V_{DS} are less than zero, the Fermi level of channel starts shifting upward or the conduction band of the channel starts shifting downward [38]. And this bias condition makes the barrier very narrow, so the electrons available in the conduction band of the source find a direct path and narrow potential barrier by which they can tunnel to the valance band of the channel as shown in Figure 9.6b. The increase in the bias voltage results in a narrower barrier and increases the tunneling probability of the electrons, which renders a large current [39].

CNT diameter is one of the most important parameters of the CNTFETs since CNTs are used as the channel material for CNTFETs and also the bandgap of the CNTs is inversely proportional to the diameter of the CNTs. Figure 9.7 shows the effect of the diameter variation on the off-state and on-state current, which corresponds with the reported results in ref. [40]. It can be observed that as the diameter of the CNT decreases the current capability of the CNTFET increases, it also can be seen that the effect on the off-state current is more significant as compared to the on-state current. Therefore, to remove this ambiguity the adjustment of the polarity gate is necessary.

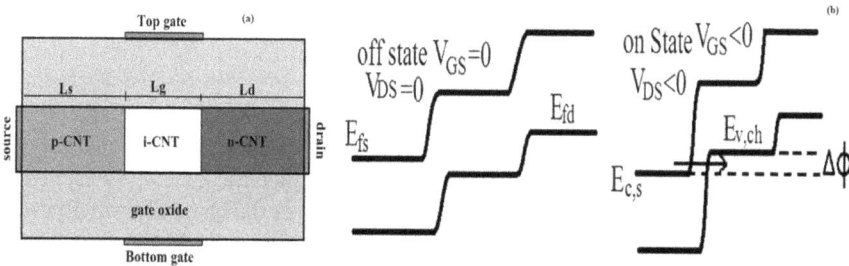

FIGURE 9.6 (a) Device structure. (b) Energy band diagram for the OFF and ON states of p-type Tunnel CNTFET [36].

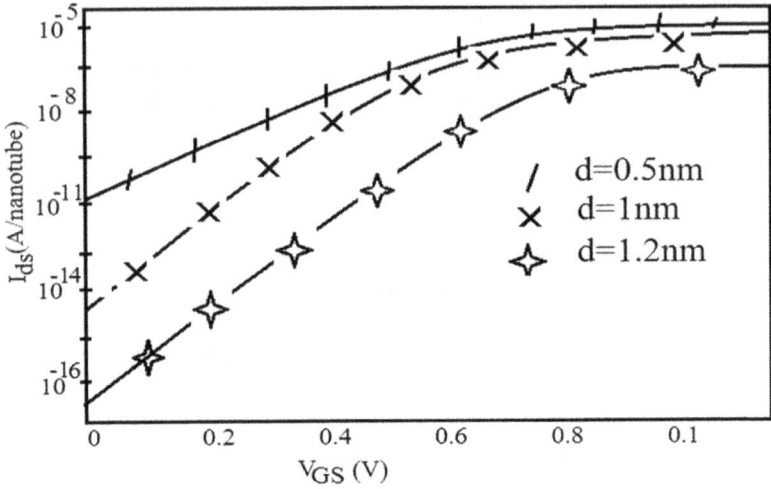

FIGURE 9.7 Impact of the diameter variation of the CNT on the drain current of the CNTFETs.

9.4.3 PERFORMANCE METRICS OF CNTFETS

The CNTFETs offer the sub-threshold slope in the range of 65mv/dec to 100mv/dec for varying the structural configuration of the CNTFETs [41]. The tunnel CNTFETs have the maximum SS whereas the conventional CNTFETs have the minimum SS. The conventional CNTFETs offer Ion/Ioff ratio in the orders of four while the tunnel CNTFETs offer Ion/Ioff ration in the orders of two. Apart from this, the trans-conductance ranges from 5 to 15, and the tunnel has the lowest whereas the partially gated CNTFETs have the highest trans-conductance value. The drive current offered by the single CNT is in the μA range and to obtain the current in mA hundreds of such CNTs are required beneath the gate [42]. The cutoff frequency f_T is 10 GHz reported at the gate length of 450 nm [43]. As the gate length is reduced, the pads' complete metallization involved in the cutoff frequency raises as far as 80GHz. Due to the limitation of the fabrication technique, most of the fabricated CNTFETs are metallic in nature, resulting in a low drain gain gd and thus poor intrinsic gain. The typical range of the gm/gd ratio reported is below unity [44].

9.5 NANOWIRE FETS (NWFETS)

Since short channel effects are a major concern in the nanoscale regime and prevent further scaling so as to avoid the SCEs, a surrounding gate usage is found to be an effective method. The surrounding gate is a way by which the gate is overlapped all around the channel and thus provides strong control over the channel from all directions [45]. It is reported that the use of the tri-gate can reduce the fabrication technology up to 22 nm [46]. A double gate gives double-inversion layers and thus increases the charge carriers or the capacity of the channel and also reduces the proportion of the leakage current. The tri-gate covers the channel from the three sides and thus gives more drive current by creating a stronger inversion layer on the underside of the gate [47]. The tri-gate technology is extended to the gate all around which is also called nanowire, hence achieving strong control of the gate over the channel and thus the drive current. It also removes the limitation of further scaling below 22 nm [48].

There are various advantages to the NWFETs over the conventional MOSFETs. First, the carries in the channel feel less scattering as compared to carriers in MOS devices since nanowire supports a one-dimensional conduction mechanism. A second benefit is that the nanowire exhibits different band structures from the planer device, since the nanowire band structure consists of some additional sub-bands near the lowest band in the energy band diagram. These sub-bands develop a multi-quantum channel and support the carrier conduction [49, 50].

9.5.1   DEVICE STRUCTURE OF NWFETs

A nanowire is a circular wire or hair-like structure, with a diameter in the range of 1 nm or below and length in the range of a few microns [51]. The materials used for the nanowire vary from application to application and these materials can be metal, semiconductor, chalcogenides, transition metal oxide, or a combination of the III-V group materials [52]. A small strip of the gate oxide and the gate electrode is overlapped over the nanowire in such a way that it does not cover both ends of the nanowire. The open end of the nanowire will work as a drain and source for the NWFETs [53]. Figure 9.8a shows double gate NWFETs, in this structure one gate is working as a control gate whereas another gate is working as a screen gate for the NWFETs. The cross-sectional view of Figure 9.8a is shown in Figure 9.8b, which shows two gate oxides one with high dielectric constant and another is with low dielectric constant. The two dielectrics are used to enhance the performance of the device [54].

9.5.2   CONDUCTION MECHANISM IN NWFETs

The conduction mechanism involves some additional quantum mechanical effects since the device is highly scaled [55]. In the nanoscale devices, the channel length of the device approaches the mean free path of the different carrier collision and the oxide thickness approaches some atomic layers' scale. Energy gets quantized near the oxide-channel interface due to the presence of a high electric field. The quantization of energy divides the continuous energy band into multiple sub-bands (Figure 9.9). The splitter energy band consists of two-dimensional (2D) density of states while the classical continuous energy band consists of three-dimensional (3D) density of states; and the charge carrier density in the 2D energy of states is less as compared to the 3D density of states. Therefore, to develop the same density of state a large gate voltage is needed [56, 57].

The quantization effects decrease the carrier density in the interface of the source and channel whereas the classical theory supports the high carrier density in the interface of the channel and source. This will also cause a large threshold voltage with respect to the conventional MOSFETs. The threshold voltage increases in order to account for the same band bend bending extra voltage required as compared to

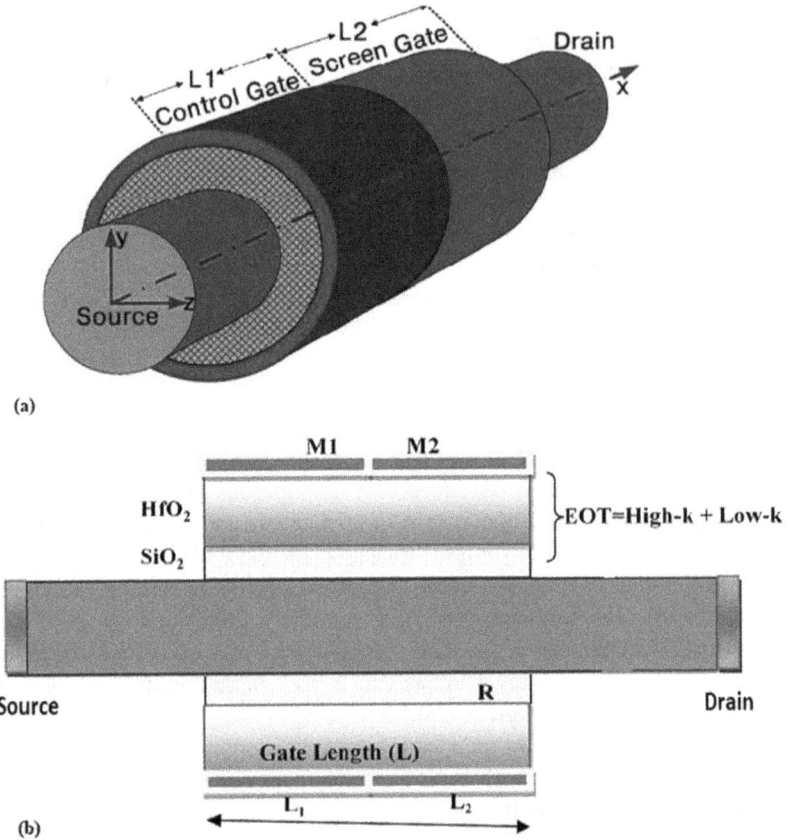

FIGURE 9.8 Device structure of the NWFETs (a) side view double gate; (b) cross-sectional view with dual gate and dual gate oxides [54].

a conventional MOS device, sine the 2D sheets have less charge carries. Due to the presence of the energy quantization effects, the behavior of the device cannot support classical physics therefore these devices need quantum physics theory [58].

9.5.3 PERFORMANCE PARAMETERS OF NWFETs

The performance parameters of the NWFETs are measured in the same ways as that of MOSFETs. Wenjun Li, et al., [59] reported various parameters of the Al_2O_3 gate-oxide based GaN NWFET, which exhibits 60mV/dec minimum SS. The same device exhibits a typical 27mV/V DIBL and a drive current of around 42 μA is observed. The Ion/Ioff ratio is large enough at 10^8, whereas the intrinsic transconductance is 27.8 μS/μm. For the Si-NW based FET, average values of various parameters like subthreshold slope range 100-130 mV/dec, Ion is in the range of μA whereas the I_{off} is in the range of nA. The transconductance ranges from (17-100) μS/μm [60]. These

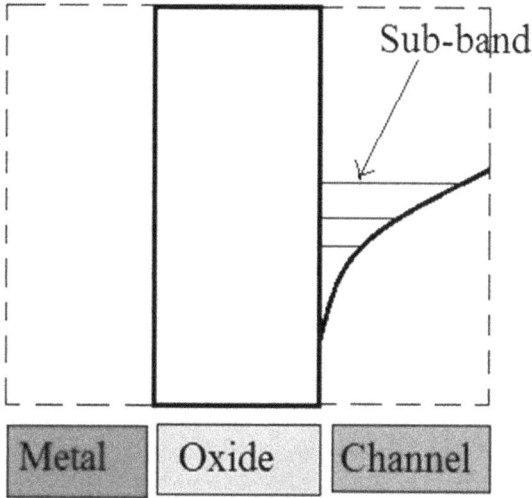

FIGURE 9.9   Multiple sub-bands due to the quantization effect.

parameters derive these GaN-based NWFETs, and have potential applications in the field of low power designs like the Internet of Things and sensors [61].

9.6 MEMRISTOR

A memristor is a nonlinear resistive device that has the ability to retain the particulars in the form of a high and low resistance state. It is observed that none of the electrical elements can give a direct relation between the electric fluxes and charge Equation 9.4 [62, 63]. The absent relation between the electric flux and electric charge was initially seen by Lion Chua in 1971. Both these parameters are connected by the memristor [64]. Since its behavior cannot be duplicated by the resistor, capacitor, and inductor, therefore it becomes the fourth fundamental component [62, 63]. Further, the concept of the memristor was generalized, and a class of memristor predicted by Kang and Chua. According to them, a device is said to be a memristor if its behavior can be characterized by two coupled Equations 9.5 and 9.6 [65, 66]. In these equations, the resistance of the device depends upon the state and the current flowing through the device. Moreover, it should form a pinched hysteresis loop between the current through the device and a periodic signal should be applied.

$$M = \frac{dq}{d\varphi} \tag{9.4}$$

$$v = R(s,i)i \tag{9.5}$$

$$\frac{ds}{dt} = f(s,i) \tag{9.6}$$

9.6.1   Device Structure of Memristor

Memristors consist of transition metal oxide (TMO) having a thickness of a few nm; this thin TMO layer is separated by two metal electrodes [62, 67]. The combined geometry of the memristor builds the metal-insulator-metal configuration. The first memristor was fabricated in 2008 with TiO2 metal oxide and platinum metal electrodes. In this structure, some engineered oxygen vacancies are created with the help of the annealing process at the one interface between the metal electrode and TMO [62]. Therefore, out of the two interfaces of the TMO with a metal electrode, one contains the TMO with less oxygen proportion, say TiO2-X, where 'x' denotes the proportion of oxygen vacancies [67, 68]. Further, Pickett modified the memristor structure and assumed a conductive filament along with a small tunneling gap inside the TMO, which is developed due to the electro-formation process and made of oxygen vacancies as shown in Fig. 9. 10 [69]. This conductive structure has the ability to realize well the dynamic behavior of the memristor. Essentially, the conductive filament is made of oxygen vacancies, which are important charge carriers and storage elements of the memristors [70].

9.6.2   Conduction Mechanism in Memristor

The transition metal oxides, which are placed in between the two metal electrodes to build a memristor of Resistive RAM (ReRAM), show complex, coupled electric-ionic motions, which involve the thermophoresis (the diffusion of the oxygen vacancies under the influence of the temperature) drift of the carriers under the applied field, diffusion of the carriers [71, 72]. Therefore, there is no single mechanism that governs the motion of the charge carriers inside the device. The hoping of the carriers is also included in the metal-oxides to enable the set and reset process [72].

Two types of conduction mechanism are generally supported by most of the ReRAMs and, depending upon these mechanisms, the RRAMs are classified into two groups [73, 74]. The first group is called the "non-filamentary" or "interfacial/electrode-limited", and the other group is known as "bulk limited". The ReRAMs fall into the electrode-limited group which supports Fowler-Nordheim tunneling, Schottky emission, and direct tunneling mechanism [74, 75]. On the other hand, the RRAMs which support Poole-Frankel emissions, Mott hoping, space charge limited conduction and trap assisted tunneling fall into the latter group [76]. In the electrode-limited conduction, the electrode interface and barrier height are the crucial parameters, while in the bulk-limited conduction, oxide electrical properties, defects, and traps inside the oxide are the key elements that are decisive in the conduction process. The TiO2 memristor shown in Figure 9.10 has a low-resistance conductive filament made of TiO2-x along with TiO2 insulating film and this was suggested by Pickett et al., [69]. This insulating film is separated by the conductive filament and platinum electrode and thus comprises a tunneling barrier. The tunneling process occurs between the metallic conductive filament and platinum electrode through the tunneling barrier of thickness considered as 's'. The tunneling width increases and decreases depending upon the nature of the applied voltage on the electrode. The positive bias repels the oxygen vacancies and thus the tunneling gap increases

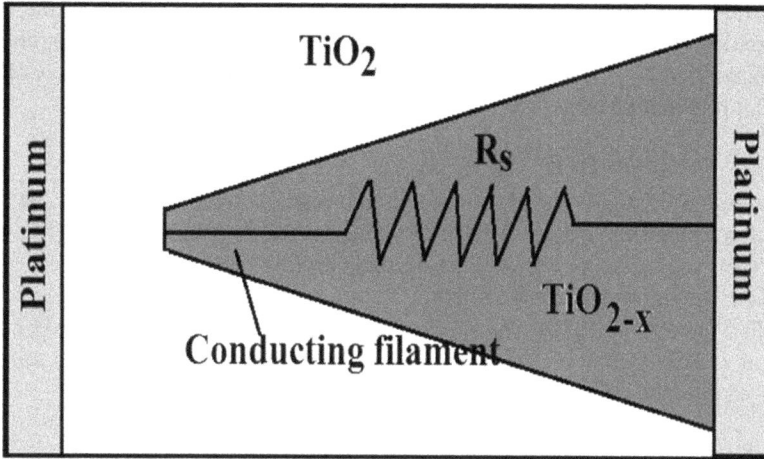

FIGURE 9.10 Memristor device structure.

while the negative voltage attracts the oxygen vacancies, hence the tunneling gap decreases.

9.6.3 PERFORMANCE PARAMETERS OF MEMRISTOR

The performance of non-volatile memories is determined by various parameters like the ratio of high resistance state (HRS) and low resistance state (LRS), endurance, retention time, read/write time, and capacity of storage. A stable and reliable memory cell is categorized on the basis of high endurance and large retention time.

9.6.3.1 ON/OFF Resistance Ratio

The HRS represents the OFF state and LRS represents the ON state of the memristors. To use memristors in switching application, the ON/OFF ratio is an important parameter and it is denoted by the proportion between resistances in HRS and LRS. The HRS and LRS are achieved in the memristor due to the rupture and formation of the conductive filament with the help of oxygen vacancies [77]. High ON/OFF resistance (>103) is often achieved in transition metal oxides for conductive filament-based memristors [78]. Spin Transfer Torque (STT) RAM and ferromagnetic RAM offer ON/OFF resistance less than 102 [79].

9.6.3.2 Endurance

The ability to retain a ON/OFF resistance state greater than a particular value for multiple reads and write operations is described as "endurance". The endurance or the time to hold the ON/ OFF state in resistive switching RAMs is an important parameter to determine metrics of the non-volatile memory. Since it is required to read and write the memory unit repeatedly, after a certain time the ON/OFF resistance state will collapse and the switching operation will abrupt, therefore it is mandatory for a switching operation to have high endurance. Endurance can also be defined as the number of successful write operations without affecting the stored information. The endurance of the Al/ZnO/Al-based memristor shows more than 250 stable switching cycles [80], whereas Pt/TaO$_x$/Ta based memristor shows 1×10^{10} open-loop switching cycles [81].

9.6.3.3 Retention Time

The retention time tells the ability of the memristor to retain the state of the memristor for a long time in the absence of applied input. It is basically the time taken by oxygen vacancies to diffuse back to their original state when the input field is removed. Al/ZnO/Al stacked structure shows 106 s retention time, whereas ITO/ZnO/ITO shows 105 s [82]. The standard retention time of 1 year is seen in emerging non-volatile memories; however, the retention time of STT-MRAM is degraded below 45 nm technology.

9.6.3.4 Read Time/Write Time

Writing to the memristor depends upon the magnitude, polarity, and the interval of the applied pulse, which has the ability to change the state of the memristor and it is given by the following equation:

$$\frac{dw}{dt} = \mu_n \frac{R_{on}}{D} i(t) \tag{9.7}$$

Where D is the dimension of the memristor, μ_n is the dopant mobility, and $i(t)$ is the current flowing the memristor. Reading from the memristor is found somehow to be a destructive operation because the applied read field can change or disturb the stored data in a memristor. The memristors like phase-change memory, and ReRAM write time varies from tens of to hundreds of picoseconds [83].

9.6.3.5 Storage Density

The storage density of the memristor-based memory goes beyond 100Gbits/cm2, which is higher than the NAND/NOR based flash memory having a storage density of 32Gbits/cm2. This is possible because of its nanoscale size [84]. It is found that memristor promises better performance than flash memory in terms of power dissipation, speed, endurance, and retention of data. Memristor-based Solid State Drive (SSD) has become more popular as compared to the flash-based SSD because of its high speed and good scalability. Memristor-based memory cells are also good at committing speed like the SRAM and density like the DRAM [85, 86].

9.7 CONCLUSION

This paper compares and analyses the structures, conduction mechanism, and static performance parameters of various advance semiconductor nanoscale devices like Tunnel FETs (TFETs), FinFETs, and Carbon Nanotube FETs (CNTFETs), Nanowire FETs, and Memristors. The TFETs use a PIN structure and incorporate BTBT and limit the off-state current which is high in the case of conventional MOSFETs, due to the thermionic emission. The FinFETs have three side gates over the channel and somehow limit the leakage since it gives more control over the channel, and also renders high on current. The CNTFETs have a planer as well as the co-axial structure, and these become one of the suitable candidates for the advances in the semiconductor industry. The NWFETs use the surrounding gate around the channel and also include quantum effects due to the very small dimension of the channel. Nanowire provides high Ion/Ioff ratio and is thus suitable for low-power sensor applications. Both CNTFETs and NWFETs support one-dimensional conduction instead of three-dimensional conduction like conventional MOSFETs. Among all the four novel FETs discussed above, the subthreshold slope of the TFETs is found to be at a minimum, although it is lacking in drive current. Memristor becomes one of the key elements in the development of advance memory and modern memory design since it integrates the speed of the SRAM, density, and storage capacity like DRAM, and non-volatility like flash memory.

REFERENCES

1. Veeraraghavan, Surya, and Jerry G. Fossum. "Short-channel effects in SOI MOSFETs." *IEEE Transactions on Electron Devices* 36(3) (1989): 522–528.
2. Kim, Yong-Bin. "Challenges for nanoscale MOSFETs and emerging nanoelectronics." *Transactions on Electrical and Electronic Materials* 11(3) (2010): 93–105.
3. Birla, Shilpi, Neeraj Kr. Shukla, Manisha Pattanaik, and R. K. Singh. "Device and circuit design challenges for low leakage SRAM for ultra-low power applications." *Canadian Journal on Electrical & Electronics Engineering* 1(7) (2010): 156–167.
4. Rana, Ashwani K., Narottam Chand, and Vinod Kapoor. "Modeling gate current of nano scale MOSFET for circuit simulation." *Multidiscipline Modeling in Materials and Structures* 7(2) (2011): 115–130.
5. Vishvakarma, S. K., A. K. Saxena, and S. Dasgupta. "Analytical modeling of inversion charge density for nanoscale dual metal gate (Hf/AlNx) and midgap symmetric double gate MOSFET." *Journal of Nanoelectronics and Optoelectronics* 4(3) (2009): 370–378.
6. Allendorf, Mark D., Adam Schwartzberg, Vitalie Stavila, and A. Alec Talin. "A roadmap to implementing metal–organic frameworks in electronic devices: Challenges and critical directions." *Chemistry–a European Journal* 17(41) (2011): 11372–11388.
7. Ionescu, Adrian M., and Heike Riel. "Tunnel field-effect transistors as energy-efficient electronic switches." *Nature* 479(7373) (2011): 329–337.
8. Zener, Clarence. "A theory of the electrical breakdown of solid dielectrics." *Proceedings of the Royal Society of London. Series A, Containing Papers of a Mathematical and Physical Character* 145(855) (1934): 523–529.
9. Kane, Evan O. "Theory of tunneling." *Journal of Applied Physics* 32(1) (1961): 83–91.
10. Raj, B., A. K. Saxena, and S. Dasgupta. "A compact drain current and threshold voltage quantum mechanical analytical modeling for FinFETs." *Journal of Nanoelectronics and Optoelectronics* 3(2) (2008): 163.
11. Choi, Woo Young, Byung-Gook Park, Jong Duk Lee, and Tsu-Jae King Liu. "Tunneling field-effect transistors (TFETs) with subthreshold swing (SS) less than 60 mV/dec." *IEEE Electron Device Letters* 28(8) (2007): 743–745.
12. Bhuwalka, Krishna Kumar, Jörg Schulze, and Ignaz Eisele. "Performance enhancement of vertical tunnel field-effect transistor with SiGe in the δp+ layer." *Japanese Journal of Applied Physics* 43(7R) (2004): 4073.
13. Colinge, J.-P. "Subthreshold slope of thin-film SOI MOSFET's." *IEEE Electron Device Letters* 7(4) (1986): 244–246.
14. Zhang, Qin, Wei Zhao, and Alan Seabaugh. "Low-subthreshold-swing tunnel transistors." *IEEE Electron Device Letters* 27(4) (2006): 297–300.

15. Raj, Balwinder, A. K. Saxena, and S. Dasgupta. "Analytical modeling for the estimation of leakage current and subthreshold swing factor of nanoscale double gate FinFET device." *Microelectronics International* 26(1) (2009): 53–63.

16. Boucart, Kathy, and Adrian Mihai Ionescu. "Length scaling of the double gate tunnel FET with a high-k gate dielectric." *Solid-State Electronics* 51(11–12) (2007): 1500–1507.

17. Taur, Yuan, Douglas A. Buchanan, Wei Chen, David J. Frank, Khalid E. Ismail, Shih-Hsien Lo, George A. Sai-Halasz et al. "CMOS scaling into the nanometer regime." *Proceedings of the IEEE* 85(4) (1997): 486–504.

18. Balestra, F., S. Cristoloveanu, F. Benachir, J. Brini, and T. Elewa. "Double-gate silicon-on-insulator transistor with volume inversion: A new device with greatly enhanced performance." *IEEE Electron Device Letters* 8(9) (1987): 410–412.

19. Khandelwal, Sourabh, Yogesh Singh Chauhan, Darsen D. Lu, Sriramkumar Venugopalan, and Muhammed Ahosan Ul Karim, Angada Bangalore Sachid, Bich-Yen Nguyen et al. "BSIM-IMG: A compact model for ultrathin-body SOI MOSFETs with back-gate control." *IEEE Transactions on Electron Devices* 59(8) (2012): 2019–2026.

20. Wind, S. J., D. J. Frank, and H.-S. Wong. "Scaling silicon MOS devices to their limits." *Microelectronic Engineering* 32(1–4) (1996): 271–282.

21. de Andrade, Maria Glória Caño, João Antonio Martino, Marc Aoulaiche, Nadine Collaert, Eddy Simoen, and Cor Claeys. "Behavior of triple-gate Bulk FinFETs with and without DTMOS operation." *Solid-State Electronics* 71 (2012): 63–68.

22. Jeong, Yongjin, In Man Kang, Seongjae Cho, Jisun Park, and Hyungsoon Shin. "Charge based current–voltage model for the silicon on insulator junctionless field-effect transistor." *Journal of Nanoscience and Nanotechnology* 20(8) (2020): 4920–4925.

23. Choi, Yunho, Kitae Lee, Kyoung Yeon Kim, Sihyun Kim, Junil Lee, Hyun-Min Kim Ryoongbin Lee, H. Kim et al. "Simulation of the effect of parasitic channel height on characteristics of stacked gate-all-around nanosheet FET." *Solid-State Electronics* 164 (2020): 107686.

24. Raj, Balwinder, A. K. Saxena, and S. Dasgupta. "Nanoscale FinFET based SRAM cell design: Analysis of performance metric, process variation, underlapped FinFET, and temperature effect." *IEEE Circuits and Systems Magazine* 11(3) (2011): 38–50.

25. Cho, Hyungki, and Changhwan Shin. "DIBL enhancement in ferroelectric-gated FinFET." *Semiconductor Science and Technology* 34(2) (2019): 025004.

26. Banerjee, Subham, and Buddhadev Pradhan. "Analytical model of subthreshold swing in triangular-shaped FinFET." In: *Devices for Integrated Circuit (DevIC)*. IEEE, Kalyani, India, 2019, pp. 42–44.

27. Narendar, Vadthiya, Richa Parihar, and Ashutosh Kumar Pandey. "Short Channel effects (SCEs) based comparative study of double-gate (DG) and gate-all-around (GAA) FinFET structures for nanoscale applications." In: *Advances in VLSI, Communication, and Signal Processing*. Springer, Singapore, 2020, pp. 673–681.

28. Jain, N., and B. Raj, 2016. "Device and circuit co-design perspective comprehensive approach on FinFET technology-a review." *Journal of Electron Devices* 23(1): 1890–1901.

29. Yao, Zhen, Henk W. Ch. Postma, Leon Balents, and Cees Dekker. "Carbon nanotube intramolecular junctions." *Nature* 402(6759) (1999): 273–276.

30. Lin, Sheng, Yong-Bin Kim, and Fabrizio Lombardi. "CNTFET-based design of ternary logic gates and arithmetic circuits." *IEEE Transactions on Nanotechnology* 10(2) (2009): 217–225.

31. Singh, Amandeep, Mamta Khosla, and Balwinder Raj. "Design and analysis of dynamically configurable electrostatic doped carbon nanotube tunnel FET." *Microelectronics Journal* 85 (2019): 17–24.

32. Singh, Amandeep, Mamta Khosla, and Balwinder Raj. "Design and analysis of electrostatic doped Schottky barrier CNTFET based low power SRAM." *AEU-International Journal of Electronics and Communications* 80 (2017): 67–72.

33. Koswatta, Siyuranga O., Dmitri E. Nikonov, and Mark S. Lundstrom. "Computational study of carbon nanotube pin tunnel FETs." In: *IEEE InternationalElectron Devices Meeting, 2005. IEDM Technical Digest.* IEEE, Washington, DC, 2005, pp. 518–521.

34. Ohno, Yutaka, Shigeru Kishimoto, Takashi Mizutani, Toshiya Okazaki, and Hisanori Shinohara. "Chirality assignment of individual single-walled carbon nanotubes in carbon nanotube field-effect transistors by micro-photocurrent spectroscopy." *Applied Physics Letters* 84(8) (2004): 1368–1370.

35. Knoch, Joachim, and Joerg Appenzeller. "Tunneling phenomena in carbon nanotube field-effect transistors." *Physica Status Solidi (A)* 205(4) (2008): 679–694.

36. Bala, Shashi, and Mamta Khosla. "Electrostatically doped tunnel CNTFET model for low-power VLSI circuit design." *Journal of Computational Electronics* 17(4) (2018): 1528–1535.

37. Zener, C. "A theory of the electrical breakdown of solid dielectrics." *Proceedings of the Royal Society of London. Series A, Containing Papers of a Mathematical and Physical Character* 145(855) (1934): 523–529.

38. Appenzeller, Joerg, Yu-Ming Lin, Joachim Knoch, Zhihong Chen, and Phaedon Avouris. "Comparing carbon nanotube transistors-the ideal choice: A novel tunneling device design." *IEEE Transactions on Electron Devices* 52(12) (2005): 2568–2576.

39. Raj, Balwinder, A. K. Saxena, and S. Dasgupta. "Quantum mechanical analytical modeling of nanoscale DG FinFET: Evaluation of potential, threshold voltage and source/drain resistance." *Materials Science in Semiconductor Processing* 16(4) (2013): 1131–1137.

40. Bala, S., and M. Khosla. "Design and analysis of electrostatic doped tunnel CNTFET for various process parameters variation." *Superlattices and Microstructures* 124 (2018): 160–167.

41. Naderi, Ali, Maryam Ghodrati, and Sobhi Baniardalani. "The use of a Gaussian doping distribution in the channel region to improve the performance of a tunneling carbon nanotube field-effect transistor." *Journal of Computational Electronics* 19(1) (2020): 283–290.

42. Sharma, Vijay Kumar, Manisha Pattanaik, and Balwinder Raj. "INDEP approach for leakage reduction in nanoscale CMOS circuits." *International Journal of Electronics* 102(2) (2015): 200–215.

43. Naderi, Ali, and Maryam Ghodrati. "Cut off frequency variation by ambient heating in tunneling pin CNTFETs." *ECS Journal of Solid-State Science and Technology* 7(2) (2018): M6–M10.

44. Shaukat, Ayesha, Rahila Umer, and Naz Islam. "Impact of dielectric material and oxide thickness on the performance of Carbon nanotube Field Effect Transistor." In: *2017 IEEE 17th International Conference on Nanotechnology (IEEE-NANO).* IEEE, Pittsburgh, PA, 2017, pp. 250–254.

45. Auth, Christopher P., and James D. Plummer. "Scaling theory for cylindrical, fully-depleted, surrounding-gate MOSFET's." *IEEE Electron Device Letters* 18(2) (1997): 74–76.

46. Jain, Aakash, Sanjeev Kumar Sharma, and Balwinder Raj. "Design and analysis of high sensitivity photosensor using cylindrical surrounding gate MOSFET for low power applications." *Engineering Science and Technology, An International Journal* 19(4) (2016): 1864–1870.

47. Knoch, Joachim, Siegfried Mantl, and J. Appenzeller. "Impact of the dimensionality on the performance of tunneling FETs: Bulk versus one-dimensional devices." *Solid-State Electronics* 51(4) (2007): 572–578.

48. Veloso, A., A. De Keersgieter, P. Matagne, N. Horiguchi, and N. Collaert. "Advances on doping strategies for triple-gate finFETs and lateral gate-all-around nanowire FETs and their impact on device performance." *Materials Science in Semiconductor Processing* 62 (2017): 2–12.

49. Khandelwal, Saurabh, Balwinder Raj, and R. D. Gupta. "Finfet based 6t sram cell design: Analysis of performance metric, process variation and temperature effect." *Journal of Computational and Theoretical Nanoscience* 12(9) (2015): 2500–2506.

50. Song, Taigon. "Opportunities and challenges in designing and utilizing vertical nanowire FET (V-NWFET) standard cells for beyond 5 nm." *IEEE Transactions on Nanotechnology* 18 (2019): 240–251.

51. Hayden, Oliver, Ritesh Agarwal, and Wei Lu. "Semiconductor nanowire devices." *Nano Today* 3(5–6) (2008): 12–22.

52. Zekentes, K., and K. Rogdakis. "SiC nanowires: Material and devices." *Journal of Physics. Part D: Applied Physics* 44(13) (2011): 133001.

53. Goh, Kian-Hui, Sachin Yadav, Kain Lu Low, Gengchiau Liang, Xiao Gong, and Yee-Chia Yeo. "Gate-all-around in 0.53 Ga 0.47 As junctionless nanowire FET with tapered source/drain structure." *IEEE Transactions on Electron Devices* 63(3) (2016): 1027–1033.

54. Sharma, Sanjeev Kumar, Balwinder Raj, and Mamta Khosla. "Subthreshold Performance of In 1–x Ga x As Based dual Metal with Gate Stack Cylindrical/Surrounding Gate nanowire MOSFET for Low Power Analog Applications." *Journal of Nanoelectronics and Optoelectronics* 12(2) (2017): 171–176.

55. Dasgupta, Avirup, Amit Agarwal, and Yogesh Singh Chauhan. "Unified compact model for nanowire transistors including quantum effects and quasi-ballistic transport." *IEEE Transactions on Electron Devices* 64(4) (2017): 1837–1845.

56. Kumar, Sunil, and Balwinder Raj. "Compact channel potential analytical modeling of DG-TFET based on Evanescent-mode approach." *Journal of Computational Electronics* 14(3) (2015): 820–827.

57. Singh, Karmjit, and Balwinder Raj. "Temperature-dependent modeling and performance evaluation of multi-walled CNT and single-walled CNT as global interconnects." *Journal of Electronic Materials* 44(12) (2015): 4825–4835.

58. Tamersit, Khalil. "Improving the performance of a junctionless carbon nanotube field-effect transistor using a split-gate." *AEU-International Journal of Electronics and Communications* 115 (2020): 153035.

59. Li, Wenjun, Matt D. Brubaker, Bryan T. Spann, Kris A. Bertness, and Patrick Fay. "GaN nanowire MOSFET with near-ideal subthreshold slope." *IEEE Electron Device Letters* 39(2) (2017): 184–187.

60. Guerfi, Y., and G. Larrieu. "Vertical silicon nanowire field effect transistors with nanoscale gate-all-around." *Nanoscale Research Letters* 11(1) (2016): 1–7.

61. Tan, Siew Li, Xingyan Zhao, Kaixiang Chen, Kenneth B. Crozier, and Yaping Dan. "High-performance silicon nanowire bipolar phototransistors." *Applied Physics Letters* 109(3) (2016): 033505.

62. Strukov, Dmitri B., Gregory S. Snider, Duncan R. Stewart, and R. Stanley Williams. "The missing memristor found." *Nature* 453(7191) (2008): 80–83.

63. Singh, Jeetendra, and Balwinder Raj. "Comparative analysis of memristor models and memories design." *Journal of Semiconductors* 39(7) (2018): 074006.

64. Chua, Leon. "Memristor-the missing circuit element." *IEEE Transactions on Circuit Theory* 18(5) (1971): 507–519.

65. Chua, Leon O., and Sung Mo Kang. "Memristive devices and systems." *Proceedings of the IEEE* 64(2) (1976): 209–223.

66. Chua, Leon. "Resistance switching memories are memristors." *Applied Physics. Part A* 102(4) (2011): 765–783.

67. Strukov, Dmitri B., and R. Stanley Williams. "Exponential ionic drift: Fast switching and low volatility ofáthin-film memristors." *Applied Physics: Part A* 94(3) (2009): 515–519.
68. Singh, Jeetendra, and Balwinder Raj. "Evaluation of nanoscale memristor device for analog and digital application." *Nanoscale Devices: Physics, Modeling, and Their Application* 393 (2018): 393–423.
69. Pickett, Matthew D., Dmitri B. Strukov, Julien L. Borghetti, J. Joshua Yang, Gregory S. Snider, Duncan R. Stewart, and R. Stanley Williams. "Switching dynamics in titanium dioxide memristive devices." *Journal of Applied Physics* 106(7) (2009): 074508.
70. Singh, Jeetendra, and Balwinder Raj. "Modeling of mean barrier height levying various image forces of metal–insulator–metal structure to enhance the performance of conductive filament based memristor model." *IEEE Transactions on Nanotechnology* 17(2) (2018): 268–275.
71. Strukov, Dmitri B., Julien L. Borghetti, and R. Stanley Williams. "Coupled ionic and electronic transport model of thin-film semiconductor memristive behavior." *Small* 5(9) (2009): 1058–1063.
72. Singh, Jeetendra, and Balwinder Raj. "Temperature dependent analytical modeling and simulations of nanoscale memristor." *Engineering Science and Technology, An International Journal* 21(5) (2018): 862–868.
73. Lim, Ee Wah, and Razali Ismail. "Conduction mechanism of valence change resistive switching memory: A survey." *Electronics* 4(3) (2015): 586–613.
74. Chiu, Fu-Chien. "A review on conduction mechanisms in dielectric films." *Advances in Materials Science and Engineering* 2014 (2014): 1–18.
75. Kumar, Suhas, Ziwen Wang, Xiaopeng Huang, Niru Kumari, Noraica Davila, John Paul Strachan, David Vine, A. L. David Kilcoyne, Yoshio Nishi, and R. Stanley Williams. "Conduction channel formation and dissolution due to oxygen thermophoresis/diffusion in hafnium oxide memristors." *ACS Nano* 10(12) (2016): 11205–11210.
76. Murakami, Katsuhisa, Mathias Rommel, Vasil Yanev, Tobias Erlbacher, Anton J. Bauer, and Lothar Frey. "A highly sensitive evaluation method for the determination of different current conduction mechanisms through dielectric layers." *Journal of Applied Physics* 110(5) (2011): 054104.
77. Janousch, Markus, G. Ingmar Meijer, Urs Staub, Bernard Delley, Siegfried F. Karg, and Björn P. Andreasson. "Role of oxygen vacancies in Cr-doped SrTiO3 for resistance-change memory." *Advanced Materials* 19(17) (2007): 2232–2235.
78. Jo, Sung Hyun, and Wei Lu. "CMOS compatible nanoscale nonvolatile resistance switching memory." *Nano Letters* 8(2) (2008): 392–397.
79. Wang, Xiaobin, Yiran Chen, Haiwen Xi, Hai Li, and Dimitar Dimitrov. "Spintronic memristor through spin-torque-induced magnetization motion." *IEEE Electron Device Letters* 30(3) (2009): 294–297.
80. Kumar, A., M. Das, V. Garg, B. S. Sengar, M. T. Htay, S. Kumar, A. Kranti, and S. Mukherjee. "Forming-free high-endurance Al/ZnO/Al memristor fabricated by dual ion beam sputtering." *Applied Physics Letters* 110(25) (2017): 253509.
81. Yang, J. Joshua, M.-X. Zhang, John Paul Strachan, Feng Miao, Matthew D. Pickett, Ronald D. Kelley, G. Medeiros-Ribeiro, and R. Stanley Williams. "High switching endurance in TaO x memristive devices." *Applied Physics Letters* 97(23) (2010): 232102.
82. Li, Wenqing, Xinqiang Liu, Yongqiang Wang, Zhigao Dai, Wei Wu, Li Cheng, Yupeng Zhang, Qi Liu, Xiangheng Xiao, and Changzhong Jiang. "Design of high-performance memristor cell using W-implanted SiO2 films." *Applied Physics Letters* 108(15) (2016): 153501.
83. Seo, Jung Won, Jae-Woo Park, Keong Su Lim, Ji-Hwan Yang, and Sang Jung Kang. "Transparent resistive random-access memory and its characteristics for nonvolatile resistive switching." *Applied Physics Letters* 93(22) (2008): 223505.

84. Hwang, Y. N., J. S. Hong, S. H. Lee, S. J. Ahn, G. T. Jeong, G. H. Koh, J. H. Oh et al. "Full integration and reliability evaluation of phase-change RAM based on 0.24/ spl mu/m-CMOS technologies." In: *2003 Symposium on VLSI Technology. Digest of Technical Papers (IEEE Cat. No. 03CH37407)*. IEEE, Kyoto, Japan, 2003, pp. 173–174.

85. Talati, N., S. Gupta, P. Mane, and S. Kvatinsky. "Logic design within memristive memories using memristor-aided loGIC (MAGIC)." *IEEE Transactions on Nanotechnology* 15(4) (2016): 635–650.

86. Singh, J., and B. Raj. "Design and investigation of 7T2M-NVSRAM With enhanced stability and temperature impact on store/restore energy." *IEEE Transactions on Very Large-Scale Integration (VLSI) Systems* 27(6) (2019): 1322–1328.

10 Nanotechnology-Mediated Strategy for the Treatment of Neuropathic Pain

A Promising Approach

Pankaj Prashar, Ankita Sood, Anamika Gautam, Pardeep Kumar Sharma, Bimlesh Kumar, Indu Melkani, Sakshi Panchal, Sachin Kumar Singh, Monica Gulati, Narendra Kumar Pandey, Linu Dash, Anupriya and Varimadugu Bhanukirankumar Reddy
Lovely Professional University, Phagwara, India

CONTENTS

10.1   NEUROPATHIC PAIN

In 1994, the definition of neuropathic pain (NP) was described by the International Association for the Study of Pain as "pain initiated or caused by a primary lesion or dysfunction in the nervous system." Then after, in 2008, a task force begun by the IASP Special Interest Group on NP (NeuSIG) decided to omit the term "dysfunction" in order to distinguish NP from nociceptive pain. Hence, a slightly modified version of definition was endorsed by a taxonomical committee of IASP: "pain caused by a lesion or disease of the somatosensory nervous system" (Taxonomy 2012).

NP (NP) is different from nociceptive pain in terms of various manners i.e. nociceptive pain requires the process of transduction to convert a non-electrical signal (noxious stimuli) to an electrical one (nociceptive signal), but NP involves direct nerve stimulation. In addition, most people with nociceptive pain (e.g. post-surgery period) recover which is quite difficult in people suffering from nerve injuries (Ciaramitaro et al. 2010). It is known to have association with multiple conditions like multiple sclerosis, diabetes, trigeminal neuralgia, carpal tunnel syndrome, Guillain-Barré syndrome, cancer-related pain, etc. (McCarberg et al. 2017). NP is identified by both positive (tingling, prickling, aching, tightening, burning, and electrical-like sensations) as well as negative symptoms (numbness, deadness, anxiety, and a feeling of wearing socks all the time).

Quantifying the prevalence of NP within the general population is very difficult due to a lack of appropriate research designs as well as diagnostic criteria. However, it has been estimated that between 3–17% of the general population is suffering from NP globally. The perceived pain in patients experiencing NP is usually spontaneous, i.e. occurring without any stimulus. This condition highly affects the quality of life of patients, including their psychological state (Cavalli et al. 2019). Hence, it has become a major concern worldwide.

Various diagnostic procedures are also available that articulate NP, including the history of the patient (any lesion that can damage somatosensory system), physical examination, and their clinical assessments (standard electromyography, quantitative sensory testing, brain or spinal cord imaging, nerve or skin biopsies) (Gilron, Baron, and Jensen 2015; Colloca et al. 2017). In addition, screening tools also have been designed that help to identify NP easily. These tools have been developed and validated in the form of questionnaires making the use of verbal pain descriptors like tingling, prickling, etc. Over the last 15 years, various screening tools have been developed, like the Michigan Neuropathy Screening Instrument, NP Scale, NP Questionnaire, NP Symptom Inventory, etc. (Didier Bouhassira and Attal 2011; Attal, Bouhassira, and Baron 2018; D Bouhassira 2019).

Despite multiple pharmacotherapy approaches, it is still very difficult to treat this distressing disorder as only 15–30% of patients receive adequate relief. Various factors like inaccurate dose, side effects, and psychological as well as associated emotional disturbances (sleep, depression, anxiety) are responsible for inadequate relief. Hence to resolve these limitations, a methodical, mechanistic, and interdisciplinary approach is urgently needed for the treatment of NP. The progress of nanotechnology and its uses in the twentieth century in the field of drugs and pharmaceuticals have contributed very well to the availability of a solution.

10.2 MOLECULAR TARGETS OF NP

In recent times, research has focused on new drug development, cell therapy, genes, and tissue regeneration-like approaches for the treatment of NP. Various novel molecular targets have been identified for the treatment of NP (Table 10.1). Sigma-1 receptors are found to be a promising target whose inhibition by Sigma-1 receptor antagonists can be used for the treatment of NP (Bravo-Caparrós et al. 2019). Furthermore, targets like ephrin b receptors as well as endoplasmic reticulum stress (ERS) receptors show anti-neuropathic activity (Y. Peng, Zang, et al. 2019; Kong et al. 2020). Inhibition of NOS1AP (nitric oxide synthase 1 adapter protein) is also one of the new approaches that develops an anti-hyperalgesic as well as anti-allodynic effect (Lee et al. 2018).

10.2.1 SIGMA RECEPTORS

Sigma receptors (Sig-Rs) are special proteins first identified in 1976 and initially combined with a different type of opioid receptors. Sigma receptors have been evaluated and cloned, no structural or biochemical resemblance to opioid receptors has been identified. Two subtypes of sigma receptors have been established in a biochemical analysis: Sig-1Rs (5-01) and Sig-2Rs (5-02). Sig-1Rs are found at the endoplasmic membrane connected with the mitochondria and are broadly spread to the respiratory, kidney, digestive, and nervous systems. Several studies have shown the function of Sig-1R in pain sensitization, and may therefore aid in the treatment of pain, particularly NP. Research has shown that the amount of Sig-1Rs in the damaged and adjacent neurons is increased as a consequence of nerve injury (Bruna and Velasco 2018; T. Song et al. 2017). The essential function of spinal Sig-1Rs in mechanical allodynia was shown by the stimulation of spinal NMDA receptors in the chronic constriction injury model of rats. Mechanical allodynia was mitigated by intrathecal delivery of Sig-1R antagonist, BD1047, two times every day in the induction period i.e. 0–5 days of NP. Rather, BD1047 did not attenuate mechanical allodynia in the maintenance phase (15–20 days). Such results indicate Sig-1R antagonists could be

TABLE 10.1

Recent Molecular Targets for Treatment of NP

S. No.	Molecular Targets	Mechanisms	Treatments and Evidence	Chemical Structures	References
1	Sigma-1 receptors	• Activates p38 MAPK (mitogen-activated protein kinase) pathway via Ca^{2+} cascades/NO (nitric oxide) signaling pathways • Increases TNF-α (tumor necrotic factor) in hippocampus • Increases reactive oxygen species production through increased NADPH oxidase 2 (Nox2) expression • Increase Ca^{2+} dependent ERK (extracellular kinase) phosphorylation	• Intrathecal administration of Sigma-1 receptor antagonist BD1047 twice daily for postoperative days 0 to 5 attenuated the mechanical allodynia	BD1047	(Huang et al. 2017; Martuscello et al. 2012; Choi et al. 2013; Z. Chen et al. 2010; Bravo-Caparrós et al. 2019; Ishikawa and Hashimoto 2010)
2	Eph (ephrin) B1 receptor	• Increases the excitability of nociceptive related neurons and synaptic plasticity at spinal level, leading to induction of NP • Activation of MAPK pathways through NMDA receptors to induce pain behaviors	• Administration of ephrinB2 siRNA led to reduction in the expression of ephrinB2 and attenuated nerve injury-induced mechanical allodynia	–	(X.J. Song et al. 2008; Kobayashi et al. 2007; Y. Peng, Zang, et al. 2019)
3	Endoplasmic reticulum stress (ERS) receptors	• Continuous and persistent activation of ER stress response causes neuroinflammation through various mechanisms like reactive oxygen species, calcium, nuclear factor-κB (NK-κB), and mitogen-activated protein kinase (MAPK)	• Salubrinal (ERS modulator) into TG (trigeminal ganglion) for 5 days following CFA (Complete Freund's Adjuvant) injection, attenuated heat hyperalgesia	Salubrinal	(Hu et al. 2006; Yang et al. 2014; Kong et al. 2020; S. Chen et al. 2017)
4	Wnt/β-catenin pathway	• Increased β-catenin binds to TCF4 (T cell 4 factor) and activates Wnt target genes like (TNF-α, IL-18 and BDNF (brain derived neurotropic factor) that are responsible for NP	Intraperitoneal injections of crocin at doses of 50 and 100mg/kg attenuated the hyperalgesia	Crocin	(Zhang et al. 2013; J.-F. Wang et al. 2020; Ahmad Dar et al. 2012)

(Continued)

TABLE 10.1 (CONTINUED)

Recent Molecular Targets for Treatment of NP

S. No.	Molecular Targets	Mechanisms	Treatments and Evidence	Chemical Structures	References
5	Histone deacetylase enzymes	• Nerve injury produces epigenetic changes due to activation of histone deacetylase enzymes, decrease histone acetylation to induce NP	• Oral doses (25 mg/kg and 50 mg/kg) of sodium valproate was found to have modulatory effect on NP and inflammatory reactions	Sodium valproate	(Khangura et al. 2017; Elsherbiny et al. 2019; Alsarra, Al-Omar, and Belal 2005)
6	Mitochondrial ATPase	• Formation of peroxynitrite due to nerve injury damages the mitochondria and decrease the production of ATP, which may contribute to induction of NP • The decreased energy production associated with mitochondrial dysfunction reduces Na⁺/K⁺ATPase activity is responsible for membrane hyper-excitability that causes NP	• Intrathecal ATP injection reduced thermal and mechanical hyperalgesia after nerve injury	–	(Lim et al. 2015; K.-H. Chen et al. 2014; Areti et al. 2016)
7	Tetrahydrobiopterin pathway (BH4)	• Sepiapterin reductase (SPR) catalyzes process of production of BH4 Induces macrophage dependent inflammatory cascade led to NP	• Tranilast treated NP by Inhibition of Sepiapterin Reductase in the BH4 Pathway • SPRi3 administered at 100, 200, or 300 mg/kg (i.p.) produced a dose-dependent anti-tactile allodynic effect in mice in the SNI model of NP	Tranilast	(Latremoliere et al. 2015; Moore et al. 2019; Fernandes et al. 2018; Khan et al. 2019)

effective as a preventive analgesic in chronic NP. The anti-allodynic effect of the Sig-1R antagonist at an induction period was associated with blocking the expression and phosphorylation of the NMDA receptor subunit NR1, the key receptor present in the spinal cord (Sánchez-Fernández et al. 2017).

10.2.2 EPH RECEPTOR

Eph receptor tyrosine kinases and their ligands include ephrins in many areas of growth, such as patterning of the tissue, angiogenesis, and formation of synapses. The laminae I-III and the tiny and mid-sized DRG nerves are found in many Eph and ephrin-receptor proteins. EphB receptors and ephrins amplify the neuronal behavior of the spinal cord, thus inducing sensory disturbances of NMDA-dependent pain disorders indicating a significant role of ephrin in spinal cords physiologic and pathological pain regulation. Eph B2 and Eph B1 isoforms in neurons of the spinal cord are increased in spinal nerve damage (Li-Na Yu et al. 2017). Administration of Eph B2 siRNA decreased production of Eph B2 and inhibited mechanical allodynia caused by nerve damage. Activation of the EphB1 and ephrin B2 signaling pathway may also be suspected in NP. There is a link between the production of hyperalgesia in CCI and dorsal rhizotomy (DR) models that is due to an increase in the expression of Eph B1 and EphB receptor proteins in DRG neurons and the dorsal horn. After nerve damage, NP is caused by the stimulation and transmission of EphB receptors in DRG and dorsal horn (Khangura et al. 2019). EphB-receptor antagonists' intrathecal administration has also been shown to prevent the induction and maintenance of a mechanical allodynia and thermal hyperalgesia due to nerve injury. Also, EphB antagonists blocked hyperactivity of nociceptive small DRG neurons and dorsal horn neurons. In addition, intrathecal injection of EphB activator in non-injured animals caused thermal hypersensitivity and decreased the long-term potentiation (LTP) threshold. Therefore, the increased regulation of ephrinB1 and EphB1 receptor proteins after nerve injury will enhance the excitability and plasticity of the neurons at spinal level that contribute to NP induction. Increased EphrinB1/EphB signaling results in increased PKCμ, NMDA, MAPK, P13 K, and p-AKT phosphorylation. All of these leads to enhanced excitability of nociceptive neurons and synaptic plasticity that are basic pathways of NP induction (Lombardi 2017).

10.2.3 ERS (ENDOPLASMIC RETICULUM STRESS) RECEPTORS

In many pathological conditions, including glucose deficiency, calcium exhaustion, free radical exposure, and the buildup of unfolded proteins, the normal working of ER has been impaired, inducing ER stress, which severely inhibits the protein folding cycle. Cells respond to stress by molecular chaperone expression like immunoglobulin protein binding, which starts a defensive mechanism known as unfold protein response (UPR). Immunoglobulin protein binding activation is used as a UPR initiation marker (Kong et al. 2020; Khangura et al. 2017). Three ER stress receptors facilitate the UPR: ER kinase PKR-like (PERK), the enzyme inositol requirement (IRE1), and the active transcription-factor (ATF6). Studies indicate that ER stress response is also a significant reaction, involving neuroinflammation, in inflammatory diseases (Sunderhaus, Law, and Kretzschmar 2019). The signaling process among ER stress and neuroinflammatory reaction is related by various mechanisms, such as reactive oxygen species, mitogen-activated protein kinase (MAPK), nuclear factor-kB, and calcium. The pro-inflammatory agents cause ER stress reaction, which in turn leads to induction of NP (Khangura et al. 2019).

10.2.4 β-CATENIN, WNT/RYK AND WNT/β-CATENIN

β-catenin is a cadherin protein complex subunit located at synaptic associations and is related with the development of synapse and neuronal network assembly (Resham et al. 2020). Wnts glycoprotein acts in several aspects of CNS growth (neural inductions, cell proliferations, synaptogenesis axon instructions, neuronal migrations, and dendritic arborization). For a normal axon guidance and neural differentiation, Ryk,a receptor of tyrosine kinase binds to Wnt for modulating its function (Z. Peng, Zha, et al. 2019; Resham and Sharma 2019). Menin (MEN I), is a tumor-suppressing gene located within the nucleus and essential for the creation of synapse among neurons. Menin also plays an important role in signaling Wnt/β-catenin through histone methylation of downstream target gene promoters. The Wnt/β-catenin route, a traditional Wnt signaling pathway, plays an important role in the formation of neurons, axon guidance, NP relief, and neuronal survival. The receptors of β-catenin, Wnt/Ryk, and Wnt/β-catenin are regarded as clinical targets for the management of NP(Zhang et al. 2013; Miranpuri et al. 2016).

10.2.5 HISTONE DEACETYLASE

Histone deacetylase enzymes (HDAC) together with histone acetyltransferase (HAT) regulate the acetylation cycle of histone lysine residues (like lysine residue9), that further amplify transcription, expression and also facilitate transcriptional elongation. Histone acetylation is among the epigenetic pathways which are believed to trigger NP, the reasons for which are mainly focused on impaired transcriptional function (Khangura et al. 2017). Histone deacetylase 4 is gradually released as a consequence of spinal nerve ligation in rats. Histone deacetylase 4 inhibition, utilizing LMK 235, stopped allodynia. A rise in histone deacetylase 1 and reduction in histone (H3) acetylation has been shown in the spinal nerve ligation model (Elsherbiny et al. 2019; Zhao and Wu 2018). Treatment with baicalin (anti-inflammatory flavonoid) greatly decreased production of deacetylase 1 and reverse pain response. Nerve disruption can be speculated to contribute to epigenetic modifications attributable to the stimulation of histone-deacetylase enzymes. This can reduce histone acetylation induction, and inhibitors of histone-deacetylase can be used to prevent NP (Z. Li et al. 2019; Van Helleputte 2018).

10.2.6 MITOCHONDRIAL ATPASE

Mitochondria, a cell powerhouse, are essential for the development of energy, apoptosis, ROS, cellular signaling and homeostasis of calcium. Mitochondrial and bioenergy deficiency are known to play a significant part in NP induction. The chemotherapeutic drugs, such as paclitaxel and oxaliplatin, have been found to form peroxynitrite in the mitochondria. The development of peroxynitrite during nerve injury affects the mitochondria and decreases ATP output, leading to NP. Research indicates the function of ATP mitochondrial synthase in adult rats following sciatic nerve injury (Rumora et al. 2019). Studies have demonstrated a decline in expression in unmyelined C-fibers and myelined A Ś-fibers of mitochondrial ATP synthase, leading to a decrease in ATP output. So, intrathecal ATP injection decreased thermal and mechanical hyperalgesia following nerve damage has been the most canonical finding. It is necessary to note that ATP is not an energy drug, but instead can be used to modulate the pain propagation on different membrane receptors (Bagli et al. 2017). It is also found that increased oxygen uptake and glycolysis contributes to hypoxia and acidosis that cause certain mitochondrial and bioenergetic dysfunction defects after nerve damage. Reduced energy generation due to mitochondrial dysfunction decreases activation of the Na+/K+ATPase accountable for hyper-excitability of the membrane. It is therefore necessary to protect mitochondria to relieve NP symptoms (van Hameren and Tricaud 2018).

10.3 NANOTECHNOLOGY

The transformation of fundamental science work into sophisticated nanosystems provides massive diagnostic and therapeutic potential to tackle chronic and acute pain (Bhushan 2017; Chakravarthy, Boehm, and Christo 2018). Nanotechnology involves materials from various compositions that show nanometrically unique attributes. Advanced nanosystems are used for targeted drug delivery to increase visualization and diagnostic testing (Moradkhani, Karimi, and Negahdari 2018). Nanomedicine is prepared to achieve a significant effect on pain management in both hospital and outpatient environments. Physicians are responsible for making themselves more aware of the capacity and limitations of emerging nanomedical technology, as an increasing number of new nanodrugs and formulations enter clinical practice worldwide (Keskinbora and Jameel 2018).

The usage of nanotechnology has increased in our everyday lives, raising an urgent need for the creation of new nano-scaled approaches (~100 nm) Nanotechnology and nano-based materials are broadly used across a number of industries, including health care, electronics, cosmetics, farming, and continuing development. Advancement in this area is seen as having exceptional potential to significantly enhance nerve injury diagnosis and treatment (Beiranvand and Sorori 2019). Therefore, the use of specifically designed nanomaterials will now be seen at all levels of disease: treatment, evaluation, and rehabilitation, and even neurological pathologies. The greatest benefit of medicinal nanomaterials is their capacity to communicate with biological structures at a molecular level with strong precision. This ensures that such materials can move across cell membranes, improved solubility, flexibility, and bioavailability of biomolecules, thus improving their distribution abilities. Therefore, apart from integrating nanomaterials into scaffolds with a view to enhancing their properties, nanomaterials alone may be used to bioimage and introduce desirable particular molecules into cells in a controlled manner (Kuthati et al. 2020). Nanomaterials therefore replicate whatever occurs at a nano-scale stage in the human body and are thereby known to be the ultimate biomimetic substance.

10.3.1 Different Types of Nanotechnological Approaches Used to Treat NP

Peripheral nerve injuries account for severe neuropathic conditions, affecting 1 in 1000 individuals per year (Zhu et al. 2014). Huge amounts of money (~$150 billion per annum are spent in the USA alone) due to concerns of nerve injuries developing NP (Grinsell and Keating 2014). In addition, it has been reported that current therapies for the treatment of NP reduces quality of life in patients due to various factors like loss of motor, sensory, or cognitive function (Nejati-Koshki et al. 2017). This situation tends to worsen with the growing world population and respective average lifespan. Therefore, medical sciences have drawn considerable interest from nanotechnology. Recently, various nanotechnology-based devices, diagnostic techniques, as well as treatment approaches for NP have been developed, including nanofibers, carbon nanomaterials, nanoimaging, and nano-topographic cues, etc., as mentioned in Figure 10.1.

10.3.1.1 Carbon Nanomaterials

For several industrial and technological uses, carbon nanomaterials have been seen as an attractive option. With such intense growth, worldwide business interest in carbon nanomaterials is expressed every year. Nanomaterials are nowadays composed of diverse commercial items, including batteries, motorized parts, and sportive goods, some form of structural content, and even agricultural products. This is because of their excellent properties, which were considered to be beneficial in conjunction with other biomaterials in therapeutic effect approaches (Rauti et al. 2019). Carbon nanostructures were suggested for the creation of neural scaffolds as viable candidates. Electrical stimulation and the sufficient electrical conductivity of the implemented biomaterials is considered to be one of our major benefits for neuronal regeneration as natural PNS can readily transmit electrical impulses. In comparison, carbon products are comprised mostly of basic factors that occur in all facets, e.g. carbon (Baldrighi et al. 2016). The most common carbohydrate nanostructures currently fall into four separate categories: carbon nanotubes (CNT), graphene, carbon nanostructures (CNFs), and nano-diamonds (ND) (Oprych et al. 2016).

10.3.1.2 Carbon Nanotubes

One of the most investigated carbon allotropes is carbon nanotubes (CNTs). CNTs are graphite sheets moving in cylindrical channels, which are made of a hexagonal atomic structure of nano-scale dimensions. These can primarily be classified into two groups, according to the size of the shells: the SWCNTs (one-wall carbon nanotubes), with diameters varying from 0.8 nm to 2.0 nm and multi-wall carbon nanotubes. Several CNT-processing strategies include electrical discharge, chemical vapor deposition, and laser ablation (Min and Yoon 2017). The serious mechanical, thermal, magnetic, visual, chemical, and electrical properties of CNTs are some of

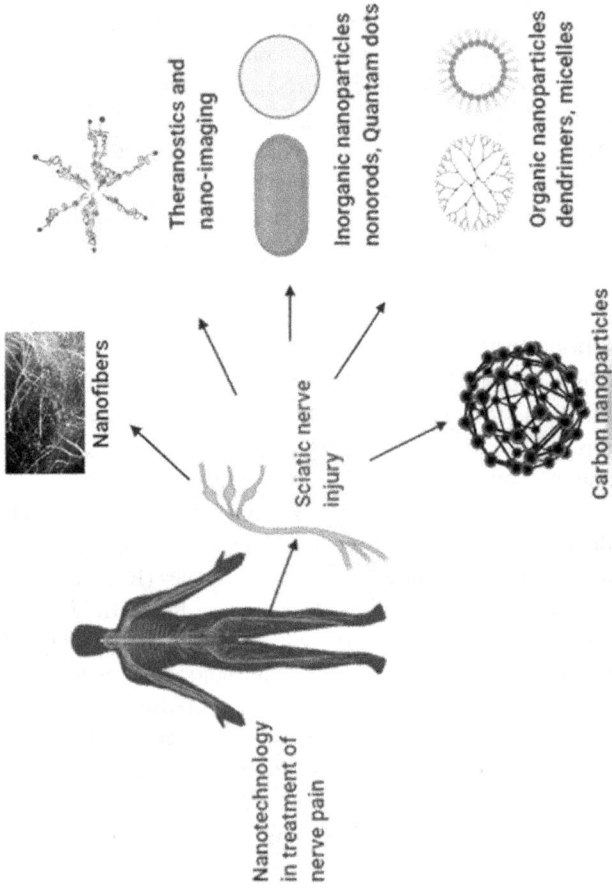

FIGURE 10.1 Nanotechnological approaches for treatment of neuropathic pain.

the striking characteristics. The combination of these features expands the spectrum of future biomedical applications where they may be utilized. As described above, the most desirable aspect of such products, apart from their nanoscale, relies on the capacity to show metallic and superconductive electron transportation characteristics (Banderas et al. 2018). The initial CNTs do not, however, have the requisite solubility for biomedical application. Two key approaches were established to allow the implementation of covalent, non-covalent functional CNTs under physiological conditions. It is therefore essential to render CNTs not only more soluble but also incorporate them in many organic, inorganic, and biological systems and applications. CNTs may be used in numerous biomedical systems, such as medication distribution, diagnostics, biosensors, biological imaging, and regenerative and cellular medicine with these modifications (Assaf et al. 2017).

10.3.1.3   Graphene

Graphene is among the allotropic carbon types. This two-dimensional planar monolayer nanoparticle, consisting of an SP2 bound carbon atom, is standardly structured as a honeycomb with an interatomic length from carbon to carbon and has many impressive mechanical, optical, and conductive characteristics (Tour and Sikkema 2019). The work on biological uses of graphene and its variants is focused on its numerous desirable properties, including large different surface areas' exceptional electric conductivity, and mechanical strength. Graphene may also be used in its elementary shape to support nerve repair. However, owing to the absence of functional groups on the graphene sheet, solvent dissolution is complicated and can be agglomerated quickly. Graphene oxidized (GO) is also used in other works, because its surface is abundant in functional groups that carry oxygen (Askari, Askari, and Shafieipour 2019).

10.3.1.4   Carbon Nanofibers

Carbon nanofibers (CNFs) are some other kinds of carbon nanoparticles, such as moisture-produced carbon nanofibers and polymer-based nanofibers, developed by using various methods. Such nanofibers are defined as high-surface (100–200 m2/g), high purity, and tunable surface chemistry non-microporous graphical products (Zhou et al. 2019; Demir et al. 2018). Due to their specific physical characteristics, such as high efficiency, metal permeability, low density, tunable composition, chemical, and environmental durability, and compliance with organo-chemical shifts, several uses have been studied (Bonferoni et al. 2018; Faccendini et al. 2017).

10.3.1.5   Nanoparticles

Researchers also suggested the usage of NPs in neuronal regeneration strategies in recent years. NPs offer a range of advantages, including their smaller size, and physical characteristics which vary from bulk, surface functionality, and chemical stability (Harris 2017). They also vary in their electrical and magnetic properties from certain nano-sized objects. The main aim of these nano-sized nanomaterials is to provide drugs and other biological agents both regulated and monitored, such as growth factors, both in vitro and in vivo (Whitehead 2016; Najafi et al. 2017). Several forms of molecules that may be connected to NPs, such as medication distribution or cell tracking, include labeling sensors, hydrophobic or hydrophilic molecules, DNA, and oligonucleotides. This polyvalence is largely attributed to the wide variety of solutions to the implementation of NPs. In addition, NPs are used extensively in numerous TE applications to boost the properties of scaffolds, such as mechanical properties and deterioration levels (Ganugula et al. 2019).

10.3.1.6   Inorganic Nanoparticles

Inorganic nanomaterials are commonly used for the reconstruction of nerves. The fields of gold (Au), silver (Ag), zinc (Zn), and silica NPs (SiO2-NPs) gained significant interest (Distasi et al. 2019). For different medicinal applications, magnetic components on the nanometer scale (e.g. nickel, cobalt, iron and their oxides) are used. The magnetism of iron is a feature that is changed when it is converted to a nano-scale material. Reduction in the size of iron leads to nano-paramagnetic behavior. Super paramagnetic iron NPs are non-magnetic particles unless subjected to a heavy magnetic field. As the magnetic field is reduced, it moves again to a non-magnetic state. In previous years, many methods have used such magnetic NPs (MNPs). The key benefits are that the loaded MNP cells may be guided at a specific location, in response to an outward gradient of the magnetic field (Baskaran et al. 2017; Baldrighi et al. 2016; Raffa 2020).

10.3.1.7   Organic Nanoparticles

10.3.1.7.1   Polymeric Nanoparticles

Because of their bulk physical properties, tunable structural and construction design, and biodegradability, polymers have been deemed desirable products. There are numerous functional polymer synthesis techniques and the polymer chains permit a large variety of molecules to work. Within this case, various formulations and characteristics may be produced of the final polymer materials, envisioning a vast variety of applications and strategies (Berrocoso et al. 2017; Jeong et al. 2009).

10.3.1.8   Biologically Derived Nanoparticles

Intercellular transfers through vesicles of macromolecules, known as exosomes, have become increasingly important as an innovative way of intercellular crosstalk. One type of NP may also be called exosomes. In terms of their growth, the exosomes are formed by the internal growing of endosomes into multi-vesicular structures that merge into the surrounding region with the plasma membranes (Shiue et al. 2019). Depending on the cell forms, they include a number of elements, including proteins, mRNA, and miRNA. Such components are transmitted through exosomes, may be considered a "cargo" and are either distributed to cells around them or rendered to function in distant cells. It is thus reasonable why various anomalies will arise in the receiver cells, including the reprogrammed DNA, based on the cargo material. Not only does the cargo affect receptor cells, it also includes proteins that serve as distributors on the surface membrane of the exosomes. Therefore, exosomes are known as modern intercellular connectivity between cell-based elements, albeit without the anticipated direct interaction from cell to cell. After recognizing the functioning of exosomes, the capacity of these NPs for treatment and other therapy can be believed (Jean-Toussaint et al. 2020; X. Yu et al. 2020).

10.3.1.9 Nanofibers

Nanofibers are fibers one or two classes smaller than standard fibers that precisely resemble the ECM. Such fibers have an unusually high surface-to-mass ratio and typically can be generated with electrospinning and self-assembly. Electrospinning is an easy and flexible electrodynamic method in which a polymer solution can be dispersed through a high potential electric field for obtaining nano-large fibers, which implies that nanofibers are consistently removed from a viscoelastic fluid by depending on the electrostatic repulsion of the surface charges. As exposed to the ambient electric field, the fluid in the metallic needle is induced to lengthen at the tip of the needle, assuming the shape of a cone that is stretched like a plane (Nune et al. 2019). The typical electrospinning equipment consists of 4 key elements: I) a high voltage source, II) a capillary, III) a syringe pump, and IV) a conductive collector, which is typically a basic surface or revolving mandrel. An enormous number of components, like polymers, small molecules and their variations, or ceramics may be used for electrospinning. In addition to traditional biomaterials, the ultimate smooth gel of nanofibers with a variety of secondary structures like transparent, pit, or core-sheath arrangements can also be adapted. However, nanofibers may be obtained by controlling their orientation, piling or folding in structured arrays or systemic structures. Therefore, nanofibrous scaffolds should be able to provide tremendous promise in nerve regenerative medicine at least for a decade, as they can mimic native tissue tubular structures like axons, microtubules, and ion channels (Mammadov et al. 2016; Lu et al. 2018).

10.3.1.10 Topographical Cues

There is ample proof that artificial spatial cues like topography may have a direct impact on the activity of the neuronal cells. However, a fairly recent hypothesis notes that the standards for growth cones obey the "substrate-cytoskeleton coupling" pattern. "Growth cones will only travel forward because they are able to bind intracellular motility signals with a specified extra-cellular translocation substrate." (Thomson et al. 2017). There are various methods available in the context of topography for the creation of patterned biomaterial surfaces, varying from basic manual scraper to highly regulated processing methods. Contact scanning, microfluidic patterning, electrospinning, and lithography are several methods to obtain more accurate physical indications at nanoscale. The usage of nano-scaled topographical markers in PN regeneration reveals the strongest outcomes of the topographical and neuronal research (Gosling et al. 2017; Porto, Porto, and Brotto 2016).

10.3.1.11 Nanotheranostics and Imaging

Combining therapy and diagnostics in one group, the definition of theranostics is centered. It is focused on picture-oriented therapy and therefore determines the result of the procedure at an early stage. Theranostics should envision and monitor photographic elements, not just for the detection of an issue, but can even determine the bio-distribution of an efficacious medication or a particular molecular target (Janjic and Gorantla 2018). Till now, when damage to PN has been verified in the surgical setting, conventional diagnostics usually focus on the interpretation of specific findings from surgical evaluations by the doctors. Consequently, diagnostic precision is sometimes decreased because this method does not have sufficient details on the requisite surgical reparation, which contributes to inadequate care, bad medical results, and lifelong impairment (Wright et al. 2018).

10.4 NANOTECHNOLOGY-BASED EVIDENCE IN THE TREATMENT OF NP

Monotherapy for acute and chronic medical conditions is not quite so successful due to limited time and insufficient transmission, since a mixture of two or three medications is used for synergistic effect. A modern drug delivery method focused on nanocarriers has been developed as a promising approach in the distribution of drugs (Table 10.2). The two popular medicines in the treatment of NP are lidocaine and thalidomide. Lidocaine works on the peripheral afferent fibers by nonspecifically inhibiting the sodium channels, although the mode of action of the thalidomides is still not known but is expected to have anti-inflammatory, immunomodulatory, and antiangiogenic effects. Nonetheless, both medications have drawbacks, such as the fact that lidocaine metabolizes very easily, and that repeated administration is required and thalidomide has low stability and aqueous solubility. Hence, for usage such limitations must first be addressed by utilizing a separate formulation approach. For its synergistic influence, Song and colleagues produced a sustained release formulation of graphene oxide nanoparticles, consisting of two medications: lidocaine and thalidomide. Due to the numerous functionality that allows it to be an effective carrier for hydrophobic medicines, graphene oxide has been thoroughly studied. These medications may be easily immobilized on the surface of graphene oxide. Tests from in vitro and in vivo research showed that this may be a successful potential approach for creating a new formulation for the treatment of NP. Furthermore, amine-functionalized nanodiamonds and phosphorous dendrimers are also modern drug delivery approaches to treat NP. The correct treatment of pain is still a significant medical problem. Opioid treatment is the gold standard medication to resolve chronic pain and achieve a long-term analgesia. Nanoparticles, liposomes and nanocapsules, micelles, dendrimers, and nanotubes are well-known forms of product for delivery of drugs using nanotechnology. However, not all are focused on nanotechnology-based formulation due to issues of biocompatibility in the formulation of therapeutic agents encapsulated in nanocarriers (specifically in the range of 10–200 nanometers) (Table 10.3).

The use of nanotechnology to build nanocarriers as vehicles for drug delivery can be seen as a positive path to enhancing the supply of pharmacologically active drugs. For analgesia, nanostructures, the loaded medication is secured from deterioration, delayed therapeutic release, and pharmacological length and thus decreases adverse reactions and toxicity. Nanoparticles help in the supply of drugs and the visualization of pharmaceutical and pharmacodynamic effects that enable the transmission of nanoparticles to different tissue cells at the site of the disease can be developed. They are an effective therapeutic tool for treating various diseases because of their special mechanical, electrical, chemical, optical, and biological properties. Some

TABLE 10.2

Evidence for Nanotechnological Approaches to Treat NP

Nanotechnology Approach	Strategy Adopted	Outcomes for Treatment of NP	References
Carbon nanotubes (CNTs)	CNT-interfaced Phosphate Glass Fibers	Carbon nanotubes were interfaced effectively with nerve guidance phosphate glass fibers and then incorporated in a 3D scaffold with physicochemical integrity with strong cellular viability and neuronal activity.	(Nawrotek et al. 2016)
	Chitosan-multiwalled CNTs	MWCNTs can be incorporated in peripheral nerve with an electrodeposited phenomenon and cytocompatibility was proved between these materials and cell	(Ahn et al. 2015)
	Chitin/CNTs nanofibrillar structure	Three different concentrations of CNTs were used, being the one with 5 wt% the most successful one in nerve regeneration.	(Wu et al. 2017)
Graphene	Graphene oxidized combining with decellularized rat sciatic nerve defect (allograft)	GO nanomaterial could be paired with decellularized allergenic sciatic nerve scaffolds for faster nerve regeneration	(Q. Wang et al. 2017)
Nanodiamonds (NDs)	Amine functionalized ND	The amine-terminated surface NDs have been a good substratum because they have facilitated neuronal adhesion, proliferation and neuritis development	(Hopper et al. 2014)
Nanoparticles (NPs)	Magnetic NP (Fe_3O_4)–chitosan mixture was fabricated into membranes or scaffolds	10% MNP magnet nanocomposites may aid binding and disperse SCs under magnetic field exposure. MNP production, gene expression, and BDNF, GDNF, NT-3 and VEGF protein secretion improved	(Z. Liu, Huang, et al. 2015)
	Nanohydroxyapatite (n-HA) coated Fe_3O_4 magnetic NP	Fe_3O_4 coated (n-HA) can successfully increase axonal elongation and cell viability. Netrin-1 axonal guidelines also rise significantly after n-HA-coated Fe_3O_4 treatment.	(M. Liu, Zhou, et al. 2015)

(Continued)

TABLE 10.2 (CONTINUED)
Evidence for Nanotechnological Approaches to Treat NP

Nanotechnology Approach	Strategy Adopted	Outcomes for Treatment of NP	References
	Polyethylenimine -coated gold-NPs (AuNPs)	Pulsed current stimulation-induced neurite outgrowth of PC12 cells to the AuNPs coated surfaces, thus proving the potential of AuNPs as an electrically conductive matrix for nerve regeneration	(Adel et al. 2017)
	SiO2-NPs encapsulated with Tetrodotoxin	The proven continuous release characteristics of hollow SiO2-NPs have helped expand the nerve block and improve safety through slow release	(Q. Liu et al. 2018)
Dendrimers	Phosphorus dendrimers	Cell-based GDNF (glial cell-derived neurotrophic factor) therapy has been shown to improve the level of axonic regeneration, whereas regulated GDNF deactivation effectively prevented the capture of GDNF-enriched regenerative axons.	(Shakhbazau et al. 2013)

nanoparticles have antioxidant characteristics and act on free radicals in order to treat NP (Moradkhani, Karimi, and Negahdari 2018; Beiranvand and Sorori 2019; Cerna, Eckschlager, and Stiborova 2016). Liposomes involve lipids such as phospholipids that are combined with other compounds such as the cholesterol to make a hollow sphere that includes the bilayer phospholipid. The exterior layer is hydrophilic, and the inner layer is hydrophobic, and the liposome is less secure. However, liposomes can improve the solubility and permeability of BCS II and Class IV drugs. Liposomal formulations are investigated not only for NP but also for various disorders like cancer, liver, viral infections, and fungal infections (Bulbake et al. 2017).

10.5 NANOTECHNOLOGY-BASED CELL THERAPY IN THE TREATMENT OF NP

The current clinical plan targets a small number of people, and innovative strategies are needed to alleviate NP. The research now centers on the usage of stem cells of various backgrounds to relieve NP. The reason for the usage of stem cells is dependent on the capacity of stem cells to substitute the damaged or defective neuronal cells with totipotent cells. Numerous studies have shown that stem cell treatment is not only restricted to regenerative effect but also provides an antinociceptive effect (Chawla, Chawla, and Jaggi 2016). Two primary forms of stem cells are used for NP management. These are neuronal and mesenchymal stem cells. Because of the

TABLE 10.3
Nanotechnological Approaches in Development of Formulations to Treat NP

Treatment Approach	Dose	Animal Species	Animal Model	Molecular Target	Kind of Pain	References
Nanoparticle-encapsulated curcumin	16 mg/kg	Male Sprague Dawley rats	Streptozotocin-induced neuropathy (32 mg/kg)	P2Y12 (purinergic receptors) receptor-mediated Akt activation in the DRGs (dorsal root ganglions)	Thermal hyperalgesia and mechanical allodynia	(Jia et al. 2018)
Nanoparticle-encapsulated emodin	5 mg/ml	Male Sprague Dawley rats	Streptozotocin-induced neuropathy (30 mg/kg)	P2X3 (purinergic receptors) receptor in DRGs	Thermal hyperalgesia and mechanical allodynia	(L. Li et al. 2017)
Cerium oxide nanoparticles	65 mg/kg 85 mg/kg	Male Wistar rats	Streptozotocin-induced neuropathy (65 mg/kg)	Radical oxygen species induced oxidative stress	Thermal hyperalgesia	(Najafi et al. 2017)
Magnesium oxide and zinc oxide nanoparticles	5 mg/kg	Adult male Wistar rats	Stress-induced NP (using semi-cylinder plexiglass tube)	Alteration of Glutamate level	Thermal hyperalgesia	(Torabi et al. 2020)
Celecexib loaded nanoemulsion	0.24 mg/kg	Male Sprague Dawley rats	CCI (chronic constriction injury) induced NP	macrophage production of PGE2 (prostaglandin E2) at the site of injury	Tactile allodynia	(Stevens et al. 2019)
CX001 (mixture of innovative polymers and lidocaine)- a novel topical cream preparation	2 puffs of CX001 spray	Adult male Wistar rats	-	-	Thermal, cold hyperalgesia and mechanical allodynia	(Luca et al. 2019)

(Continued)

TABLE 10.3 (CONTINUED)
Nanotechnological Approaches in Development of Formulations to Treat NP

Treatment Approach	Dose	Animal Species	Animal Model	Molecular Target	Kind of Pain	References
3,3′-diindolylmethane (DIM)-loaded nanocapsules (NCs) suspension	10 ml/kg	Male adult Swiss mice	Formalin induced NP	Inhibition of cyclooxygenase-2, matrix metalloproteinases and inducible nitric oxide synthase, interleukin-8 levels and increase in interferon-gamma production	Thermal hyperalgesia	(Mattiazzi et al. 2019)
Liposome-encapsulated clodronate	30 μg/kg	Male adult and male and female 1-day-old neonatal Wistar rats	Spinal nerve ligation	Depletion of spinal microglia	Allodynia to mechanical stimulation	(Y.-R. Wang et al. 2018)
Nefopam hydrochloride loaded sustained release microspheres	20 mg/kg	Wistar rats	Plantar incision model	-	Mechanical allodynia and thermal hyperalgesia	(N. Sharma, Arora, and Madan 2018)
Polyamidoamine dendrimer-conjugated triamcinolone acetonide	Single injection of reaction mixture	Male C57BL/6 mice	Spinal nerve ligation	Inhibition of spinal microglia activation	Mechanical allodynia	(H. Kim, Choi, et al. 2017)
Brucine-loaded liposome	30 mg/kg/ day	8-to-10week old males in a C57Bl/6 background (wild type mice)	CCI (chronic constriction injury) induced NP	Inhibition of sodium channels	Mechanical allodynia and thermal hyperalgesia	(G. Yu et al. 2019)

ethical issue and tumor-related threat embryonic stem cells are never used. Neural stem cells seem to be the most acceptable cells and can be differentiated into nerves, oligodendrocytes, and astrocytes (Carvalho et al. 2019; Nejati-Koshki et al. 2017). In 2011, Silvia Franchi and researchers became the first to utilize murine neural stem cells for NP management. This group has been able to explain the biochemical function of neural stem cell therapy and indicated that all these cells communicate bi-directionally with immune cells and modulatory variables triggered by nervous lesions. One of the key issues with this treatment was that the patients experienced a compromised spinal cord and have a poor survival rate. This problem was then solved by employing a hybrid technique. Co-transplantation of neural stem cells with olfactory unsheathing cells enhances the sensory functions linked to unpleasant behavior. The combinatorial technique has increased the axonal stability and plasticity of the up and down routes. Neural stem cells, derived from both mice and humans, can be used for cell therapy. Mononuclear cells originating from bone marrow can also be used to relieve the NP (Sari et al. 2018). Mesenchymal stem cells constitute a highly diverse sub-set of stromal cells that can be derived from a specific source, such as bone marrow, lung, and fetal liver. Bone marrow and mesenchymal stem cells originated from adipose tissue were more commonly utilized for stem cell therapy. Bone marrow includes a hematopoietic and vascular element. The vascular portion includes multipotent, non-hematopoietic stem cells, which can be divided into osteoblast, adipocytes, myocytes, and chondrocytes. The mesenchymal stem cells originating from human and rodent bone marrow may be used for NP therapy. Stem cells derived from adipose tissue provides a substantial advantage over all other forms of mesenchymal stem cells owing to their low immunogenicity and strong immunomodulative properties, rendering them ideal for the treatment of diseases like NP. Such cells can effectively be used for autologous transplantation (Kelly et al. 2017). It has been proposed that a cell reprogramming will undo the minimal trans-differentiation capacity of adult stem cells. Moreover, some studies found that stem cells are extremely reactive to the microenvironment and may affect cell activity and functionality. The key objective in the area of tissue engineering was to get a signal that effectively directs stem cells to divide into the particular lines of existing cells or tissues, hence maintaining the undifferentiated form (Alvarado-Vazquez et al. 2017; Kwon et al. 2019).

10.6   NANOTECHNOLOGY-BASED GENE THERAPY FOR THE MANAGEMENT OF NP

This provides a modern and focused solution to chronic NP treatment. In this method, both viral and non-viral vectors are used to carry genes that encode applicants such as natural opioids (ENK, POMC), anti-sense gene RNA to specific sites, neurotransmitters (GAD, Glut1), immunomodulatory factor (IL-10, TNFα, IL-2, and IL-4) and neurotrophins. The usage of viral vectors for gene therapy offers a highly effective transduction system for achieving high rates of genetic expression and thus achieve a more antinociceptive effect (Bernal et al. 2017). Many problems are often connected with the usage of viral and non-viral vectors, such as I) toxicity, II) immunogenicity, (III) development of tumors, (IV) secondary immune reaction.

Naked plasmid DNA is one of the simplest approaches for inducing intrinsic gene expression in cells. The possibilities of host immune system stimulation are reduced since no foreign material is involved. Nevertheless, other problems with naked DNA vectors include short-term transgenic expression, poor stability, and fast DNA reaction. Physical strategies have been developed for transmitting naked DNA, such as liposomes, niosomes, gene gun, electroporation, and shockwaves have increased in vitro transduction capacity to primary DRG neurons. TNFα is a crucial mediator for NP pathogenesis and is increased in the sites of nerve damage during initial production of NP, so pain relief may be provided by drugs or genes that inhibit TNFα activity. In 2015, Gerard and his colleagues produced gold nanoparticles with changed surface characteristics which function as a SiRNA nano-carrier and given a stable and regulated release in the brain, subsequently used for the purposes of the prolonged delivery of small interfering RNA (siRNA). In vivo studies showed that siRNA-loaded gold nanoparticles would reduce NP (Cardoso et al. 2016). IL-10 is an anti-inflammatory agent that has proven very important to NP therapy. Several transmission methods were utilized for extended release, utilizing both plasmid and virus vectors such as HSV (Herpes Simples virus-based vector). In 2010 Soderquist and colleagues devised nanoparticles for the supply of the IL-10 encoded in plasmid DNA or long-term relief from NP (NP). In vitro and in vivo studies showed that these micro and nanoparticles facilitated cell invasion and enhanced p-DNA production in meningeal tissues in the spinal cord. This technique that prolongs the release using micro and nanoparticles may also be beneficial in the treatment of many other neuro-inflammatory disorders (Shin et al. 2019).

10.7 NANOTECHNOLOGY: SUPERIOR TO CONVENTIONAL TREATMENTS

Widely accepted medication for the control of NP is the prescribing of analgesics, the implanting of intrathecal devices, stimulators of the spinal cord, and physical therapy. The greatest unaddressed need of NP is insufficient pain relief. First line treatments widely used in NP include tricyclic antidepressants (TCAs) such as duloxetine and nortriptyline, as well as anticonvulsants such as gabapentin, pregabalin, and carbamazepine. TCA is considered to be cardiotoxic in NP patients. Anticonvulsants can increase the chance of suicide (Agarwal and Kansal 2018; Kapural et al. 2018). While localized solutions for peripheral care like topical lidocaine patches provide temporary pain relief, they are believed to cause allergies and are ineffective at treating long-term pain. Non-traditional drugs approved for the treatment of certain conditions including lamotrigine, NMDA, cannabinoids, capsaicin, and serotonin inhibitors. They are among the last line of treatment methods. Indeed, all these drugs lead to severe adverse effects in the case of long-term use. Implantable intrathecal drug delivery systems (IDDs) have gained significant interest in long-term NP control in patients who reject traditional treatment approaches (Didier Bouhassira and Attal 2018; Fornasari 2017).

BBB is a daunting impediment for the transmission of drugs to the central nervous system (CNS). This prevents the movement of most substrates by selectively transporting the substrates necessary for homeostasis of the brain. More than 95 percent of drugs for CNS disease are estimated to have been impeded by their ineffi-ciency in reaching BBB, a major problem in the area of CNS drug delivery (D Skaper and Stephen 2016). The opioids are some of the main medicines used for chronic pain and rely on the drug's ability to cross the blood–brain barrier, which suggests its key function, through the mechanism by which opioids cause analgesic results. BBB is believed to play a key role in opiate tolerance by means of BBB receptors of P-129 glycoprotein (P-gp), which help to control the efflux of opiate from brain to blood. Long-term exposure to opiates is shown to enhance BBB expression of P-gp contrib-uting to opiate tolerance. The most effective method of administering brain medi-cines is to administer analgesics by intrathecal catheters, by offering adequate dose with no adverse effects (D. Kim, Min, et al. 2017).However, the invasiveness of the technique and the preservation of appropriate drug concentrations over a longer time are significant challenges. The non-surgical routes to the supply of drugs include the disturbance of the permeability of BBB through the destruction of endothelial tight interstices, either through osmotic pressure or microblasting, which improves the drug's entry via BBB. Formulating a reliable drug delivery system crosses the BBB is an important move in facilitating the widespread use of CNS therapies (Cruccu and Truini 2017). Recent nanotechnological developments have shown that nanoma-terials with an effective design can penetrate the BBB efficiently. Selective function-alization for a number of biomedical applications included complex moieties such as diagnostic imaging agents, fluorescent dyes, photosensitive agents, antibiotics, anticancer agents, and analgesics, as well as within and on the surface. The spe-cific properties of nanoparticles such as small size, switchable surface charges, high biocompatibility, and enhanced circulation have now been utilized for the develop-ment of new platforms to supply drugs through BBB (De Marco and Janecka 2016; Puthenveedu, Shiwarski, and Pradhan 2017).

MMP-12 Matrix metalloproteinases are operational magnetic nanoparticles capable of identifying NP specifically. MMP-12 is considered to play a crucial role in peripheral neuroinflammation and there are increased levels of MMP-12 in NP (Husain et al. 2019). In addition, histology studies have confirmed the presence of iron nanoparticles at the nervous site in particular. The strategy developed can be useful in pain imaging and other proteins can be used for pain imaging apart from MMP-12, which upregulate during nerve damage. For instance, aquaporin-4 and interlecin-1 receptors are two additional proteins which may be used to conju-gate unique antibodies with nanocarriers to represent the NP site. Delivering and maintaining analgesics in the spinal cord for long periods is a significant obstacle in the treatment of NP and other neuro-inflammatory disorders, including can-cer (Kuthati et al. 2020; Miranpuri et al. 2016). So, a magnetic nanoparticle is preferred to treat spinal cord glioblastoma (GBM) for the specific transmission of an anti-cancer product. A similar method may be used to treat NP for the maintenance of analgesic substances in the spinal cord. Metal nanoparticles are conjugated with doxorubicin and delivered by intrathecal injection in this method. The magnetic field has been used to direct the magnetic nanoparticles filled with doxorubicin to the tumor site. Directed analgesic delivery with a similar strategy will minimize CNS risk at the spinal cord level and increased drug concentration (S. Sharma et al. 2016; BANDERAS et al. 2018). All these studies show that the nanoparticles produce better results with fewer side effects. We can say therefore that nanotechnology is superior to conventional treatment.

10.8 APPLICATIONS OF NANOTECHNOLOGY IN NP

10.8.1 IN DRUG DELIVERY

Among various lipid nanoparticles, the classical example is liposomes that mainly consist of phospholipids and have already been extensively examined by various lipid-based nanoparticles (Bidve et al. 2020). Liposomes are vesicles with a two-dimensional lipid membrane composed of an aqueous film. The amphiphilic fragments used in the liposome synthesis are the same as biological membranes that enhance the effectiveness and protection of different medicines. Liposomes are effective transport agents for both hydrophilic and hydrophobic pharmaceuticals. The hydrophilic drugs may be inserted into the aqueous matrix of liposomes, and hydrophobic drugs are attached to the lipid membrane. The liposomal production could be customized based on the physical and chemical characteristics of the drug (Hasanzadeh-Kiabi 2018). Liposomes can typically be formulated in water by sonication of amphiphilic lipids such as phospholipids. However, other biological molecules are known to be affected by the sonication process. Extrusion and Mozafari techniques are the most popular techniques of liposomal formulations developed for human use. Liposome-dependent medicines have been of great interest in recent years and are one of the most effective carriers of medicines with FDA approval. Their increased biocompatibility and biodegradability along with their small size and their simple development process make them suitable for different drugs formulation (Isacchi et al. 2017; Beltrán-Gracia et al. 2019) (Table 10.4).

In contrast with free medicines at the same dosage, the encapsulation of medicines like zoledronic acid and clodronate have greatly improved the biodistribution of mechanically disposed allodynic in animal models (Y.-R. Wang et al. 2018; Caraglia et al. 2013). This indicates that the structure of nanocarriers is useful (Figure 10.2). Problems with a low diffusion period are currently addressed by the creation of injectable long-acting nanoformulations.

Liposomal medication structure is considered to extend product diffusion. For example, the liposomal combination of saxitoxin and dexamethasone is found to be effective in NP with only one single injection (Lin, Paul, and Kuo 2018). This is also noteworthy that well-known analgesics such as opioids have been shown to reduce NP over one week with one single liposomal injection. Liposomes are also considered for gene therapy (Alvarado-Vazquez et al. 2017). Recent research shows that carbon monoxide (CO) may reduce inflammation. Furthermore, work suggests that CO administration inhibits the development of lipopolysaccharide (LPS)-mediated cytokine and instead demonstrates significant cytoprotective and anti-inflammatory results. The release molecule 2 (CORM-2) of carbon monoxide is among the well-established CO producers with an effectiveness in NP relief. Its hydrophobic existence and minimal half-life, however, impede its therapeutic use. CORM-2 encapsulated strong lipid nanoparticles which were effective in the reduction of NP caused by CCI in contrast to free medicines (Gabriel and Ilfeld 2019; Ismailova et al. 2018).

TABLE 10.4

Liposomal Formulation of Drugs Used in Treatment of NP

Drug	Effect	Challenge	Application of Liposomes	Reference
Zolendronic acid	Inhibit ras dependent Erk pathway	Bio-distribution is limited to bones	Liposomes cross the blood–brain barrier	(Caraglia et al. 2013)
Clodronate	Inhibit vesicular ATP release from neurons	Does not produce specific action	Liposomal preparation depletes spinal microglia and reduces initial NP	(Y.-R. Wang et al. 2018)
Saxitoxin and Dexamethasone	sodium channel blocker, and the glucocorticoid agonist	-	Liposomal preparation delay the NP	(Shankarappa and Kohane 2016)
Vincristine and posaconazole	Microtubules damaging agent and lanosterol-14alpha-demethylase inhibitor	Increased vincristine toxicity on co-administration with posaconazole		(Lin, Paul, and Kuo 2018)
Verbascoside	Antioxidant	Poor chemical stability	Liposomes ameliorate drug solubility, stability, and bioavailability	(Isacchi et al. 2017)

Polymeric nanoparticles constitute enticing nanocarriers, which can dissolve or capture the therapeutic medication or encapsulate it into an integrated polymer matrix. Polymer nanoparticles are either produced from natural sources (e.g. chitosans) or from synthetic (e.g. polylactides). Double emulsion, emulsification, and emulsion-solvent evaporation are widely used for the development of polymer nanoparticles. Synthesis with nanoprecipitation and supercritical fluid technology are other popular methods. High biocompatibility, biodegradability, and hydrophobic drug solubility are the key benefits of polymeric nanomaterials. PLGA is also one of the biodegradable polymers widely used for the distribution of medications. Many of the other devices that have been utilized successfully in the diagnosis of NP include metal, silicon, and hybrid nanoparticles (Nigam et al. 2019; De Freitas et al. 2018). In recent years stem cell therapy has acquired significant popularity, and viral vectors are widely used for gene transfection. However, viral vectors, in view of their good efficiency, often activate the immune system of the host, thus reducing the effectiveness of gene distribution.

Metallic gold nanoparticles have also been studied extensively for their possible application in the transmission of medicines and genes. Gold nanoparticles have

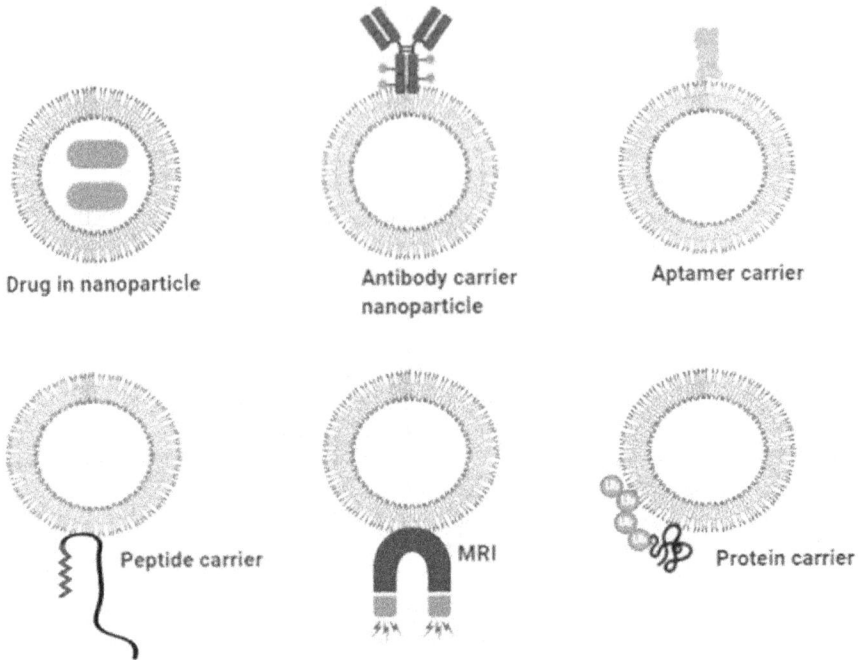

FIGURE 10.2   Role of nanocarriers in capturing therapeutic medications.

numerous benefits as therapeutic carriers, for example fast adsorption by biological molecules, such as enzymes, peptides and DNA, strong biocompatibility, and stability (Harris 2017). Gold nano-rods are also used to inhibit the tumor necrosis factor (TNF) by powerful, non-viral vector to deliver siRNA, which relieves NP. Recently, catalytic nanoparticles with natural antioxidant properties were produced as emergent synthetic enzymes (Ren et al. 2018). Nanomaterials have extraordinary physiochemical properties because of their relatively limited size and their large volume/surface ratio.

10.8.2   IN DIAGNOSIS OF NP

NP is widely recognized as a pain syndrome which is both difficult to diagnose and treat. However, there is no gold standard for diagnosing NP at present. The only way of diagnosing NP is through the use of patient history and clinical examinations and various screening tools like Pain Detect, Leeds Assessment of Neuropathic Symptoms and Sign sand Douleur Neuropathique etc. (Mistry et al. 2019). Imaging techniques to diagnose any nerve injury like electromyography and nerve conduction studies have several limitations like poor localization, variations of reading with the operator, and ambient temperature. All these assessment methods lack sensitivity and objectivity (Jones, Eisenberg, and Jia 2016; Scheib and Höke 2013). To overcome this, nanotechnology has devised reliable, reproducible, and non-invasive strategies to evaluate any nerve injury. Although only a limited number of examples are reported for nanotechnology-based diagnosis of NP (Figure 10.3), these approaches have advantages like highly dynamic processes, limited time, and spatial resolution (Rangavajla et al. 2014; Gustafson et al. 2012). Recently, Zheng et al. have reported the use of MRI nanoimaging probes for nerve-specific myelin-associated protein zero (P0); and peripheral myelin protein 22 kDa (PMP22) proteins have been suggested to monitor the progress of nerve regeneration in living subjects (Zheng et al. 2014). Live magnetic resonance images of multiple cell types can contribute both to a quantitative understanding of complex pathology and a comprehensive understanding of therapeutic strategies for NP (Hitchens et al. 2015; Zhong et al. 2015). Monocytes/macrophages are essential characters in neuroinflammation and neuro-regeneration and are a highly accessible cellular focus for nanoimaging due to their ability to carry nanoparticles into the blood stream. This allows their noninvasive and dynamic in vivo tracking using NIRF, PET, and MRI. Schwann cells produce inflammatory mediators after axon damage to attract nerve macrophages. A process of Wallerian degeneration that is marked by an increased presence of macrophages is activated by monocyte chemoattractant-1 (MCP-1), interleukins, and a factor in nerve growth. For this process to complete, nanoparticles specifically undergo labeling the in vivo macrophages, which are used to monitor neuro-inflammation and longitudinal neuro-regeneration as such. Interest in the study of in vivo macrophage and inflammation of nanoparticles as molecular image agents and sensors has grown dramatically in recent years. In one study, ultra-small iron oxide nanoparticles (USPIOs) were successfully used in the MRI monitoring of peripheral neuronal

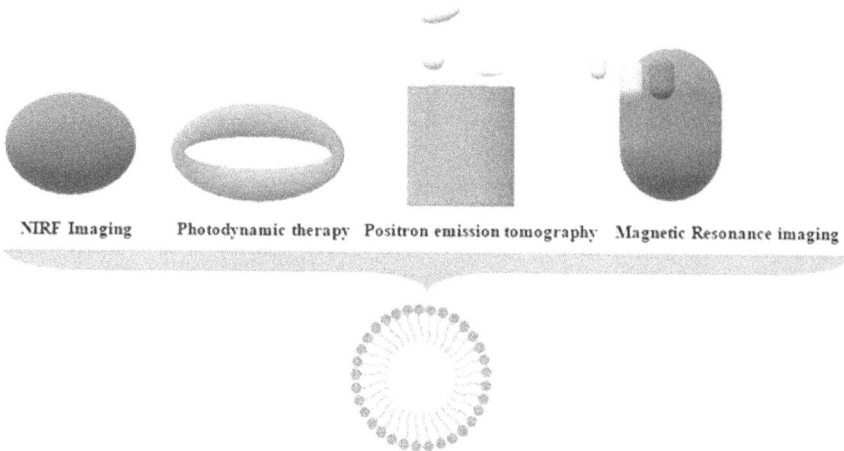

NIRF Imaging   Photodynamic therapy   Positron emission tomography   Magnetic Resonance imaging

FIGURE 10.3   Different nanotechnology-based diagnosis techniques for NP.

injury patterns in live animals in response to minocycline, tetracycline antibiotic (Ghanouni et al. 2012). Vasudeva et al. showed near-infrared imaginary fluorescence (NIRF) imagery of macrophages infiltrating the site of neuronal injury as a surrogate marker for neuroinflammation with intravenously injected NIRF-labelled PFC nanoemulsions in live animals. PFC nanoemulsions carried by infiltrating macrophages accumulated in the chronic compressed nerve (CCI) and produced a signal to the damaged nerve that NIRF could image (Vasudeva et al. 2014). Taken together, nanoparticles and nanoimaging probes inspire and encourage further research and development in the fields of NP medicine and surgery.

10.9 CHALLENGES AND FUTURE ASPECTS

Although nanomedicines have proven potential advantages in the use of NP treatment over traditional drugs, thorough testing should be conducted for safety, potency, and prophecy after in vivo administration. Organic nanomaterials such as polymer and lipid-based nanoparticles are considered to degrade quickly in elemental form and are not cytotoxic. Several FDA-approved polymer-based nanomedicines are also an important concept for future research. Nevertheless, the toxic profile of inorganic nanomaterials composed of transparent, solid, metallic, and metal oxide nanomaterials is very complicated to determine. Recent experiments have shown that metal nanoparticles are at risk of neurotoxicity by growing ROS inducing neuroinflammation and cell death (Crisponi et al. 2017; Ahmad et al. 2016). Moreover, several studies have suggested that anti-oxidant pre-treatment will remove or reduce metal-inducing ROS elevation that suppresses inflammatory responses (Brand et al. 2017). Nonetheless, the knowledge surrounding the neurotoxicity of inorganic nanomaterials, which impede risk evaluations, is significantly lacking. The physiochemical characteristics of inorganic nanomaterials such as scale, form, porosity, surface region, charge, and chemical composition have a direct effect on their cytotoxicity. In order to fully understand the impact of physicochemical substances on a toxicity profile, researchers must specify techniques to monitor the physiochemical characteristics and the effect of these variables on toxicity. Certain underestimated problems are the long-term consequences of nanomaterials. The short-term consequences of nanomaterials that cannot be extrapolated over the long term have been shown in several experiments. To understand biocompatibility, scientists should extensively record nanoparticles' physiochemical characteristics to better analyze, replicate, and compare toxicity results (Khorrami et al. 2019; De Matteis 2017; Malaviya, Shukal, and Vasavada 2019).

10.10 CONCLUSION

Nanotechnology is the advanced technology for immediate relief from NP. Many traditional medical formulations have been modified with the use of nanotechnology. A mediated therapeutic approach and conventionallyprepared synthetic drugs are currently available as nanotubes or in dot form to avoid pharmaceutical challenges. Sometimes self-nanoemulsifying drug delivery systems have also been adopted in combination (e.g. duloxetine and curcumin), to avoid pharmaceutical challenges and so make it available for the treatment of NP. Currently, the superiority of nanotechnology over other formulations provides a new avenue in pharmaceutical fields and the health sector. Nevertheless, scientists must consider the challenges associated with nanodelivery for such drugs in the treatment of NP.

REFERENCES

Adel, Moein, Masoumeh Zahmatkeshan, Behrooz Johari, Sharmin Kharrazi, Mehdi Mehdizadeh, Bahram Bolouri, and Seyed Mahdi Rezayat. 2017. "Investigating the effects of electrical stimulation via gold nanoparticles on in vitro neurite outgrowth: Perspective to nerve regeneration." *Microelectronic Engineering* 173: 1–5.

Agarwal, Ruchika, and Saurabh Kansal. 2018. "Antinociceptive evaluation of conventional anticonvulsant with conventional analgesics on pain model of albino rats and mice." *International Journal of Research in Medical Sciences* 6(3): 790.

Ahmad, Mohammad Zaki, Basel A Abdel-Wahab, Afroze Alam, Sobiya Zafar, Javed Ahmad, Farhan Jalees Ahmad, Patrick Midoux, Chantal Pichon, and Sohail Akhter. 2016. "Toxicity of inorganic nanoparticles used in targeted drug delivery and other biomedical application: An updated account on concern of biomedical nanotoxicology." *Journal of Nanoscience and Nanotechnology* 16(8): 7873–7897.

Ahmad Dar, Riyaz, Pradeep Kumar Brahaman, Sweety Tiwari, and Krishna Sadashiv Pitre. 2012. "Indirect electrochemical analysis of crocin in phytochemical sample." *E-Journal of Chemistry* 9(2): 918–925.

Ahn, Hong-Sun, Ji-Young Hwang, Min Soo Kim, Ja-Yeon Lee, Jong-Wan Kim, Hyun-Soo Kim, Ueon Sang Shin, Jonathan C Knowles, Hae-Won Kim, and Jung Keun Hyun. 2015. "Carbon-nanotube-interfaced glass fiber scaffold for regeneration of transected sciatic nerve." *Acta Biomaterialia* 13: 324–334.

Alsarra, Ibrahim A, M Al-Omar, and F Belal. 2005. "Valproic acid and sodium valproate: Comprehensive profile." *Profiles of Drug Substances, Excipients and Related Methodology* 32: 209–240.

Alvarado-Vazquez, Perla Abigail, Laura Bernal, Candler A Paige, Rachel L Grosick, Carolina Moracho Vilrriales, David Wilson Ferreira, Cristina Ulecia-Morón, and EA Romero-Sandoval. 2017. "Macrophage-specific nanotechnology-driven CD163 overexpression in human macrophages results in an M2 phenotype under inflammatory conditions." *Immunobiology* 222(8–9): 900–912.

Areti, Aparna, Veera Ganesh Yerra, Prashanth Komirishetty, and Ashutosh Kumar. 2016. "Potential therapeutic benefits of maintaining mitochondrial health in peripheral neuropathies." *Current Neuropharmacology* 14(6): 593–609.

Askari, Nahid, Mohammad Bagher Askari, and Ali Shafieipour. 2019. "Investigation the molecular structure of novel graphene hybrid scaffold in nerve regeneration." *Journal of Molecular Structure* 1186: 393–403.

Assaf, Kyl, Claudenete Vieira Leal, Mariana Silveira Derami, Eliana Aparecida de Rezende Duek, Helder Jose Ceragioli, and Alexandre Leite Rodrigues de Oliveira. 2017. "Sciatic nerve repair using poly (ε-caprolactone) tubular prosthesis associated with nanoparticles of carbon and graphene." *Brain and Behavior* 7(8): e00755.

Attal, Nadine, Didier Bouhassira, and Ralf Baron. 2018. "Diagnosis and assessment of NP through questionnaires." *The Lancet Neurology* 17(5): 456–466.

Bagli, Eleni, Anastasia K Zikou, Niki Agnantis, and Georgios Kitsos. 2017. "Mitochondrial membrane dynamics and inherited optic neuropathies." *in vivo* 31(4): 511–525.

Baldrighi, Michele, Massimo Trusel, Raffaella Tonini, and Silvia Giordani. 2016. "Carbon nanomaterials interfacing with neurons: An in vivo perspective." *Frontiers in Neuroscience* 10: 250.

Banderas, Lucía Martín, Mercedes Fernández Arévalo, Esther Berrocoso Domínguez, and Juan Antonio Mico Segura. 2018. "Method for producing a pharmaceutical composition of polymeric nanoparticles for treating NP caused by peripheral nerve compression." Google Patents.

Baskaran, Mrudhula, Padmamalini Baskaran, Navamoney Arulsamy, and Baskaran Thyagarajan. 2017. "Preparation and evaluation of PLGA-coated capsaicin magnetic nanoparticles." *Pharmaceutical Research* 34(6): 1255–1263.

Beiranvand, Siavash, and Mohamad Masud Sorori. 2019. "Pain management using nanotechnology approaches." *Artificial Cells, Nanomedicine, and Biotechnology* 47(1): 462–468.

Beltrán-Gracia, Esteban, Adolfo López-Camacho, Inocencio Higuera-Ciapara, Jesús B Velázquez-Fernández, and Alba A Vallejo-Cardona. 2019. "Nanomedicine review: Clinical developments in liposomal applications." *Cancer Nanotechnology* 10(1): 11.

Bernal, Laura, Abigail Alvarado-Vázquez, David Wilson Ferreira, Candler A Paige, Cristina Ulecia-Morón, Bailey Hill, Marina Caesar, and EA Romero-Sandoval. 2017. "Evaluation of a nanotechnology-based approach to induce gene-expression in human THP-1 macrophages under inflammatory conditions." *Immunobiology* 222(2): 399–408.

Berrocoso, Esther, Raquel Rey-Brea, Mercedes Fernández-Arévalo, Juan Antonio Micó, and Lucía Martín-Banderas. 2017. "Single oral dose of cannabinoid derivate loaded PLGA nanocarriers relieves NP for eleven days." *Nanomedicine: Nanotechnology, Biology and Medicine* 13(8): 2623–2632.

Bhushan, Bharat. 2017. "Introduction to nanotechnology." In: *Springer Handbook of Nanotechnology*, 1–19. Springer Nature Switzerland AG.

Bidve, Pankaj, Namrata Prajapati, Kiran Kalia, Rakesh Tekade, and Vinod Tiwari. 2020. "Emerging role of nanomedicine in the treatment of NP." *Journal of Drug Targeting* 28(1): 11–22.

Bonferoni, MC, F Riva, A Invernizzi, E Dellera, G Sandri, S Rossi, G Marrubini, G Bruni, B Vigani, C Caramella, and F Ferrari. 2018. "Alpha tocopherol loaded chitosan oleate nanoemulsions for wound healing. Evaluation on cell lines and ex vivo human biopsies, and stabilization in spray dried Trojan microparticles." *European Journal of Pharmaceutics and Biopharmaceutics* 123: 31–41.

Bouhassira, D. 2019. "NP: Definition, assessment and epidemiology." *Revue Neurologique* 175(1–2): 16–25.

Bouhassira, Didier, and Nadine Attal. 2011. "Diagnosis and assessment of NP: The saga of clinical tools." *Pain* 152(3): S74–S83.

Bouhassira, Didier, and Nadine Attal. 2018. "Emerging therapies for NP: New molecules or new indications for old treatments?" *Pain* 159(3): 576–582.

Brand, Walter, Cornelle W Noorlander, Christina Giannakou, Wim H De Jong, Myrna W Kooi, Margriet VDZ Park, Rob J Vandebriel, Irene EM Bosselaers, Joep HG Scholl, and Robert E Geertsma. 2017. "Nanomedicinal products: A survey on specific toxicity and side effects." *International Journal of Nanomedicine* 12: 6107.

Bravo-Caparrós, Inmaculada, Gloria Perazzoli, Sandra Yeste, Domagoj Cikes, José Manuel Baeyens, Enrique José Cobos, and Francisco Rafael Nieto. 2019. "Sigma-1 receptor inhibition reduces NP induced by partial sciatic nerve transection in mice by opioid-dependent and-independent mechanisms." *Frontiers in Pharmacology* 10: 613.

Bruna, Jordi, and Roser Velasco. 2018. "Sigma-1 receptor: A new player in neuroprotection against chemotherapy-induced peripheral neuropathy." *Neural Regeneration Research* 13(5): 775.

Bulbake, Upendra, Sindhu Doppalapudi, Nagavendra Kommineni, and Wahid Khan. 2017. "Liposomal formulations in clinical use: An updated review." *Pharmaceutics* 9(2): 12.

Caraglia, Michele, Livio Luongo, Giuseppina Salzano, Silvia Zappavigna, Monica Marra, Francesca Guida, Sara Lusa, Catia Giordano, Vito De Novellis, and Francesco Rossi 2013. "Stealth liposomes encapsulating zoledronic acid: A new opportunity to treat NP." *Molecular Pharmaceutics* 10(3): 1111–1118.

Cardoso, AM, JR Guedes, AL Cardoso, C Morais, P Cunha, AT Viegas, R Costa, A Jurado, and MC Pedroso de Lima. 2016. "Recent trends in nanotechnology toward CNS diseases: Lipid-based nanoparticles and exosomes for targeted therapeutic delivery." In: Khuloud T. Al-Jamal (ed.), *International Review of Neurobiology*, 1–40. Elsevier.

Carvalho, Cristiana R, Joana Silva-Correia, Joaquim M Oliveira, and Rui L Reis. 2019. "Nanotechnology in peripheral nerve repair and reconstruction." *Advanced Drug Delivery Reviews* 148: 308–343.

Cavalli, Eugenio, Santa Mammana, Ferdinando Nicoletti, Placido Bramanti, and Emanuela Mazzon. 2019. *The NP: An Overview of the Current Treatment and Future Therapeutic Approaches.* SAGE Publications Sage UK: London, England.

Cerna, Tereza, Tomas Eckschlager, and Marie Stiborova. 2016. "Targeted nanoparticles- a promising opportunity in cancer therapy-Review." *Journal of Metallomics and Nanotechnologies* 4: 6–11.

Chakravarthy, Krishnan V, Frank J Boehm, and Paul J Christo. 2018. "Nanotechnology: A promising new paradigm for the control of pain." *Pain Medicine* 19(2): 232–243.

Chawla, Aastha, Rajeev Chawla, and Shalini Jaggi. 2016. "Microvasular and macrovascular complications in diabetes mellitus: Distinct or continuum?" *Indian Journal of Endocrinology and Metabolism* 20(4): 546.

Chen, Kuan-Hung, Chung-Ren Lin, Jiin-Tsuey Cheng, Jen-Kun Cheng, Wen-Tzu Liao, and Chien-Hui Yang. 2014. "Altered mitochondrial ATP synthase expression in the rat dorsal root ganglion after sciatic nerve injury and analgesic effects of intrathecal ATP." *Cellular and Molecular Neurobiology* 34(1): 51–59.

Chen, Shuyi, Chunli Sun, Huawei Gu, Haiying Wang, Shan Li, Yi Ma, and Jufang Wang. 2017. "Salubrinal protects against Clostridium difficile toxin B-induced CT26 cell death." *Acta biochimica et biophysica sinica* 49(3): 228–237.

Chen, Zhoumou, Carolina Muscoli, Tim Doyle, L Bryant, Salvatore Cuzzocrea, Vincenzo Mollace, Rosanna Mastroianni, Emanuela Masini, and Daniela Salvemini. 2010. "NMDA-receptor activation and nitroxidative regulation of the glutamatergic pathway during nociceptive processing." *Pain®* 149(1): 100–106.

Choi, Sheu-Ran, Dae-Hyun Roh, Seo-Yeon Yoon, Suk-Yun Kang, Ji-Young Moon, Soon-Gu Kwon, Hoon-Seong Choi, Ho-Jae Han, Alvin J Beitz, and Seog-Bae Oh. 2013. "Spinal sigma-1 receptors activate NADPH oxidase 2 leading to the induction of pain hypersensitivity in mice and mechanical allodynia in neuropathic rats." *Pharmacological Research* 74: 56–67.

Ciaramitaro, Palma, Mauro Mondelli, Francesco Logullo, Serena Grimaldi, Bruno Battiston, Arman Sard, Cecilia Scarinzi, Giuseppe Migliaretti, Giuliano Faccani, and Dario Cocito. 2010. "Traumatic peripheral nerve injuries: Epidemiological findings, NP and quality of life in 158 patients." *Journal of the Peripheral Nervous System* 15(2): 120–127.

Colloca, Luana, Taylor Ludman, Didier Bouhassira, Ralf Baron, Anthony H Dickenson, David Yarnitsky, Roy Freeman, Andrea Truini, Nadine Attal, and Nanna B Finnerup. 2017. "NP." *Nature Reviews Disease Primers* 3: 17002.

Crisponi, Guido, Valeria M Nurchi, Joanna I Lachowicz, Massimiliano Peana, Serenella Medici, and Maria Antomietta Zoroddu. 2017. "Toxicity of nanoparticles: Etiology and mechanisms." In: Alexandru Mihai Grumezescu (ed.), *Antimicrobial Nanoarchitectonics*, 511–546. Elsevier.

Cruccu, Giorgio, and Andrea Truini. 2017. *Neuropathic Pain: The Scope of the Problem.* Springer Nature Switzerland AG.

D Skaper, Stephen. 2016. "Mast cell–glia dialogue in chronic pain and NP: Blood-brain barrier implications." *CNS and Neurological Disorders-Drug Targets (Formerly Current Drug Targets-CNS & Neurological Disorders)* 15(9): 1072–1078.

De Freitas, Guilherme BL, Durinézio J De Almeida, Emerson Carraro, Ivo I Kerppers, Guilherme AG Martins, Rubiana M Mainardes, Najeh M Khalil, and Iara JT Messias-Reason. 2018. "Formulation, characterization, and in vitro/in vivo studies of capsaicin-loaded albumin nanoparticles." *Materials Science and Engineering: Part C* 93: 70–79.

De Marco, Rossella, and Anna Janecka. 2016. "Strategies to improve bioavailability and in vivo efficacy of the endogenous opioid peptides endomorphin-1 and endomorphin-2." *Current Topics in Medicinal Chemistry* 16(2): 141–155.

De Matteis, Valeria. 2017. "Exposure to inorganic nanoparticles: Routes of entry, immune response, biodistribution and in vitro/in vivo toxicity evaluation." *Toxics* 5(4): 29.

Demir, Ulku, R Shahbazi, S Calamak, S Ozturk, M Gultekinoglu, and K Ulubayram. 2018. "Gold nano-decorated aligned polyurethane nanofibers for enhancement of neurite outgrowth and elongation." *Journal of Biomedical Materials Research Part A* 106(6): 1604–1613.

Distasi, Carla, Marianna Dionisi, Federico Alessandro Ruffinatti, Alessandra Gilardino, Roberta Bardini, Susanna Antoniotti, Federico Catalano, Eleonora Bassino, Luca Munaron, and Gianmario Martra. 2019. "The interaction of SiO2 nanoparticles with the neuronal cell membrane: Activation of ionic channels and calcium influx." *Nanomedicine* 14(5): 575–594.

Elsherbiny, Nehal M, Eman Ahmed, Ghada Abdel Kader, Yousra Abdel-mottaleb, Mohamed H ElSayed, Amal M Youssef, and Sawsan A Zaitone. 2019. "Inhibitory effect of valproate sodium on pain behavior in diabetic mice involves suppression of spinal histone deacetylase 1 and inflammatory mediators." *International Immunopharmacology* 70: 16–27.

Faccendini, Angela, Barbara Vigani, Silvia Rossi, Giuseppina Sandri, Maria Cristina Bonferoni, Carla Marcella Caramella, and Franca Ferrari. 2017. "Nanofiber scaffolds as drug delivery systems to bridge spinal cord injury." *Pharmaceuticals* 10(3): 63.

Fernandes, Valencia, Dilip Sharma, Shivani Vaidya, PA Shantanu, Yun Guan, Kiran Kalia, and Vinod Tiwari. 2018. "Cellular and molecular mechanisms driving NP: Recent advancements and challenges." *Expert Opinion on Therapeutic Targets* 22(2): 131–142.

Fornasari, Diego. 2017. "Pharmacotherapy for NP: A review." *Pain and Therapy* 6(1): 25–33.

Gabriel, Rodney A, and Brian M Ilfeld. 2019. "Peripheral nerve blocks for postoperative analgesia: From traditional unencapsulated local anesthetic to liposomes, cryo-neurolysis and peripheral nerve stimulation." *Best Practice and Research Clinical Anaesthesiology* 33(3): 293–302.

Ganugula, R, M Deng, M Arora, H-L Pan, and MNV Ravi Kumar. 2019. "Polyester nanoparticle encapsulation mitigates paclitaxel-induced peripheral neuropathy." *ACS Chemical Neuroscience* 10(3): 1801–1812.

Ghanouni, Pejman, Deepak Behera, Jin Xie, Xiaoyuan Chen, Michael Moseley, and Sandip Biswal. 2012. "In vivo USPIO magnetic resonance imaging shows that minocycline mitigates macrophage recruitment to a peripheral nerve injury." *Molecular Pain* 8: 1744-8069-8-49.

Gilron, Ian, Ralf Baron, and Troels Jensen. 2015. "Neuropathic pain: Principles of diagnosis and treatment." In: *Mayo Clinic Proceedings* 90(4): 532–545.

Gosling, Artur Padao, Maria Dias Torres Kenedi, AJLA Da Cunha, FJJD Reis, VLR De Castro Halfoun, and Llcia Margarida De Vilhena Saadi. 2017. "Characteristics of NP after multidrug therapy in a tertiary referral centre for leprosy: A cross-sectional study in Rio de Janeiro, Brazil." *Leprosy Review* 88(1): 109–121.

Grinsell, D, and CP Keating. 2014. "Peripheral nerve reconstruction after injury: A review of clinical and experimental therapies." *BioMed Research International* 2014: 1–13.

Gustafson, Tiffany P, Ying Yan, Piyaraj Newton, Daniel A Hunter, Samuel Achilefu, Walter J Akers, Susan E Mackinnon, Philip J Johnson, and Mikhail Y Berezin. 2012. "A NIR dye for development of peripheral nerve targeted probes." *Medicinal ChemistryComm* 3(6): 685–690.

Harris, Chelsea. 2017. "Evaluation of nanoparticles for the treatment of chronic NP".

Hasanzadeh-Kiabi, Farshad. 2018. "Nano-drug for pain medicine." *Drug Research* 68(05): 245–249.

Hitchens, T Kevin, Li Liu, Lesley M Foley, Virgil Simplaceanu, Eric T Ahrens, and Chien Ho. 2015. "Combining perfluorocarbon and superparamagnetic iron-oxide cell labeling for improved and expanded applications of cellular MRI." *Magnetic Resonance in Medicine* 73(1): 367–375.

Hopper, AP, JM Dugan, AA Gill, OJL Fox, PW May, JW Haycock, and F Claeyssens, A A Gill, O J L Fox, and P W May. 2014. "Amine functionalized nanodiamond promotes cellular adhesion, proliferation and neurite outgrowth." *Biomedical Materials* 9(4): 045009.

Hu, Ping, Zhang Han, Anthony D Couvillon, Randal J Kaufman, and John H Exton. 2006. "Autocrine tumor necrosis factor alpha links endoplasmic reticulum stress to the membrane death receptor pathway through IRE1α-mediated NF-κB activation and down-regulation of TRAF2 expression." *Molecular and Cellular Biology* 26(8): 3071–3084.

Huang, Qian, Xiao-Fang Mao, Hai-Yun Wu, Hao Liu, Ming-Li Sun, Xiao Wang, and Yong-Xiang Wang. 2017. "Cynandione A attenuates NP through p38β MAPK-mediated spinal microglial expression of β-endorphin." *Brain, Behavior, and Immunity* 62: 64–77.

Husain, Syeda Fabeha, Raymond WM Lam, Tao Hu, Michael WF Ng, ZQG Liau, Keiji Nagata, Sanjay Khanna, Yulin Lam, Kishore Bhakoo, and Roger Ho. 2019. "Locating the site of NP in vivo using MMP-12-targeted magnetic nanoparticles." *Pain Research and Management* 2019.

Isacchi, Benedetta, Maria Camilla Bergonzi, Romina Iacopi, Carla Ghelardini, Nicoletta Galeotti, and Anna Rita Bilia. 2017. "Liposomal formulation to increase stability and prolong antineuropathic activity of verbascoside." *Planta Medica* 83(05): 412–419.

Ishikawa, Masatomo, and Kenji Hashimoto. 2010. "The role of sigma-1 receptors in the pathophysiology of neuropsychiatric diseases." *Journal of Receptor, Ligand and Channel Research* 3: 25–36.

Ismailova, Aiten, David Kuter, D Scott Bohle, and Ian S Butler. 2018. "An overview of the potential therapeutic applications of CO-releasing molecules." *Bioinorganic Chemistry and Applications* 2018: 1–24.

Janjic, Jelena M, and Vijay S Gorantla. 2018. "Clinically translatable nanotheranostic platforms for peripheral nerve regeneration: Design with outcome in mind." *Reporters, Markers, Dyes, Nanoparticles, and Molecular Probes for Biomedical Applications X.*

Jean-Toussaint, Renee, Yuzhen Tian, Amrita Datta Chaudhuri, Norman J Haughey, Ahmet Sacan, and Seena K Ajit. 2020. "Proteome characterization of small extracellular vesicles from spared nerve injury model of NP." *Journal of Proteomics* 211: 103540.

Jeong, Injae, Beom-Soo Kim, Hyejung Lee, Kang-Min Lee, Insop Shim, Sung-Keel Kang, Chang-Shick Yin, and Dae-Hyun Hahm. 2009. "Prolonged analgesic effect of PLGA-encapsulated bee venom on formalin-induced pain in rats." *International Journal of Pharmaceutics* 380(1–2): 62–66.

Jia, Tianyu, Jingan Rao, Lifang Zou, Shanhong Zhao, Zhihua Yi, Bing Wu, Lin Li, Huilong Yuan, Liran Shi, and Chunping Zhang. 2018. "Nanoparticle-encapsulated curcumin inhibits diabetic NP involving the P2Y12 receptor in the dorsal root ganglia." *Frontiers in Neuroscience* 11: 755.

Jones, Salazar, Howard M Eisenberg, and Xiaofeng Jia. 2016. "Advances and future applications of augmented peripheral nerve regeneration." *International Journal of Molecular Sciences* 17(9): 1494.

Kapural, L, C Yu, MW Doust, L Kapural, C Yu, MW Doust, CDC National Diabetes, KE Schmader, R Pop-Busui, and AJM Boulton. 2018. "Pharmacotherapy for NP in adults: A systematic review and meta-analysis." *Postgraduate Medicine* 130(supl): 1–91.

Kelly, Jessica M, Allison Bradbury, Douglas R Martin, and Mark E Byrne. 2017. "Emerging therapies for neuropathic lysosomal storage disorders." *Progress in Neurobiology* 152: 166–180.

Keskinbora, KH, and MA Jameel. 2018. "Nanotechnology applications and approaches in medicine: A review." *Journal of Nanoscience and Nanotechnology Research* 2(2): 1–5.

Khan, Md Imran, Deepu Dowarha, Revansiddha Katte, Ruey-Hwang Chou, Anna Filipek, and Chin Yu. 2019. "Lysozyme as the anti-proliferative agent to block the interaction between S100A6 and the RAGE V domain." *PLoS One* 14(5).

Khangura, Ravneet, A Bali, AS Jaggi, and N Singh Kaur, Anjana Bali. 2017. "Histone acetylation and histone deacetylation in NP: An unresolved puzzle?" *European Journal of Pharmacology* 795: 36–42.

Khangura, Ravneet, J Sharma, A Bali, N Singh, and AS Jaggi Kaur, Jasmine Sharma, Anjana Bali. 2019. "An integrated review on new targets in the treatment of NP." *The Korean Journal of Physiology and Pharmacology* 23(1): 1–20.

Khorrami, Mohammad Bagher, Hamid Reza Sadeghnia, Alireza Pasdar, Majid Ghayour-Mobarhan, Bamdad Riahi-Zanjani, Alireza Hashemzadeh, Mohammad Zare, and Majid Darroudi. 2019. "Antioxidant and toxicity studies of biosynthesized cerium oxide nanoparticles in rats." *International Journal of Nanomedicine* 14: 2915.

Kim, Danbi, Hansol Min, Kyoungsu Jeon, Sangdong Kim, Sanghyun Ahn, Raehyung Ryu, Junsik Kim, Mun Han, Choongyong Kim, and Juyoung Park. 2017. "The kinetics of P-glycoprotein after blood-brain barrier disruption in rat brain by MRI-guided focused ultrasound." 한국실험동물학회학술발표대회논문집 56: 99–99.

Kim, Hwisung, Boomin Choi, Hyoungsub Lim, Hyunjung Min, Jae Hoon Oh, Sunghyun Choi, Joung Goo Cho, Jong-Sang Park, and Sung Joong Lee. 2017. "Polyamidoamine dendrimer-conjugated triamcinolone acetonide attenuates nerve injury-induced spinal cord microglia activation and mechanical allodynia." *Molecular Pain* 13: 1744806917697006.

Kobayashi, Hideo, Takuya Kitamura, Miho Sekiguchi, Miwako K Homma, Yukihito Kabuyama, Shin-ichi Konno, Shin-ichi Kikuchi, and Yoshimi Homma. 2007. "Involvement of EphB1 receptor/EphrinB2 ligand in NP." *Spine* 32(15): 1592–1598.

Kong, Dawei, Zhuangli Guo, Wenqiang Yang, Qi Wang, Yu Yanbing, and Li Zhang. 2020. "Tanshinone II A affects diabetic peripheral NP via spinal dorsal horn neuronal circuitry by modulating endoplasmic reticulum stress pathways." *Experimental and Clinical Endocrinology and Diabetes* 128(01): 59–65.

Kuthati, Yaswanth, Vaikar Navakanth Rao, Prabhakar Busa, Srikrishna Tummala, Goutham Davuluri Venkata Naga, and Chih Shung Wong. 2020. "Scope and applications of nanomedicines for the management of neuropathic pain." *Molecular Pharmaceutics* 17(4): 1015–1027.

Kwon, Song, Kwai Han Yoo, Sun Jin Sym, and Dongwoo Khang. 2019. "Mesenchymal stem cell therapy assisted by nanotechnology: A possible combinational treatment for brain tumor and central nerve regeneration." *International Journal of Nanomedicine* 14: 5925.

Latremoliere, Alban, Alexandra Latini, Nick Andrews, Shane J Cronin, Masahide Fujita, Katarzyna Gorska, Ruud Hovius, Carla Romero, Surawee Chuaiphichai, and Michio Painter 2015. "Reduction of neuropathic and inflammatory pain through inhibition of the tetrahydrobiopterin pathway." *Neuron* 86(6): 1393–1406.

Lee, Wan-Hung, Li-Li Li, Aarti Chawla, Andy Hudmon, Yvonne Y Lai, Michael J Courtney, and Andrea G Hohmann. 2018. "Disruption of nNOS-NOS1AP protein-protein interactions suppresses NP in mice." *Pain* 159(5): 849.

Li, Lin, Xuan Sheng, Shanhong Zhao, Lifang Zou, Xinyao Han, Yingxin Gong, Huilong Yuan, Liran Shi, Lili Guo, and Tianyu Jia 2017. "Nanoparticle-encapsulated emodin decreases diabetic NP probably via a mechanism involving P2X3 receptor in the dorsal root ganglia." *Purinergic Signalling* 13(4): 559–568.

Li, Zhihua, Yanyan Guo, Xiuhua Ren, Lina Rong, Minjie Huang, Jing Cao, and Weidong Zang. 2019. "HDAC2, but not HDAC1, regulates Kv1. 2 Expression to mediate NP in CCI rats." *Neuroscience* 408: 339–348.

Li-Na Yu, MD, and MS Li-Hong Sun, MD Min Wang, MS Lie-Ju Wang, MS Ying Wu, MS Jing Yu, MD Wen-Na Wang, MD Feng-Jiang Zhang, MD Xue Li, and MD Min Yan. 2017. "EphrinB-EphB signaling induces hyperalgesia through ERK5/CREB pathway in rats." *Pain Physician* 20: E563–E574.

Lim, Tony KY, Malena B Rone, Seunghwan Lee, Jack P Antel, and Ji Zhang. 2015. "Mitochondrial and bioenergetic dysfunction in trauma-induced painful peripheral neuropathy." *Molecular Pain* 11: s12990-015-0057-7.

Lin, Mark John-Yung, Megan Rose Paul, and Dennis John Kuo. 2018. "Severe NP with concomitant administration of vincristine and posaconazole." *The Journal of Pediatric Pharmacology and Therapeutics* 23(5): 417–420.

Liu, Meili, Gang Zhou, Yongzhao Hou, Gang Kuang, Zhengtai Jia, Ping Li, and Yubo Fan. 2015. "Effect of nano-hydroxyapatite-coated magnetic nanoparticles on axonal guidance growth of rat dorsal root ganglion neurons." *Journal of Biomedical Materials Research – Part A* 103(9): 3066–3071.

Liu, Qian, Claudia M Santamaria, Tuo Wei, Chao Zhao, Tianjiao Ji, Tianshe Yang, Andre Shomorony, Bruce Y Wang, and Daniel S Kohane. 2018. "Hollow silica nanoparticles penetrate the peripheral nerve and enhance the nerve blockade from tetrodotoxin." *Nano Letters* 18(1): 32–37.

Liu, Zhongyang, Liangliang Huang, Liang Liu, Beier Luo, Miaomiao Liang, Zhen Sun, Shu Zhu, Xin Quan, Yafeng Yang, and Teng Ma 2015. "Activation of Schwann cells in vitro by magnetic nanocomposites via applied magnetic field." *International Journal of Nanomedicine* 10: 43.

Lombardi, Sara. 2017. "Eph receptors as a target to develop novel therapies to control neurodegeneration and neuropathic diseases." Alma.

Lu, Jiaju, Xun Sun, Heyong Yin, Xuezhen Shen, Shuhui Yang, Yu Wang, Wenli Jiang, Yue Sun, Lingyun Zhao, and Xiaodan Sun 2018. "A neurotrophic peptide-functionalized self-assembling peptide nanofiber hydrogel enhances rat sciatic nerve regeneration." *Nano Research* 11(9): 4599–4613.

Luca, Andrei, Cosmin-Teodor Mihai, Gabriela-Dumitriţa Stanciu, Veronica Bild, Elena Cojocaru, Robert Ancuceanu, Valeria Harabagiu, Cristian Peptu, Catalina Anişoara Peptu, and Maria Leonconstantin. 2019. "In-vivo safety and efficacy evaluation of a novel polymeric based lidocaine formulation for topical analgesia." *Pain* 9: 10.

Malaviya, D Shukal, and AR Vasavada. 2019. "Nanotechnology-based drug delivery, metabolism and toxicity." *Current Drug Metabolism* 20(14): 1167–1190.

Mammadov, Busra, Melike Sever, Mevhibe Gecer, Fatih Zor, Sinan Ozturk, Hakan Akgun, Umit H Ulas, Zeynep Orhan, Mustafa O Guler, and Ayse B Tekinay. 2016. "Sciatic nerve regeneration induced by glycosaminoglycan and laminin mimetic peptide nanofiber gels." *RSC Advances* 6(112): 110535–110547.

Martuscello, Regina T, Robert N Spengler, Adela C Bonoiu, Bruce A Davidson, Jadwiga Helinski, Hong Ding, Supriya Mahajan, Rajiv Kumar, Earl J Bergey, and Paul R Knight 2012. "Increasing TNF levels solely in the rat hippocampus produces persistent pain-like symptoms." *Pain®* 153(9): 1871–1882.

Mattiazzi, Juliane, Marcel Henrique Marcondes Sari, Taíne de Bastos Brum, Paulo César Oliveira Araújo, Jéssica Mendes Nadal, Paulo Vítor Farago, Cristina Wayne Nogueira, and Letícia Cruz. 2019. "3, 3'-Diindolylmethane nanoencapsulation improves its antinociceptive action: Physicochemical and behavioral studies." *Colloids and Surfaces, Part B: Biointerfaces* 181: 295–304.

McCarberg, Bill, Yvonne D'Arcy, Bruce Parsons, Alesia Sadosky, Andrew Thorpe, and Regina Behar. 2017. " NP: A narrative review of etiology, assessment, diagnosis, and treatment for primary care providers." *Current Medical Research and Opinion* 33(8): 1361–1369.

Min, Joo-Ok, and Bo-Eun Yoon. 2017. "Glia and gliotransmitters on carbon nanotubes." *Nano Reviews and Experiments* 8(1): 1323853.

Miranpuri, Gurwattan S, Dominic T Schomberg, Bahauddeen Alrfaei, Kevin C King, Bryan Rynearson, Vishwas S Wesley, Nayab Khan, Kristen Obiakor, Umadevi V Wesley, and Daniel K Resnick. 2016. "Role of matrix metalloproteinases 2 in spinal cord injury-induced NP." *Annals of Neurosciences* 23(1): 25–32.

Mistry, Jai, Nicola R Heneghan, Timothy Noblet, Deborah Falla, and Alison Rushton. 2019. "Diagnostic utility of patient history, clinical examination and screening tool data to identify NP in low back-related leg pain: Protocol for a systematic review." *BMJ Open* 9(11).

Moore, Benjamin JR, Barira Islam, Sean Ward, Olivia Jackson, Rebecca Armitage, Jack Blackburn, Shozeb Haider, and Patrick C McHugh. 2019. "Repurposing of tranilast for potential NP treatment by inhibition of sepiapterin reductase in the BH4 pathway." *ACS Omega* 4(7): 11960–11972.

Moradkhani, Mahmoud Reza, Arash Karimi, and Babak Negahdari. 2018. "Nanotechnology application for pain therapy." *Artificial Cells, Nanomedicine, and Biotechnology* 46(2): 368–373.

Najafi, Rezvan, Asieh Hosseini, Habib Ghaznavi, Saeed Mehrzadi, and Ali M Sharifi. 2017. "Neuroprotective effect of cerium oxide nanoparticles in a rat model of experimental diabetic neuropathy." *Brain Research Bulletin* 131: 117–122.

Nawrotek, Katarzyna, Michał Tylman, Karolina Rudnicka, Justyna Gatkowska, and Jacek Balcerzak. 2016. "Tubular electrodeposition of chitosan–carbon nanotube implants enriched with calcium ions." *Journal of the Mechanical Behavior of Biomedical Materials* 60: 256–266.

Nejati-Koshki, Kazem, Yousef Mortazavi, Younes Pilehvar-Soltanahmadi, Sumit Sheoran, and Nosratollah Zarghami. 2017. "An update on application of nanotechnology and stem cells in spinal cord injury regeneration." *Biomedicine and Pharmacotherapy = Biomedecine and Pharmacotherapie* 90: 85–92.

Nigam, Kuldeep, Atinderpal Kaur, Amit Tyagi, Kailash Manda, Reema Gabrani, and Shweta Dang. 2019. "Baclofen-loaded poly (d, l-lactide-co-glycolic acid) nanoparticles for NP management: In vitro and in vivo evaluation." *Rejuvenation Research* 22(3): 235–245.

Nune, Manasa, Anuradha Subramanian, Uma Maheswari Krishnan, and Swaminathan Sethuraman. 2019. "Peptide nanostructures on nanofibers for peripheral nerve regeneration." *Journal of Tissue Engineering and Regenerative Medicine* 13(6): 1059–1070.

Oprych, Karen M, Raymond LD Whitby, Sergey V Mikhalovsky, Paul Tomlins, and Jimi Adu. 2016. "Repairing peripheral nerves: Is there a role for carbon nanotubes?." *Advanced Healthcare Materials* 5(11): 1253–1271.

Peng, Yunan, Ting Zang, Luyang Zhou, Kun Ni, and Xuelong Zhou. 2019. "COX-2 contributed to the remifentanil-induced hyperalgesia related to ephrinB/EphB signaling." *Neurological Research* 41(6): 519–527.

Peng, Zhiyou, Leiqiong Zha, Meijuan Yang, Yunze Li, Xuejiao Guo, and Zhiying Feng. 2019. "Effects of ghrelin on pGSK-3β and β-catenin expression when protects against NP behavior in rats challenged with chronic constriction injury." *Scientific Reports* 9(1): 1–9.

Porto, Fábio Henrique de Gobbi, Gislaine Cristina Lopes Machado Porto, and Mario Wilson Lervolino Brotto. 2016. "Additional tests to investigate NP. The value of electroneuromyography for NP." *Revista Dor* 17: 23–26.

Puthenveedu, Manojkumar A, Daniel J Shiwarski, and Amynah Pradhan. 2017. "Method to increase bioavailability of the delta-opioid receptor for management of pain and neuropsychiatric disorders." Google Patents.

Raffa, Vittoria. 2020. "Engineering magnetic nanoparticles for repairing nerve injuries." In: Vincenzo Guarino, Maria Letizia Focarete, and Dario Pisignano (eds.), *Advances in Nanostructured Materials and Nanopatterning Technologies*, 167–200. Elsevier.

Rangavajla, Gautam, Nassir Mokarram, Nazanin Masoodzadehgan, S Balakrishna Pai, and Ravi V Bellamkonda. 2014. "Noninvasive imaging of peripheral nerves." *Cells, Tissues, Organs* 200(1): 69–77.

Rauti, Rossana, Mattia Musto, Susanna Bosi, Maurizio Prato, and Laura Ballerini. 2019. "Properties and behavior of carbon nanomaterials when interfacing neuronal cells: How far have we come?" *Carbon* 143: 430–446.

Ren, Jianmin, Fufu Ma, Yujing Zhou, Anting Xu, Jianjian Zhang, Rong Ma, and Xiaoyan Xiao. 2018. "Hearing impairment in type 2 diabetics and patients with early diabetic nephropathy." *Journal of Diabetes and Its Complications* 32(6): 575–579.

Resham, Kahkashan, Pragyanshu Khare, Mahendra Bishnoi, and Shyam S Sharma. 2020. "Neuroprotective effects of isoquercitrin in diabetic neuropathy via Wnt/β-catenin signaling pathway inhibition." *BioFactors* 46(3): 411–420.

Resham, Kahkashan, and Shyam S Sharma. 2019. "Pharmacological interventions targeting Wnt/β-catenin signaling pathway attenuate paclitaxel-induced peripheral neuropathy." *European Journal of Pharmacology* 864: 172714.

Rumora, Amy E, Masha G Savelieff, Stacey A Sakowski, and Eva L Feldman. 2019. "Disorders of mitochondrial dynamics in peripheral neuropathy: Clues from hereditary neuropathy and diabetes." *International Review of Neurobiology* 145: 127–176.

Sánchez-Fernández, Cristina, José Manuel Entrena, José Manuel Baeyens, and Enrique José Cobos. 2017. "Sigma-1 receptor antagonists: A new class of neuromodulatory analgesics." In: Sylvia B. Smith and Tsung-Ping Su (eds.), *Sigma Receptors: Their Role in Disease and as Therapeutic Targets*, 109–132. Springer Nature Switzerland AG.

Sari, Marcel Henrique Marcondes, Vanessa Angonesi Zborowski, Luana Mota Ferreira, Natália da Silva Jardim, Paulo Cesar Oliveira Araujo, César Augusto Brüning, Letícia Cruz, and Cristina Wayne Nogueira. 2018. Enhanced pharmacological actions of p, p-methoxyl-diphenyl diselenide-loaded polymeric nanocapsules in a mouse model of NP: Behavioral and molecular insights." *Journal of Trace Elements in Medicine and Biology* 46: 17–25.

Scheib, Jami, and Ahmet Höke. 2013. "Advances in peripheral nerve regeneration." *Nature Reviews. Neurology* 9(12): 668.

Shakhbazau, Antos, Chandan Mohanty, Dzmitry Shcharbin, Maria Bryszewska, Anne-Marie Caminade, Jean-Pierre Majoral, Jacob Alant, and Rajiv Midha. 2013. "Doxycycline-regulated GDNF expression promotes axonal regeneration and functional recovery in transected peripheral nerve." *Journal of Controlled Release* 172(3): 841–851.

Shankarappa, Sahadev Aramanethalgur, and Daniel S Kohane. 2016. "Formulations and methods for delaying onset of chronic NP." Google Patents.

Sharma, Neelam, Sandeep Arora, and Jitender Madan. 2018. "Nefopam hydrochloride loaded microspheres for post-operative pain management: Synthesis, physicochemical characterization and in-vivo evaluation." *Artificial Cells, Nanomedicine, and Biotechnology* 46(1): 138–146.

Sharma, Shweta, Ashwni Verma, Jyotsana Singh, B Venkatesh Teja, Naresh Mittapelly, Gitu Pandey, Sandeep Urandur, Ravi P Shukla, Rituraj Konwar, and Prabhat Ranjan Mishra. 2016. "Vitamin B6 tethered endosomal pH responsive lipid nanoparticles for triggered intracellular release of doxorubicin." *ACS Applied Materials and Interfaces* 8(44): 30407–30421.

Shin, Juhee, Yuhua Yin, Do Kyung Kim, Sun Yeul Lee, Wonhyung Lee, Joon Won Kang, Dong Woon Kim, and Jinpyo Hong. 2019. "Foxp3 plasmid-encapsulated PLGA nanoparticles attenuate pain behavior in rats with spinal nerve ligation." *Nanomedicine: Nanotechnology, Biology and Medicine* 18: 90–100.

Shiue, Sheng-Jie, Ruey-Horng Rau, Han-Shiang Shiue, Yi-Wei Hung, Zhi-Xiang Li, Kuender D Yang, and Jen-Kun Cheng. 2019. "Mesenchymal stem cell exosomes as a cell-free therapy for nerve injury–induced pain in rats." *Pain* 160(1): 210–223.

Song, Tieying, Jianhui Zhao, Xiaojing Ma, Zaiwang Zhang, Bo Jiang, and Yunliang Yang. 2017. "Role of sigma 1 receptor in high fat diet-induced peripheral neuropathy." *Biological Chemistry* 398(10): 1141–1149.

Song, Xue-Jun, Jun-Li Cao, Hao-Chuan Li, Ji-Hong Zheng, Xue-Song Song, and Li-Ze Xiong. 2008. "Upregulation and redistribution of ephrinB and EphB receptor in dorsal root ganglion and spinal dorsal horn neurons after peripheral nerve injury and dorsal rhizotomy." *European Journal of Pain* 12(8): 1031–1039.

Stevens, Andrea M, Lu Liu, Dylan Bertovich, Jelena M Janjic, and John A Pollock. 2019. "Differential expression of neuroinflammatory mRNAs in the rat sciatic nerve following chronic constriction injury and pain-relieving nanoemulsion NSAID delivery to infiltrating macrophages." *International Journal of Molecular Sciences* 20(21): 5269.

Sunderhaus, Elizabeth R, Alexander D Law, and Doris Kretzschmar. 2019. "ER responses play a key role in Swiss-Cheese/Neuropathy Target esterase-associated neurodegeneration." *Neurobiology of Disease* 130: 104520.

Taxonomy, IASP IASP. 2012. Available Online: http://www.iasp-pain.org/Education/Content .aspx.ItemNumber=1698&navItemNumber576.

Thomson, Suzanne E, Chloe Charalambous, Carol-Anne Smith, Penelope M Tsimbouri, Theophile Déjardin, Paul J Kingham, Andrew M Hart, and Mathis O Riehle. 2017. "Microtopographical cues promote peripheral nerve regeneration via transient mTORC2 activation." *Acta Biomaterialia* 60: 220–231.

Torabi, Mozhgan, Mahnaz Kesmati, Hamid Galehdari, Hossein Najafzadeh Varzi, and Nahid Pourreza. 2020. "MgO and ZnO nanoparticles anti-nociceptive effect modulated by glutamate level and NMDA receptor expression in the hippocampus of stressed and non-stressed rats." *Physiology and Behavior* 214: 112727.

Tour, James M, and William Sikkema. 2019. "Neuronal scaffold-water soluble graphene for treatment of severed spinal cords and neuronal repair." Google Patents.

Van Hameren, Gerben, and Nicolas Tricaud. 2018. "Mitochondrial hydrogen peroxide levels in healthy peripheral nerves and peripheral neuropathies." *Free Radical Biology and Medicine* 120: S24.

Van Helleputte, Lawrence. 2018. "Inhibition of histone deacetylase 6 (HDAC6) as a therapeutic strategy in chemotherapy-induced peripheral neuropathies." *Journal of the Peripheral Nervous System* 20(2).

Vasudeva, Kiran, Karl Andersen, Bree Zeyzus-Johns, T Kevin Hitchens, Sravan Kumar Patel, Anthony Balducci, Jelena M Janjic, and John A Pollock. 2014. "Imaging neuroinflammation in vivo in a neuropathic pain rat model with near-infrared fluorescence and 19F magnetic resonance." *PLoS One* 9(2): e90589.

Wang, Jin-Feng, Hai-Jun Xu, Zhao-Long He, Qin Yin, and Wei Cheng. 2020. "Crocin alleviates pain hyperalgesia in AIA rats by inhibiting the spinal Wnt5a/β-catenin signaling pathway and glial activation." *Neural Plasticity* 2020: 1–11.

Wang, Qiaoling, Jinlong Chen, Qingfei Niu, Xiumei Fu, Xiaohong Sun, and Xiaojie Tong. 2017. "The application of graphene oxidized combining with decellularized scaffold to repair of sciatic nerve injury in rats." *Saudi Pharmaceutical Journal* 25(4): 469–476.

Wang, Yi-Rui, Xiao-Fang Mao, Hai-Yun Wu, and Yong-Xiang Wang. 2018. "Liposome-encapsulated clodronate specifically depletes spinal microglia and reduces initial NP." *Biochemical and Biophysical Research Communications* 499(3): 499–505.

Whitehead, Kathryn A. 2016. "A one-two punch for pain control." *Science Translational Medicine* 8(336): 336ec69-336ec69.

Wright, M, M Centelles, W Gedroyc, and M Thanou. 2018. "Image guided focused ultrasound as a new method of targeted drug delivery."

Wu, Shuangquan, Bo Duan, Ang Lu, Yanfeng Wang, Qifa Ye, and Lina Zhang. 2017. "Biocompatible chitin/carbon nanotubes composite hydrogels as neuronal growth substrates." *Carbohydrate Polymers* 174: 830–840.

Yang, Fang, John Whang, William T Derry, Daniel Vardeh, and Joachim Scholz. 2014. "Analgesic treatment with pregabalin does not prevent persistent pain after peripheral nerve injury in the rat." *Pain®* 155(2): 356–366.

Yu, Guang, Linnan Qian, Yu Juanjuan, Min Tang, Changming Wang, Yuan Zhou, Xiao Geng, Chan Zhu, Yan Yang, and Yang Pan. 2019. "Brucine alleviates NP in mice via reducing the current of the sodium channel." *Journal of Ethnopharmacology* 233: 56–63.

Yu, Xiaolu, Mannan Abdul, Bing-Qian Fan, Lilu Zhang, Xing Lin, Yan Wu, Hui Fu, Qisi Lin, and Hao Meng. 2020. "The release of exosomes in the medial prefrontal cortex and nucleus accumbens brain regions of chronic constriction injury (CCI) model mice could elevate the pain sensation." *Neuroscience Letters* 723: 134774.

Zhang, Yan-Kai, Zhi-Jiang Huang, Su Liu, Yue-Peng Liu, Angela A Song, and Xue-Jun Song. 2013. "WNT signaling underlies the pathogenesis of NP in rodents." *The Journal of Clinical Investigation* 123(5): 2268–2286.

Zhao, Yingbo, and Tianjian Wu. 2018. "Histone deacetylase inhibition inhibits brachial plexus avulsion-induced NP." *Muscle and Nerve* 58(3): 434–440.

Zheng, Linfeng, Kangan Li, Yuedong Han, Wei Wei, Sujuan Zheng, and Guixiang Zhang. 2014. "In vivo targeted peripheral nerve imaging with a nerve-specific nanoscale magnetic resonance probe." *Medical Hypotheses* 83(5): 588–592.

Zhong, Jia, Kazim Narsinh, Penelope A Morel, Hongyan Xu, and Eric T Ahrens. 2015. "In vivo quantification of inflammation in experimental autoimmune encephalomyelitis rats using fluorine-19 magnetic resonance imaging reveals immune cell recruitment outside the nervous system." *PLoS One* 10(10).

Zhou, Xijie, Bin Zhao, Keshav Poonit, Weidong Weng, Chenglun Yao, Chao Sun, and Hede Yan. 2019. "An aligned nanofiber nerve conduit that inhibits painful traumatic neuroma formation through regulation of the RhoA/ROCK signaling pathway." *Journal of Neurologicalsurgery* 1(aop): 1–10.

Zhu, Shu, Jun Ge, Yuqing Wang, Fengyu Qi, Teng Ma, Meng Wang, Yafeng Yang, Zhongyang Liu, Jinghui Huang, and Zhuojing Luo. 2014. "A synthetic oxygen carrier-olfactory ensheathing cell composition system for the promotion of sciatic nerve regeneration." *Biomaterials* 35(5): 1450–1461.

11 Self-Nanoemulsifying Drug Delivery System
A Unique Platform for Overcoming Bioavailability of Lipophilic and Gastrointestinal Labile Drugs

Narendra Kumar Pandey, Amica Panja, Sachin Kumar Singh, Rubiya Khursheed, Bimlesh Kumar, Monica Gulati, Saurabh Singh, Dileep Singh Baghel, Rajan Kumar, Ankit Awasthi, Rajneesh Kumar Gupta, Ankit Kumar, Leander Corrie and Pooja Patni
Lovely Professional University, Phagwara, India

CONTENTS

11.1   INTRODUCTION

Many medicines – mainly BCS (Biopharmaceutical Classification System) Class II
drugs – show poor permeability and poor solubility. In the past few years many traditional techniques have been used to overcome this problem. Some such techniques
include reduction in particle size, salt form of the drug, complexation etc. These techniques show some limitations, like in the case of salt formation, when salts forms do
not present in their original form, so they may be converted into acid or base in GIT.
In the case of reduction in particle size, they may cause problems like poor wettability.
So, because of certain limitations, some lipid-based techniques are used to increase
bioavailability and solubility of the drugs, one technique being SNEDDS. SNEDDS
are isotropic mixtures containing oil, surfactant, co-surfactant, and the drug. These
are clear or translucent emulsions having a droplet size 100 nm, and can form stable
O/W type of nanoemulsions (Savale 2015). They dissolve larger quantities of lipophilic
drugs and protect the drug from hydrolysis. In the case of parental preparations, they
prevent enzymatic degradation of the drug. These nanoemulsions are thermodynamically stable. They are optimized by a pseudoternary phase diagram. Now SNEDDS
are applied in different aspects of drug delivery systems like in cancer therapy, cosmetics, parental drug delivery, transdermal drug delivery and ophthalmic drug delivery. In
recent years, SNEDDS, SMEDDS and SEDDS have been used to improve the hydrophilicity of poorly hydrophilic drugs (Basha, Rao, and Vedantham 2013) Partitioning
of the drug between lipophilic as well as hydrophilic phase can be easily done by this
system. SNEDDS are the type of systems which are thermodynamically stable and
transparent in nature. The stability of these nanoemulsions can be increased with the
addition of surfactants and co-surfactants (Savale 2015; Garg et al. 2016).

Types of self-nanoemulsifying drug delivery systems:

a) *Water in oil (W/O) nanoemulsion*: These are the type of emulsions in which
   a water droplet is dissolved in an oily phase.
b) *Oil in water (O/W)*: These are the type of emulsions in which an oil globule
   is dissolved in a water phase.
c) *Bi-continuous*: In these types of emulsions, surfactant is soluble in both
   types of solvents (oily as well as aqueous).

The comparison between SEDDS, SMEDDS, and SNEDDS is shown in Table 11.1.

11.1.1   COMPOSITION OF SNEDDS

The composition of SNEDDS is shown in Figure 11.1.

Oil: In SNEDDS, the selection of an oily phase is an essential parameter, as in
this case mainly the O/W type of emulsions are used. For nanoemulsion formulation,

TABLE 11.1

Comparison in SEDDS, SMEDDS, and SNEDDS

S. No.	SEDDS	SMEDDS	SNEDDS
1.	Self-emulsifying drug delivery system.	Self-micro-emulsifying drug delivery system	Self-nano-emulsifying drug delivery system
2.	Globule size: 500 micrometers	Globule size: Less than 100 micrometers	Globule size: Less than 100 nanometers
3.	Mix of drug, oil, surfactant	Mixture of drug, oil, surfactant, and co-surfactant	Mixture of drug, oil, surfactant, and co-surfactant
4.	Turbid appearance	Translucent in nature	Transparent in nature
5.	Thermodynamically unstable	Thermodynamically stable	Thermodynamically and kinetically stable

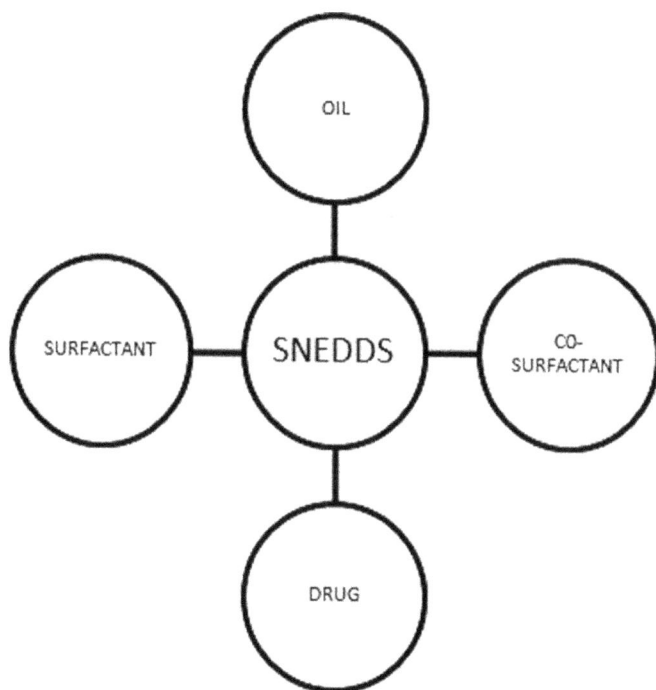

FIGURE 11.1 Composition of SNEDDS.

the oily phase should be able to solubilize drugs. Naturally as well as synthetically occurring oil, fats, and triglycerides can be used in nanoemulsions. Triglycerides are used to decrease the degree of saturation and to prevent oxidation degradation. The addition of oil is chief to increase friction for transportation of the drug into the intracellular membrane to increase the hydrophilicity of poorly soluble drugs (Singh et al. 2013; Savale 2015). Commonly used oils/lipids are listed in Table 11.2.

TABLE 11.2
Oils/Lipids Used for SEDDS

Lipid/Oil	Chemical Name	Reference
Bean phospholipids	-	(Lv et al. 2012)
Capmul MCM EP	Glycerylcaprylate/caprate	-
Caprylic/capric triglyceride	Caprylic/capric triglyceride	-
Capryol 90	Propylene glycol monocaprylate (type II) NF	(Inugala et al. 2015)
Captex 355	Glyceryltricaprylate/tricaprate	(Inugala et al. 2015)
Capmum MCM C8	Glycerylmonocaprylate	(Inugala et al. 2015)
Castor oil	Castor oil	-
Chuanxiong oil		
Cinnamon oil	Cinnamon oil	(P. Zhang et al. 2008)
Cotton seed oil	Cotton seed oil	(Kang et al. 2012)
Corn oil	Corn oil	(Kang et al. 2012)
Cradamol GTCC	Caprylic/capric triglyceride	(Yao, Lu, and Zhou 2008)
Cremophor EL castor oil	Macrogolglycerol ricinoleate, polyoxyl 35 castor oil USP	(Inugala et al. 2015)
Ethyl oleate	Ethyl oleate	(Cui et al. 2005)
Gelcire 44/14	Lauroyl macrogol-32 glycerides EP Lauroyl polyoxyl-32 glycerides NF	(Mandawgade et al. 2008)
Isopropyl myristate	Myristic acid isopropyl ester	(Wang et al. 2009)
Labrafac PG	Propylene glycol dicaprylocaprate EP Propylene glycol dicaprylate/dicaprate NF	(Setthacheewakul et al. 2010)
Lauroglycol FCC	Propylene glycol mono laurate	(Rao and Shao 2008; Rao, Yajurvedi, and Shao 2008)
Labrafac CC	Caprylic/capric triglyceride	(Inugala et al. 2015; Kang et al. 2012)
Mineral oil	Higher alkanes from mineral source	(Kang et al. 2012)
Maisine oil	Glycerylmonolinoleate	(N. Parmar et al. 2011; B. Zhang, Lu, and Chen 2008)
Miglyol 812	Liquid lipids/C8/C10 triglycerides	(Ma et al. 2012)
Myvacet 9-45	Myvacet 9-45K NF	(Kommuru et al. 2001)
Methyl decanoate	Decanoic acid methyl ester	(Wang et al. 2009)
Methyl oleate	Oleic acid methyl ester	(Wang et al. 2009)
Oleic acid	Octadececenoic acid	(Qi et al. 2011; Rao and Shao 2008; Rao, Yajurvedi, and Shao 2008)
Olive oil	Olive oil	(Qi et al. 2011)
Peanut oil	Peanut oil	(Kang et al. 2012)
Peceol	Glycerol monooleate	
Phosal 53 MCT	Lecithin in caprylic/capric triglycerides, alcohol, glyceryl stearate, oleic acid and ascorbylpalmitate	(Shanmugam, Baskaran, et al. 2011; Shanmugam, Park, et al. 2011)

(Continued)

TABLE 11.2 (CONTINUED)
Oils/Lipids Used for SEDDS

Lipid/Oil	Chemical Name	Reference
Polyoxyethylene castor oil	Polyoxyethylene castor oil	(Mekjaruskul et al. 2013)
Sesame oil	Sesame oil	(Kang et al. 2012)
Sunflower oil	Sunflower oil	(Kang et al. 2012)
Soybean oil	Soybean oil	(Qi et al. 2011)
Trilaurin	Glycerol trilaurate	(Elgart et al. 2013)

Surfactant: Surfactants are those type of compounds which lower the surface tension/interfacial tension between two immiscible liquids, between a gas/liquid, or between a liquid/solid. These may act as detergents, wetting agents, and emulsifying agents. They can be self-nanoemulsifying, self-emulsifying, and microemulsifying, and have the ability to solubilized poorly water-soluble drugs. Surfactant should be used in an optimum amount; if used in a larger amount it may cause toxicity, so to ensure the safety of emulsions it must be used in an appropriate amount. Non-ionic surfactants are more stable as compared to ionic surfactants as these are non-toxic as well as thermodynamically stable. For the preparation of emulsions or nanoemulsions, the concentration of surfactant is based upon the size of a droplet or globule. As the concentration of surfactant increases, so the globule size ultimately increases (Garg et al. 2016; Balakumar, Raghavan, and Abdu 2013; Koga et al. 2006).

Types of surfactants used:

a) Anionic surfactants: Negatively charged
b) Cationic surfactants: Positively charged
c) Ampholytic surfactants: Positively and negatively charged
d) Non-ionic surfactants: No charge
   - Anionic surfactants: These are the type of surfactants in which the hydrophilic group has negative charge such as carboxyl, e.g. potassium laurate, SLS
   - Cationic surfactants: These are the type of surfactants in which the hydrophilic group has positive charge, e.g. quaternary ammonium compound
   - Ampholytic surfactants: These are surfactants which carry both positive as well as negative charge, e.g. sulfobetaines
   - Non-ionic surfactants: These are the type of surfactants which do not carry any charge over hydrophilic phase but have solubility in water, as strong polar functional groups are present in them, e.g. polysorbates (Savale 2015)

The surfactants used for SEDDS are shown in Table 11.3.

Co-surfactant: Interfacial fluidity can be prevented by adding co-surfactants. It has a similar function to that of surfactant. It is added along with surfactant to

TABLE 11.3
Surfactants Used for SEDDS (Kang et al. 2012)

Surfactant	Chemical Name	Reference
Capmul	mono-diglyceride of medium chain fatty acids	(Basalious, Shawky, and Badr-Eldin 2010)
Cremophor RH40	PEG-40 hydrogenated castor oil	(Rao and Shao 2008)
Cremophor-EL	PEG-35 castor oil	(N. Parmar et al. 2011)
Labrafil M 2125 CS	PEG-6 corn oil	(Inugala et al. 2015; Kang et al. 2012)
Labrafil M1944CS	PEG-6 apricot kernel oil	(Inugala et al. 2015; Kang et al. 2012)
Labrasol	Caprylocaprylmacrogol glycerides	(Inugala et al. 2015; N. Parmar et al. 2011; Rao and Shao 2008; Rao, Yajurvedi, and Shao 2008)
Polysorbate 80	Polyoxy ethylene 20 sorbitan mono oleate	(Rao and Shao 2008)
Polysorbate 20	Polyoxy ethylene 20 sorbitan mono laurate	(Rao and Shao 2008)
Polyoxamer 407	Poly(ethylene glycol)-block-poly(propylene glycol)-block-poly(ethylene glycol)	(Date and Nagarsenker 2007)
Polyoxamer 188	Pluronic F-68 solution	(Date and Nagarsenker 2007)
Solutol HS 15	Macrogol (15)-hydroxystearate	(Date and Nagarsenker 2007)
Span 20	Sorbitanmonolaurate	(Kang et al. 2012; Qi et al. 2011)
Span 80	Sorbitanmonooleate	
Span 85	Sorbitantrioleate	(Qi et al. 2011)
Tween20	PEG-20 sorbitanmonolaurate	(Date and Nagarsenker 2007)
Tween-80	PEG-20 sorbitanmonooleate	(Date and Nagarsenker 2007; Qi et al. 2011; Singh et al. 2009)
Tween-85	PEG-20 sorbitantrioleate	(Singh et al. 2009)

increase the capacity of surfactant hence improving the hydrophilicity of drugs with low solubility. The most important application of co-surfactant is that they prevent interfacial tension in the oil and water phase.

Examples: Ethanol, Methanol, Pentanol, Glycol, Propylene glycol (Savale 2015). Commonly used surfactants are listed in Table 11.4.

Drug: As these formulations are used to enhance the bioavailability of the drugs so here those drugs are used which are hydrophobic in nature (Savale 2015).

11.1.1.1 Mechanism of SNEDDS

The free energy of the emulsion is described by:

$$\Delta G = \Sigma N \Pi r 2 \sigma.$$

where

ΔG = free energy

N = number of droplets

TABLE 11.4
Co-Surfactants Used for SEDDS

Co-Surfactant	Chemical Name	HLB	Reference
1,2 octane diol	1,2 octane diol		(Wang et al. 2009)
Akoline MCM	Caprylic/capric glycerides	5–6	(Date and Nagarsenker 2007)
Akomed	Oil containing triacylglycerols of caprylic and capric acid		(Date and Nagarsenker 2007)
Capmul MCM-C8	Glycerylcaprylate	5–6	(Singh et al. 2009)
Caproyl 90	Propylene glycol mono caprylate	6	(Kang et al. 2012; N. Parmar et al. 2011)
HCO-60	PEG-60 hydrogenated castor oil	14	(Singh et al. 2009)
Imwitor 742	Caprylic/capric glycerides	4	(Date and Nagarsenker 2007)
Labrafil 1944 CS	PEG-6 apricot kernel oil	4	(Date and Nagarsenker 2007)
Lauroglycol 90	Propylene glycol monolaurate	5	(Inugala et al. 2015)
Lauroglycol FCC	Propylene glycol monolaurate	4	(Rao, Yajurvedi, and Shao 2008)
Lutrol F127	PolyoxamersPh Eur., polyoxamers		(Beg et al. 2015)
PEG 400	Polyethylene glycol 400	11.6	(Rao and Shao 2008)
PG	Propylene glycol		(Date and Nagarsenker 2007; Shahba, Mohsin, and Alanazi 2012)
PlurolDioleique CC 497	Polyglyceryl-3 dioleate NF Polyglyceryl-3 oleate (USA FDA IIG)	3	(Date and Nagarsenker 2007)
Transcutol P	Diethylene glycol mono ethyl ether	-	(Date and Nagarsenker 2007; N. Parmar et al. 2011)

r = radius of droplets
σ = interfacial energy

From the above, it is found that free energy is directly proportional to interfacial energy decrease. When the energy required in the dispersion is higher than the energy involved in the formation of droplets, then self-emulsification occurs.

In the case of conventional formulations, free energy is very high because to make a new surface between two immiscible phases, high energy is required. As in the case high energy is used, so the emulsions may not be stable and the two immiscible phases may tend to separate. But in the case of SNEEDS, forms instantaneously as the system has very low free energy or sometimes negligible free energy due to the presence of a flexible interface. On combining the mixture of oil phase and surface-active agent in an aqueous phase, on gentle stirring, an interface is developed between 2 phases. Further, water permeates through the interface and is solubilized within the oil up to the solubilization limit. After water permeation, a liquid crystalline phase will form. The amount of liquid crystalline phase depends on the concentration of surface active agents (Basha, Rao, and Vedantham 2013).

11.1.1.2 Categorization of SEDDS

11.1.1.2.1 Liquid SEDDS

These are self-emulsifying isotropic mixtures of oil, surfactant, and cosolvent in liquid state. These offer the advantages of enhanced solubility of drugs and their increased lymphatic absorption. However, due to their liquid state, they are difficult to be dispensed as dosage form. To make the dosage form more convenient, they need to be incorporated into soft gelatin capsules. This, in turn, adds to the cost of formulation (Garg et al. 2016; Singh et al. 2009).

11.1.1.2.2 Super Saturable SEDDS

The concentration of surfactants in the SEDDS formulation is usually in the range of 20–60%. From a safety point of view, the use of such a high concentration of surfactant becomes a concern for the formulator, as their higher concentration is known to lead to some adverse effects (Garg et al. 2016). To overcome this problem, the concept of super saturable SEDDS was created. In these, the concentration of surfactants is reduced by the inclusion of water-soluble polymeric precipitation inhibitor (PPI). These formulations maintain a super saturable metastable state *in vivo* by reducing the drug precipitation using PPI. Hydroxypropyl methylcellulose (HPMC) of different grades of viscosity have been widely reported to prevent crystallization as PPI in super-saturable SEDDS (Garg et al. 2016; Gao and Morozowich 2006; Gao et al. 2003; Raghavan et al. 2000).

11.1.1.2.3 Solid SEDDS (S-SEDDS)

Self-emulsifying drug delivery systems were initially developed in liquid form. However, these liquid SEDDS faced the difficulty of stability, formation of unit dosage form, high production costs, low stability and portability, low drug loading, and few choices of dosage forms. Irreversible drugs/excipient precipitation may also be problematic (B. Tang, Cheng, et al. 2008). S-SEDDS come as a superior alternative to the L-SEDDS. S-SEDDS, along with the advantages of liquid SEDDS, provide better stability, ease of handling, and ease of conventional dosage forms like tablets and capsules (Mohsin, Shahba, and Alanazi 2012). Solid self-emulsifying compositions are preferred over liquid ones due to their ability to extend the drug release, higher stability, and ease of handling. Solid SEDDS, as the name suggests, are solid dosage forms which have the ability to self-emulsify when in contact with GI media (Cho et al. 2013). S-SEDDS are available in different forms like powders, granules, pellets, tablets, and self-emulsifying dispersions (Tarate, Chavan, and K Bansal 2014).

11.2 TECHNIQUES USED FOR SOLIDIFICATION OF SEDDS

11.2.1 Physical Adsorption

Physical adsorption of L-SEDDS on the solid carriers is one of the simplest techniques of solidification. In this process, L-SEDDS are added to the solid carrier and mixed either via physical blending by hand or motor pestle on the lab scale or via the use of blenders. The loading factor is calculated as the amount of solid carrier required for adsorption of L-SEDDS so that homogenous powder is obtained. After this, the weighed amount of both L-SEDDS and carrier are mixed together until a homogenous solid powder is formed via the adsorption of L-SEDDS over solid carriers. This powder should be passed through sieves to break up any lumps, if present. The resultant powder can be directly filled into capsules or can be compressed into tablets via the addition of some other excipients used for tableting (Chavan, Modi, and Bansal 2015; Milović et al. 2014; B. Tang, Cheng, et al. 2008; Zidan et al. 2015). Several carriers like silicon dioxide or syloid have the capacity to adsorb large amounts of L SEDDS (Tarate, Chavan, and K Bansal 2014).

The hydrophilic/hydrophobic nature of the carrier on which L-SEDDS have to be adsorbed affect the properties of the drug e.g. L-SNEDDS of ezetimibe were prepared with Capryol 90, Lauroglycol FCC, ethyl laurate, Cremophor EL and Transcutol P and adsorbed on hydrophobic colloidal silicon dioxide to form self-nano emulsifying granules (SNEG). X-ray diffraction (XRD) indicates that the drug is in its amorphous form, but when the same SNEDDS were loaded on magnesium stearate a eutectic mixture is resulted (Dixit and Nagarsenker 2008).

Quantities of ingredients per unit dose can be calculated for S-SEDDS. The following equation was used to calculate the amount of carrier materials (Abdelbary, Amin, and Salah 2013):

$$L = \frac{W}{Qf} \tag{11.1}$$

L is the liquid loading factor; W is the liquid medication weight; Q is the carrier material weight. The excipient ratio (R) is the ratio between the carrier (Q) and coating material (q) as presented by the following equation (Abdelbary, Amin, and Salah 2013):

$$R = \frac{Q}{q} \tag{11.2}$$

11.2.2 MELT GRANULATION

Melt granulation is a method in which S-SEDDS are prepared in a single step. In this method there is no need to prepare L-SEDDS and then adsorb on the solid carrier. In this method, oil, e.g. goat fat, or surfactant which is solid at room temperature is used. In this method, all the mixture of oil and surfactant is taken in the desired quantity and heated above the melting point. Into this melted mixture drug is added and mixed to form a homogenous mixture (Attama and Mpamaugo 2006).

When this molten mixture is added dropwise with a beaker containing cold water at 4° C at 1000 rpm, it leads to the formation of solid lipid spheres (Attama and Mpamaugo 2006). The granulation process is controlled by the parameters such as impeller speed, mixing time, binder particle size, and viscosity of the molten binder (Tarate, Chavan, and K Bansal 2014). These can be filtered out and dried. Attama et al. reported the formation of self-emulsifying liposphere using this method using goat fat and Tween 65 (Attama and Mpamaugo 2006).

11.2.3 Pour Moulding Method

Self-emulsifying suppositories and tablets can be prepared via the pour molding method. In this method, oil and surfactant are taken and heated together until they homogenize completely. Drug is added into this homogenous mixture and stirred thoroughly. This mixture is now poured into molds and allowed to settle at a temperature of 4°C. The tablets or suppositories with self-emulsifying ability are taken out from the mold and stored in a cool place (Attama et al. 2003). Self-emulsifying tablets were prepared using this method (Attama et al. 2003). Though this method is easy to execute, industry-friendly, and reproducible. the chances of degradation at a higher temperature exist. Moreover, the selection of lipids and surfactant is very important for the stability of these formulations as only those excipients which are solid at room temperature should be selected (B. Tang, Cheng, et al. 2008; J. Tang, Sun, et al. 2008).

11.2.4 Spray Congealing

Self-emulsifying microparticles can be produced with spray-congealing technology. Fluidized bed equipment is utilized for this purpose. It uses two fluid atomizers with a wide orifice opening i.e. pneumatic nozzle. External mixing of fluid and air or gas occurs outside the nozzle orifice, thus atomization can be varied by changing the air pressure without affecting the liquid flow rate, enabling the spraying of high concentration or viscous products. The temperature of the feed tank containing molten fluid should be kept higher than the melting temperature. The congealing chamber should be cooled using a refrigerator system for the solidification of droplets. A nozzle sprays the molten fluid in the form of fine droplets. These molten drops harden because of the low temperature of the chamber and are collected at the bottom of the congealing chamber (Albertini et al. 2015). Factors that affect the size include the orifice size of pneumatic nozzles, temperature of feed and congealing chamber, rate of atomization, and air pressure (Albertini et al. 2015). This method bypasses the use of traditional solvents of spray drying like water and alcohol and relies on the excipients present in typical self-emulsifying formulations (Tarate, Chavan, and K Bansal 2014).

11.2.5 Spray Drying

Spray drying is one of the commonly used techniques in the formation of S-SEDDS. A spray dryer consists of the following components viz. feed delivery system, atomizer, heated air supply, drying chamber, solid-gas separator, and product collection system. In this technique, the drug, L-SEDDS, and carrier are dissolved or suspended in a solvent to form a homogenous system. This solution is now atomized to produce liquid droplets with the help of a spray nozzle in the spray dryer. These atomized droplets come in contact with hot air in the drying chamber and are converted into fine powder which becomes separated and is collected in the cyclone and collecting container. The product is self-emulsifying in nature. Both hydrophobic and hydrophilic carriers can be used in this process. The atomizer, the temperature, the most suitable air flow pattern, and the drying chamber design are important variables affecting product characteristics (Alinaghi et al. 2015; Balakrishnan et al. 2009; Czajkowska-Kośnik et al. 2015; Tarate, Chavan, and K Bansal 2014).

11.2.6 EXTRUSION-SPHERONIZATION

S-SEDDS can also be formulated in the form of pellets via extrusion-spheroniza-tion. This process includes the wet granulation of L-SEDDS with solid excipients, followed by extrusion of the wet mass, spheronization of extrudates, drying of the spheroids, sizing, and optional coating of the pellets. The extruder consists of a die through which material is forced with the help of a single or twin screw, and shaped into cylinders of uniform length. This process is used to form uniformly sized pellets. Spheronizer is equipped with a bowl having fixed side walls and a rapidly rotating bottom plate. The bottom plate is grooved to provide the equipment-particle interactions for rounding the cylindrical pellets (Abdalla, Klein, and Mäder 2008; Tarate, Chavan, and K Bansal 2014).

Lyophilization can also be used for formulating S-SEDDS. In this process, water is evaporated directly via sublimation. It includes several steps i.e. freezing, primary drying, and secondary drying. In this process, both carrier and L-SEDDS are dissolved in a common solvent followed by freezing and sublimation process. This method gives a solid product (Tarate, Chavan, and K Bansal 2014). Jain et al., 2014 prepared S-SNEDDS using lyophilization technique. SNEDDS were diluted in a minimum quantity of deionized water and thoroughly mixed with Aerosil® 200. Lyophilization was performed after 15 minutes of equilibration (A.K. Jain, Thanki, and Jain 2014). It is a method of choice for theromolabile formulations, proteins, peptides, and vaccines, although cost and time is a constraint (Tarate, Chavan, and K Bansal 2014).

11.2.7 USE OF POROUS BEADS

Porous beads or porous tablets with a large surface area may also be used for loading L-SEDDS. Porous polystyrene beads with a surface area of 153.12 m^2/g were used by Patil and Paradkar. Results show good loading efficiency as well as drug content (S. Patel, Jani, and Patel 2012; Patil and Paradkar 2006). Christiansen et al. prepared porous tablet cores for the loading of L-SNEDDS using magnesium aluminometasilicate granules with Ac-Di-Sol (disintegrant) and magnesium stearate (anti-adhesive). These cores were then kept in a container along with L-SNEDDS for 2 h to ensure loading of 500 mg of SNEDDS on tablet cores. Excess L-SNEDDS were removed to obtain dry and shiny tablets loaded with L-SNEDDS (Christiansen et al. 2014).

11.2.8 SELF-EMULSIFYING SOLID DISPERSION

Self-emulsifying solid dispersions can also be prepared by the melting method. In this technique, drug, surfactant and fatty acids are homogenously mixed and slightly heated to get a melted mixture. This melted mixture is then added to a suitable adsorbent like Aerosil® 200 and kept at a cool temperature. Solid mass obtained is crushed and passed through a sieve of suitable size to obtain fine powder (Tran et al. 2013).

11.2.8.1 Drug Transport Mechanism of SEDDS

SEDDS offer bioavailability of water-insoluble drugs even through oral administration. Upon reaching the GIT, these SEDDS undergo absorption in three steps (Garg et al. 2016; Charman and Porter 1996; Stremmel 1988).

Step 1 (Digestion): The enzymes present in GIT hydrolyze the emulsion at oil-water interphase and enable SEDDS for absorption. The digestion process stops once these SEDDS form mixed micelles upon interacting with bile salts and fatty acids.

Step 2 (Absorption): During absorption, these micelles are taken up through active or passive transport by enterocyte membrane or through lymphatic circulation by chylomicrons.

Step 3 (Circulatory): Chylomicrons release the drug into the systemic circulation. The remaining lipids are utilized by the body.

11.3   PREPARATION OF SNEDDS

In this method, the drug is dissolved in the lipophilic part of the nanoemulsion and the liquid phase is combined with surfactant and co-surfactant. Then the liquid phase has to be added in the oily phase slowly with continuous stirring until the suspension becomes transparent. The desired range of dispersed globules can be achieved using an ultra-sonicator.

SNEDDS can be prepared in two ways:

(a) *Preparation of liquid SNEDDS*: In this, the ratio of surfactant/co-surfactant and oil/co-solvent has to be selected via pseudoternary phase diagram. Then, a series of formulations must be made by taking dissimilar concentrations of surfactant, co-surfactant, and oil. All of them should be weighed accurately and then the drug has to be dissolved in the mixture. Then that mixture has to be stored or kept at room temperature.

(b) *Preparation of solid SNEDDS*: In this type of SNEDDS preparation, the appropriate solvent has to be used which has to be added in a mortar pestle containing the drug. Then when damp mass is formed, it has to pass through the sieve and then dried at room temperature (Savale 2015).

11.3.1 Methods of Preparation of SNEDDS

The methods of preparation of SNEDDS are shown in Figure 11.2.

High energy approach: This method is based on the selection of composition of surfactant, co-surfactant, co-solvents in a mixture. This mixture is subjected to high energy using a mechanical process to form the desired nanoemulsion (Basha, Rao, and Vedantham 2013).

High-pressure homogenizer: By using this homogenizer, fine nanoemulsions can be prepared by applying very high shear stress. Oil is dispersed into a mixture

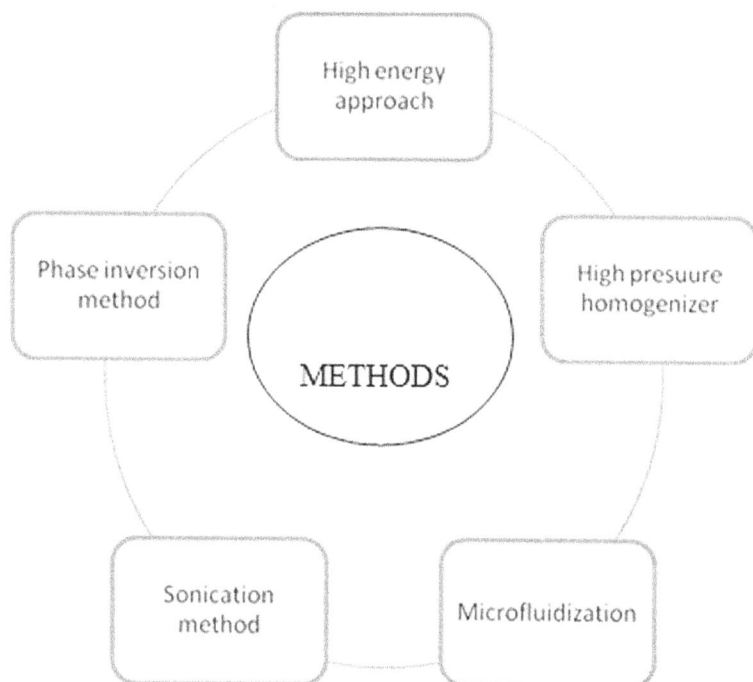

FIGURE 11.2 Methods of preparation of SNEDDS.

containing water and surfactant under very high pressure. The reduction in droplet size can be explained by a combination of two theories, i.e. turbulation and covariance (Basha, Rao, and Vedantham 2013).

When high velocity is applied to a mixture, high energy will be imparted to the liquid in the valve, which consequently generates strong turbulent eddies of the same size as MDD. When droplets diverge from the eddy current, this results in a reduction in the droplet size. Simultaneously, a pressure drop across the valve occurs (due to the cavitation), and this further generates eddies which gradually cause a disruption in droplets.

If a sufficient amount of surfactant is added, then an emulsion of less than 100 nm can be prepared using this method, as the oil-water interface will be completely covered by surfactant (Basha, Rao, and Vedantham 2013).

Microfluidization: In this method, a device named a "Microfluidizer" is used. For this high-pressure positive displacement pump, a pressure of 500-200 PSI is used, which drives the formulation through an interaction chamber which consists of small channels known as microchannels. The product flows through these channels, resulting in the formation of microparticles of a submicron range. A combination of the aqueous and oily phases is processed in the homogenizer to yield a coarse emulsion, which when introduced into the microfluidizer will result in the formation of a stable nano-emulsion (Sinha and Ganesh 2015).

Sonication method: By using this method, only small batches or lab scale batches can be prepared. With the use of the sonication method, the size of the simple conventional emulsion can be reduced (A. Patel, Shelat, and Lalwani 2014).

Phase inversion method: By changing the temperature with constant composition or by changing the composition at constant temperature, adequate phase transitions can be produced. In the case of the phase inversion temperature, this method is based on the principle that determines how the solubility of polyoxyethylene changes with a change in temperature. Therefore with an increase in temperature, the surfactant becomes lipophilic in nature (K. Jain et al. 2013).

11.4   EVALUATION OF SNEDDS

- Thermodynamic stability
- Centrifugation study
- Droplet size and polydispersibility index (PDI)
- Heating and cooling cycle
- Viscosity
- Drug content
- Freeze thaw cycle
- Dispersibility test
- Percent transmittance
- pH measurements
- Morphological study

Thermodynamic stability: The thermodynamical stability of the nanoemulsion is a crucial step which can adversely affect the emulsion by precipitating the drug in the excipient matrix. It may also lead to phase separation which not only affects the performance of the emulsion but also its appearance (Kaur, Chandel, and Harikumar 2013).

Centrifugation study: Prepared nanoemulsion can be centrifuged at 5000 rpm for 30 min in the laboratory. Then the formulation will be checked for any instability problems like creaming, cracking, or phase inversion. If any formulation is unstable, that will be excluded from the study (Kaur, Chandel, and Harikumar 2013).

Droplet size: The droplet size of SNEDDS can be checked using a zeta sizer. In this, the formulation will be diluted using water and mixed gradually using a stirrer. Then the emulsion will be subjected to a particle/droplet size analyzer using a zeta sizer. Droplet size can also be determined using an osmoviscometer (Kaur, Chandel, and Harikumar 2013).

Heating and cooling cycle: Around three heating and cooling cycles must be performed for 24 hours at a temperature of 4° C and 40° C. After these cycles, the emulsions have to be subjected to thermodynamic stability studies to know if any instability occurs like phase separation, creaming, etc. (Kaur, Chandel, and Harikumar 2013).

Viscosity: Viscosity of the self-nanoemulsifying drug delivery system is performed to determine the consistency of the nanoemulsion. The viscosity of SNEDDS can be determined using Brookfield viscometer (Vanitasagar and Subhashini 2013).

Drug content: Essentially, this is completed for the evaluation of solid self-nano emulsions. It is also carried out in order to check the content of the drug present and the purity of the nanoemulsion. Twenty tablets are taken and crushed prior to taking their average weight. Then the sample is diluted and analyzed under HPLC and the percentage of drug present can be determined using a dissolution apparatus (Vanitasagar and Subhashini 2013).

Freeze thaw cycle: This cycle determines the stability of SNEDDS. In these formulations, 3 freeze cycles are necessary for not less than 24 hours, and then thawed for 24 hours. Then, centrifugation is performed for 3000 rpm for around 5 min. Then the nanoemulsions are observed to determine if any phase separation occurs (Kaur, Chandel, and Harikumar 2013).

Dispersibility test: The efficiency of nanoemulsion can be determined using a USP type II dissolution apparatus. Nine-hundred milliliters of water is added into the apparatus, and the temperature adjusted to 37±0.5°C. The paddle rotates at 50 rpm. The in vitro performance of the formulation can be adjusted according to the following grading system:

Grade A: Nanoemulsions which rapidly form within minutes and have a transparent or bluish form.

Grade B: It also forms quickly, but is slightly less transparent and has a transparent bluish emulsion.

Grade C: It is milky in nature and is able to form in two minutes.

Grade D: It has a slightly oily appearance, is grayish-white, and dull in nature.

Grade E: Large globules are present on the surface of the emulsions. They show either low or minimal emulsification.

Out of all of the above, grade C is used as SNEDDS (Savale 2015).

Percent transmittance: The percent transmittance of SNEDDS can be determined using UV-Vis single beam spectroscopy or double beam spectroscopy at 560 nm (Kaur, Chandel, and Harikumar 2013).

pH measurements: The pH of any type of emulsion can be determined using a pH meter or potentiometer. For pH measurements, an electrode of pH meter has to be dipped in the emulsion and pH can be determined (Kaur, Chandel, and Harikumar 2013).

Morphological study: This is used for the determination of color, odor, appearance, density of nanoemulsions, or any other formulations (Kaur, Chandel, and Harikumar 2013).

11.5   APPLICATIONS OF SNEDDS

- Bioavailability of the drugs can be improved by SNEDDS.
- SNEEDS are highly stable as compared to SMEDDS and SEDDS.
- It is also used in other medicine systems, such as the Unani and Ayurvedic systems.
- Antimicrobial property: These are o/w type of emulsions and have a size range from 200–600 nm. These emulsions have a wide spectrum of activity against bacteria and viruses.
- Thermal stability enzymes can be increased by making this type of emulsion.

11.6   RECENT ADVANCES

- Usage of SNEDDS for the improvement of drugs in terms of stability.
- Controlling of the in vitro release of the drug can also be achieved using SNEDDS.
- Enhancement in drug-loading capacity (<25 mg to >2 g).
- SNEDDS is now commonly used for drug targeting, e.g. leukemia and auto-immune diseases.
- Transdermal drug delivery can be accomplished using SNEDDS.

The various SNEDDS prepared to date are shown in Table 11.5.

11.7   CONCLUSION

- SNEDDS can absorb a large number of lipophilic drugs.
- SNEDDS increase the bioavailability and solubility of poorly soluble drugs.
- Evaluation parameters of SNEDDS are also very easy to do, as they do not require highly sophisticated techniques – except for HPLC.
- SNEDDS require very simple manufacturing techniques such as mixture, homogenizer, and sonicator.
- SNEDDS are highly stable compared to SEDDS and SMEDDS.

TABLE 11.5
Various SNEDDS Prepared till Date

Drug	Composition of L-SNEDDS	Techniques of Solidification	Formulation Prepared	Carrier Used	Stage of Development	References
Loratidin	Liquid paraffin, Capriole, Span 20 and Transcutol	Extrusion Spheronization	S-SNEDDS	Aerosil	Formulation and development	(Abbaspour, Jalayer, and Makhmalzadeh 2014)
Carvedilol	Capmul MCM, Nikkol HCO 50	Congealing	S-SNEDDS	Nikkol HCO 50	Preclinical phase	(Singh et al. 2013)
Lovastatin	Capmul MCXM, Nikkol HCO-50, Lutrol F127	Melting method	S-SNEDDS	-	Preclinical phase	(Beg et al. 2015)
Nifedipine	Imwitor 742	Physical adsorption by triturate	S-SNEDDS	Aerosil 200	Formulation and development	(Weerapol et al. 2014)
Vitamin A acetate	Soyabean oil, Capmul MCM-C8, Cremophore EL	Mixing and compression into tablets	SNEDDS tablets	Avicel	Formulation and development	(Taha, Al-Suwayeh, and Anwer 2009)
Darunavir	Capmul MCM, Tween 80, Transcutol P,	Physical adsorption	S-SNEDDS	Neusilin US2	Preclinical phase	(Inugala et al. 2015)
Cilostazol	Peceol, Tween 20, Labrasol	Spray dried	S-SNEDDS	Calcium silicate	Preclinical phase	(Mustapha et al. 2017)(
Embelin	Capryol 90, Acrysol EL 135, PEG 400	Physical adsorption	S-SNEDDS	Aerosil, Neusilin US2	Formulation and development	(K. Parmar, Patel, and Sheth 2015)
Rosuvastatin calcium,	Garlic oil, olive oil, Tween-80, PEG 400	Physical mixing	Solid supersaturable SNEDDS	Maltodextrin and MCC 102	Priclinical phase	(Abo Enin and Abdel-Bar 2016)
Tacrolimus	Capryol PGMC, Transcutol HP, Labrasol	Absorption method	S-SNEDDS	Colloidal silica	Preclinical phase	(Seo et al. 2015)
Valsartan	Capmul MCM, Labrasol, Tween 20	Adsorption method	S-SNEDDS	Aerosil 200, Sylysia (350, 550, 730), Neusilin US2	Preclinical phase	(Beg et al. 2012)

(Continued)

TABLE 11.5 (CONTINUED)
Various SNEDDS Prepared till Date

Drug	Composition of L-SNEDDS	Techniques of Solidification	Formulation Prepared	Carrier Used	Stage of Development	References
Celecoxib	Capryol 90, Cremophor RH 40, Propylene glycol	-	SNEDDS	-	-	(Kaur, Chandel, and Harikumar 2013)
Rosuvastatin	Capryol 90, poloxamer 407, Transcutol P	Spray dried	S-SNEDDS	Mannitol	Formulation and development	(Kamel and Basha 2013)
Flurbiprofen	Labrafill M 1944, Labrasol, Transcutol HP	Spray dried	S-SNEDDS	Hydrophobic and hydrophilic carriers	Formulation and development	(Kang et al. 2012)
Olmesartan medoxomil	Oelic acid, Tween 80 and Transcutol HP	Surface adsorption method	S-SNEDDS	Aerosil 200, Aeroperl GT, Sylysia 550, Neusilin US2 and Fujicalin SG	Preclinical phase	(Beg et al. 2016)
Repaglinide	Olive oil, Miglyol Cremophore RH 40, Capryol 90 and Labrasol	Adsorption technique	S-SNEDDS	Neusilin US2	Formulation and development	(Reddy and Sowjanya 2015)
Erlotinib	Labrafil M2125CS, Labrasol, and Transcutol HP	Spray dried	S-SEDDS	Dextran or Aerosil	Preclinical phase	(Truong et al. 2016)
Docetaxel	Capryol 90, Cremophore EL and Transcutol HP	Absoption method	S-SNEDDS	Colloidal silica	Preclinical phase	(Quan et al. 2013)
Simvastatin	Capryol 90, Cremophore RH 40, Transcutol HP	Adsorption technique	S-SNEDDS	Crospovidone	Formulation and development	(Reddy and Sowjanya 2015)
Irbesartan	Capryol 90, Cremophor RH40 and Transcutol HP	Spray dried	S-SNEDDS	Aerosil 200	Research	(Nasr, Gardouh, and Ghorab 2016)
Glipizide	Captex 355, Solutol HS15 and Imwitor 988	Physical mixing	S-SNEDDS	Calcium carbonate	Formulation and development	(Dash et al. 2015)

REFERENCES

Abbaspour, Mohammadreza, Negar Jalayer, and Behzad Sharif Makhmalzadeh. 2014. "Development and evaluation of a solid self-nanoemulsifying drug delivery system for loratadin by extrusion-spheronization." *Advanced Pharmaceutical Bulletin* 4(2): 113.

Abdalla, Ahmed, Sandra Klein, and Karsten Mäder. 2008. "A new self-emulsifying drug delivery system (SEDDS) for poorly soluble drugs: Characterization, dissolution, in vitro digestion and incorporation into solid pellets." *European Journal of Pharmaceutical Sciences* 35(5): 457–464.

Abdelbary, Ghada, Maha Amin, and Salwa Salah. 2013. "Self nano-emulsifying simvastatin based tablets: Design and in vitro/in vivo evaluation." *Pharmaceutical Development and Technology* 18(6): 1294–1304.

Abo Enin, A Hadel, and Hend Mohamed Abdel-Bar. 2016. "Solid super saturated self-nano-emulsifying drug delivery system (sat-SNEDDS) as a promising alternative to conventional SNEDDS for improvement rosuvastatin calcium oral bioavailability." *Expert Opinion on Drug Delivery* 13(11): 1513–1521.

Albertini, Beatrice, Marcello Di Sabatino, Cecilia Melegari, and Nadia Passerini. 2015. "Formulation of spray congealed microparticles with self-emulsifying ability for enhanced glibenclamide dissolution performance." *Journal of Microencapsulation* 32(2): 181–192.

Alinaghi, Azadeh, Angel Tan, Shasha Rao, and Clive A Prestidge. 2015. "Impact of solidification on the performance of lipid-based colloidal carriers: Oil-based versus self-emulsifying systems." *Current Drug Delivery* 12(1): 16–25.

Attama, AA, and VE Mpamaugo. 2006. "Pharmacodynamics of piroxicam from self-emulsifying liDospheres formulated with homolipids extracted from Capra hircus." *Drug Delivery* 13(2): 133–137.

Attama, AA, IT Nzekwe, PO Nnamani, MU Adikwu, and CO Onugu. 2003. "The use of solid self-emulsifying systems in the delivery of diclofenac." *International Journal of Pharmaceutics* 262(1–2): 23–28.

Balakrishnan, Prabagar, Beom-Jin Lee, Dong Hoon Oh, Jong Oh Kim, Young-Im Lee, Dae-Duk Kim, Jun-Pil Jee, Yong-Bok Lee, Jong Soo Woo, and Chul Soon Yong. 2009. "Enhanced oral bioavailability of coenzyme Q10 by self-emulsifying drug delivery systems." *International Journal of Pharmaceutics* 374(1–2): 66–72.

Balakumar, Krishnamoorthy, Chellan Vijaya Raghavan, and Siyad Abdu. 2013. "Self nano-emulsifying drug delivery system (SNEDDS) of rosuvastatin calcium: Design, formulation, bioavailability and pharmacokinetic evaluation." *Colloids and Surfaces, Part B: Biointerfaces* 112: 337–343.

Basalious, Emad B, Nevine Shawky, and Shaimaa M Badr-Eldin. 2010. "SNEDDS containing bioenhancers for improvement of dissolution and oral absorption of lacidipine. I: Development and optimization." *International Journal of Pharmaceutics* 391(1–2): 203–211.

Basha, Syed Peer, Koteswara P Rao, and Chakravarthi Vedantham. 2013. "A brief introduction to methods of preparation, applications and characterization of nanoemulsion drug delivery systems." *Indian Journal of Research in Pharmacy and Biotechnology* 1(1): 25.

Beg, Sarwar, OP Katare, Sumant Saini, Babita Garg, Rajneet Kaur Khurana, and Bhupinder Singh. 2016. "Solid self-nanoemulsifying systems of olmesartan medoxomil: Formulation development, micromeritic characterization, in vitro and in vivo evaluation." *Powder Technology* 294: 93–104.

Beg, Sarwar, Premjeet Singh Sandhu, Rattandeep Singh Batra, Rajneet Kaur Khurana, and Bhupinder Singh. 2015. "QbD-based systematic development of novel optimized solid self-nanoemulsifying drug delivery systems (SNEDDS) of lovastatin with enhanced biopharmaceutical performance." *Drug Delivery* 22(6): 765–784.

Beg, Sarwar, Suryakanta Swain, Harendra Pratap Singh, Ch Niranjan Patra, and ME Bhanoji Rao. 2012. "Development, optimization, and characterization of solid self-nanoemulsifying drug delivery systems of valsartan using porous carriers." *AAPS PharmSciTech* 13(4): 1416–1427.

Charman, William N, and Christopher JH Porter. 1996. "Lipophilic prodrugs designed for intestinal lymphatic transport." *Advanced Drug Delivery Reviews* 19(2): 149–169.

Chavan, Rahul B, Sameer R Modi, and Arvind K Bansal. 2015. "Role of solid carriers in pharmaceutical performance of solid supersaturable SEDDS of celecoxib." *International Journal of Pharmaceutics* 495(1): 374–384.

Cho, Wonkyung, Min-Soo Kim, Jeong-Soo Kim, Junsung Park, Hee Jun Park, Kwang-Ho Cha, Jeong-Sook Park, and Sung-Joo Hwang. 2013. "Optimized formulation of solid self-microemulsifying sirolimus delivery systems." *International Journal of Nanomedicine* 8: 1673.

Christiansen, Martin Lau, Rene Holm, Jakob Kristensen, Mads Kreilgaard, Jette Jacobsen, Bertil Abrahamsson, and Anette Müllertz. 2014. "Cinnarizine food-effects in beagle dogs can be avoided by administration in a Self Nano Emulsifying Drug Delivery System (SNEDDS)." *European Journal of Pharmaceutical Sciences* 57: 164–172.

Cui, Shengmiao, Chunshun Zhao, Dawei Chen, and Zhonggui He. 2005. "Self-microemulsifying drug delivery systems (SMEDDS) for improving in vitro dissolution and oral absorption of Pueraria lobata isoflavone." *Drug Development and Industrial Pharmacy* 31(4–5): 349–356.

Czajkowska-Kośnik, Anna, Marta Szekalska, Aleksandra Amelian, Emilia Szymańska, and Katarzyna Winnicka. 2015. "Development and evaluation of liquid and solid self-emulsifying drug delivery systems for atorvastatin." *Molecules* 20(12): 21010–21022.

Dash, Rajendra Narayan, Habibuddin Mohammed, Touseef Humaira, and Devi Ramesh. 2015. "Design, optimization and evaluation of glipizide solid self-nanoemulsifying drug delivery for enhanced solubility and dissolution." *Saudi Pharmaceutical Journal* 23(5): 528–540.

Date, Abhijit A, and MS Nagarsenker. 2007. "Design and evaluation of self-nanoemulsifying drug delivery systems (SNEDDS) for cefpodoxime proxetil." *International Journal of Pharmaceutics* 329(1–2): 166–172.

Dixit, RP, and MS Nagarsenker. 2008. "Self-nanoemulsifying granules of ezetimibe: Design, optimization and evaluation." *European Journal of Pharmaceutical Sciences* 35(3): 183–192.

Elgart, Anna, Irina Cherniakov, Yanir Aldouby, Abraham J Domb, and Amnon Hoffman. 2013. "Improved oral bioavailability of BCS class 2 compounds by self nano-emulsifying drug delivery systems (SNEDDS): The underlying mechanisms for amiodarone and talinolol." *Pharmaceutical Research* 30(12): 3029–3044.

Gao, Ping, and Walter Morozowich. 2006. "Development of supersaturatable self-emulsifying drug delivery system formulations for improving the oral absorption of poorly soluble drugs." *Expert Opinion on Drug Delivery* 3(1): 97–110.

Gao, Ping, Bobby D Rush, William P Pfund, Tiehua Huang, Juliane M Bauer, Walter Morozowich, Ming-Shang Kuo, and Michael J Hageman. 2003. "Development of a supersaturable SEDDS (S-SEDDS) formulation of paclitaxel with improved oral bioavailability." *Journal of Pharmaceutical Sciences* 92(12): 2386–2398.

Garg, Varun, Reena Gupta, Bhupinder Kapoor, Sachin Kumar Singh, and Monica Gulati. 2016. "Application of self-emulsifying delivery systems for effective delivery of nutraceuticals." In: *Emulsions*, 479–518. Elsevier.

Inugala, Spandana, Basanth Babu Eedara, Sharath Sunkavalli, Rajeshri Dhurke, Prabhakar Kandadi, Raju Jukanti, and Suresh Bandari. 2015. "Solid self-nanoemulsifying drug delivery system (S-SNEDDS) of darunavir for improved dissolution and oral bioavailability: In vitro and in vivo evaluation." *European Journal of Pharmaceutical Sciences* 74: 1–10.

Jain, Amit K, Kaushik Thanki, and Sanyog Jain. 2014. "Solidified self-nanoemulsifying formulation for oral delivery of combinatorial therapeutic regimen: Part I. Formulation development, statistical optimization, and in vitro characterization." *Pharmaceutical Research* 31(4): 923–945.

Jain, Kunal, R Suresh Kumar, Sumeet Sood, and K Gowthamarajan. 2013. "Enhanced oral bioavailability of atorvastatin via oil-in-water nanoemulsion using aqueous titration method." *Journal of Pharmaceutical Sciences and Research* 5(1): 18.

Kamel, Rabab, and Mona Basha. 2013. "Preparation and in vitro evaluation of rutin nanostructured liquisolid delivery system." *Bulletin of Faculty of Pharmacy, Cairo University* 51(2): 261–272.

Kang, Jun Hyeok, Dong Hoon Oh, Yu-Kyoung Oh, Chul Soon Yong, and Han-Gon Choi. 2012. "Effects of solid carriers on the crystalline properties, dissolution and bioavailability of flurbiprofen in solid self-nanoemulsifying drug delivery system (solid SNEDDS)." *European Journal of Pharmaceutics and Biopharmaceutics* 80(2): 289–297.

Kaur, Gurjeet, Pankaj Chandel, and SL Harikumar. 2013. "Formulation development of self-nanoemulsifying drug delivery system (SNEDDS) of celecoxib for improvement of oral bioavailability." *Pharmacophore* 4(4): 120–133.

Koga, Kenjiro, Yoichi Kusawake, Yukako Ito, Nobuyuki Sugioka, Nobuhito Shibata, and Kanji Takada. 2006. "Enhancing mechanism of Labrasol on intestinal membrane permeability of the hydrophilic drug gentamicin sulfate." *European Journal of Pharmaceutics and Biopharmaceutics* 64(1): 82–91.

Kommuru, TR, B Gurley, MA Khan, and IK Reddy. 2001. "Self-emulsifying drug delivery systems (SEDDS) of coenzyme Q10: Formulation development and bioavailability assessment." *International Journal of Pharmaceutics* 212(2): 233–246.

Lv, Liang-Zhong, Chen-Qi Tong, Qing Lv, Xin-Jiang Tang, Li-Ming Li, Qing-Xia Fang, Yu Jia, Min Han, and Jian-Qing Gao. 2012. "Enhanced absorption of hydroxysafflor yellow A using a self-double-emulsifying drug delivery system: In vitro and in vivo studies." *International Journal of Nanomedicine* 7: 4099.

Ma, Hongda, Qingchun Zhao, Yongjun Wang, Tao Guo, Ye An, and Guobing Shi. 2012. "Design and evaluation of self-emulsifying drug delivery systems of rhizoma corydalis decumbentis extracts." *Drug Development and Industrial Pharmacy* 38(10): 1200–1206.

Mandawgade, Sagar D, Shobhona Sharma, Sulabha Pathak, and Vandana B Patravale. 2008. "Development of SMEDDS using natural lipophile: Application to β-artemether delivery." *International Journal of Pharmaceutics* 362(1–2): 179–183.

Mekjaruskul, Catheleeya, Yu-Tsai Yang, Marina GD Leed, Matthew P Sadgrove, Michael Jay, and Bungorn Sripanidkulchai. 2013. "Novel formulation strategies for enhancing oral delivery of methoxyflavones in Kaempferia parviflora by SMEDDS or complexation with 2-hydroxypropyl-β-cyclodextrin." *International Journal of Pharmaceutics* 445(1–2): 1–11.

Milović, Mladen, Spomenka Simović, Dušan Lošić, Andriy Dashevskiy, and Svetlana Ibrić. 2014. "Solid self-emulsifying phospholipid suspension (SSEPS) with diatom as a drug carrier." *European Journal of Pharmaceutical Sciences* 63: 226–232.

Mohsin, K, AA Shahba, and FK Alanazi. 2012. "Lipid based self emulsifying formulations for poorly water soluble drugs-an excellent opportunity." *Indian Journal of Pharmaceutical Education and Research* 46(2): 88–96.

Mustapha, Omer, Kyung Soo Kim, Shumaila Shafique, Dong Shik Kim, Sung Giu Jin, Youn Gee Seo, Yu Seok Youn, Kyung Taek Oh, Beom-Jin Lee, and Young Joon Park. 2017. "Development of novel cilostazol–loaded solid SNEDDS using a SPG membrane emulsification technique: Physicochemical characterization and in vivo evaluation." *Colloids and Surfaces, Part B: Biointerfaces* 150: 216–222.

Nasr, Ali, Ahmed Gardouh, and Mamdouh Ghorab. 2016. "Novel solid self-nanoemulsifying drug delivery system (S-SNEDDS) for oral delivery of olmesartan medoxomil: Design, formulation, pharmacokinetic and bioavailability evaluation." *Pharmaceutics* 8(3): 20.

Parmar, Komal, Jayvadan Patel, and Navin Sheth. 2015. "Self nano-emulsifying drug delivery system for embelin: Design, characterization and in-vitro studies." *Asian Journal of Pharmaceutical Sciences* 10(5): 396–404.

Parmar, Nitin, Neelam Singla, Saima Amin, and Kanchan Kohli. 2011. "Study of cosurfactant effect on nanoemulsifying area and development of lercanidipine loaded (SNEDDS) self nanoemulsifying drug delivery system." *Colloids and Surfaces, Part B: Biointerfaces* 86(2): 327–338.

Patel, Archita, Pragna Shelat, and Anita Lalwani. 2014. "Development and optimization of solid self-nanoemulsifying drug delivery system (S-SNEDDS) using Scheffe's design for improvement of oral bioavailability of nelfinavir mesylate." *Drug Delivery and Translational Research* 4(2): 171–186.

Patel, Swayamprakash, Girish Jani, and Mruduka Patel. 2012. "Solid self emulsifying drug delivery system: An emerging dosage form for poorly bioavailable drugs." *Inventi Rapid: NDDS* 4: 1–7.

Patil, Pradeep, and Anant Paradkar. 2006. "Porous polystyrene beads as carriers for self-emulsifying system containing loratadine." *AAPS PharmSciTech* 7(1): E199–E205.

Qi, Xiaole, Lishuang Wang, Jiabi Zhu, Zhenyi Hu, and Jie Zhang. 2011. "Self-double-emulsifying drug delivery system (SDEDDS): A new way for oral delivery of drugs with high solubility and low permeability." *International Journal of Pharmaceutics* 409(1–2): 245–251.

Quan, Qizhe, Dong-Wuk Kim, Nirmal Marasini, Dae Hwan Kim, Jin Ki Kim, Jong Oh Kim, Chul Soon Yong, and Han-Gon Choi. 2013. "Physicochemical characterization and in vivo evaluation of solid self-nanoemulsifying drug delivery system for oral administration of docetaxel." *Journal of Microencapsulation* 30(4): 307–314.

Raghavan, SL, A Trividic, AF Davis, and J Hadgraft. 2000. "Effect of cellulose polymers on supersaturation and in vitro membrane transport of hydrocortisone acetate." *International Journal of Pharmaceutics* 193(2): 231–237.

Rao, Sripriya Venkata Ramana, and Jun Shao. 2008. "Self-nanoemulsifying drug delivery systems (SNEDDS) for oral delivery of protein drugs: I. Formulation development." *International Journal of Pharmaceutics* 362(1–2): 2–9.

Rao, Sripriya Venkata Ramana, Kavya Yajurvedi, and Jun Shao. 2008. "Self-nanoemulsifying drug delivery system (SNEDDS) for oral delivery of protein drugs: III. In vivo oral absorption study." *International Journal of Pharmaceutics* 362(1–2): 16–19.

Reddy, M Sunitha, and N Sowjanya. 2015. "Formulation and in-vitro characterization of solid self nanoemulsifying drug delivery system (s-SNEDDS) of simvastatin." *Journal of Pharmaceutical Sciences and Research* 7(1): 40.

Savale, SK. 2015. "A review-self nanoemulsifying drug delivery system (snedds)." *International Journal of Research in Pharmaceutical and NANO Sciences* 4(6): 385–397.

Seo, Youn Gee, Dong-Wuk Kim, Kwan Hyung Cho, Abid Mehmood Yousaf, Dong Shik Kim, Jeong Hoon Kim, Jong Oh Kim, Chul Soon Yong, and Han-Gon Choi. 2015. "Preparation and pharmaceutical evaluation of new tacrolimus-loaded solid self-emulsifying drug delivery system." *Archives of Pharmacal Research* 38(2): 223–228.

Setthacheewakul, Saipin, Sirima Mahattanadul, Narubodee Phadoongsombut, Wiwat Pichayakorn, and Ruedeekorn Wiwattanapatapee. 2010. "Development and evaluation of self-microemulsifying liquid and pellet formulations of curcumin, and absorption studies in rats." *European Journal of Pharmaceutics and Biopharmaceutics* 76(3): 475–485.

Shahba, Ahmad Abdul-Wahhab, Kazi Mohsin, and Fars Kaed Alanazi. 2012. "Novel self-nanoemulsifying drug delivery systems (SNEDDS) for oral delivery of cinnarizine: Design, optimization, and in-vitro assessment." *AAPS PharmSciTech* 13(3): 967–977.

Shanmugam, Srinivasan, Rengarajan Baskaran, Prabagar Balakrishnan, Pritam Thapa, Chul Soon Yong, and Bong Kyu Yoo. 2011. "Solid self-nanoemulsifying drug delivery system (S-SNEDDS) containing phosphatidylcholine for enhanced bioavailability of highly lipophilic bioactive carotenoid lutein." *European Journal of Pharmaceutics and Biopharmaceutics* 79(2): 250–257.

Shanmugam, Srinivasan, Jae-Hyun Park, Kyeong Soo Kim, Zong Zhu Piao, Chul Soon Yong, Han-Gon Choi, and Jong Soo Woo. 2011. "Enhanced bioavailability and retinal accumulation of lutein from self-emulsifying phospholipid suspension (SEPS)." *International Journal of Pharmaceutics* 412(1–2): 99–105.

Singh, Bhupinder, Shantanu Bandopadhyay, Rishi Kapil, Ramandeep Singh, and Om Parkash Katare. 2009. "Self-emulsifying drug delivery systems (SEDDS): Formulation development, characterization, and applications." *Critical Reviews™ in Therapeutic Drug Carrier Systems* 26(5).

Singh, Bhupinder, Ramandeep Singh, Shantanu Bandyopadhyay, Rishi Kapil, and Babita Garg. 2013. "Optimized nanoemulsifying systems with enhanced bioavailability of carvedilol." *Colloids and Surfaces, Part B: Biointerfaces* 101: 465–474.

Sinha, MKB, and N Ganesh. 2015. "Preparation and characterization of nanoemulsion based on papaya seed oil." *Vivo Scientia* 4: 72–76.

Stremmel, Wolfgang. 1988. "Uptake of fatty acids by jejunal mucosal cells is mediated by a fatty acid binding membrane protein." *The Journal of Clinical Investigation* 82(6): 2001–2010.

Taha, Ehab I, Saleh A Al-Suwayeh, and Md Khalid Anwer. 2009. "Preparation, in vitro and in vivo evaluation of solid-state self-nanoemulsifying drug delivery system (SNEDDS) of vitamin A acetate." *Journal of Drug Targeting* 17(6): 468–473.

Tang, Bo, Gang Cheng, Jian-Chun Gu, and Cai-Hong Xu. 2008. "Development of solid self-emulsifying drug delivery systems: Preparation techniques and dosage forms." *Drug Discovery Today* 13(13–14): 606–612.

Tang, Jingling, Jin Sun, Fude Cui, Tianhong Zhang, Xiaohong Liu, and Zhonggui He. 2008. "Self-emulsifying drug delivery systems for improving oral absorption of ginkgo biloba extracts." *Drug Delivery* 15(8): 477–484.

Tarate, Bapurao, Rahul Chavan, and Arvind K Bansal. 2014. "Oral solid self-emulsifying formulations: A patent review." *Recent Patents on Drug Delivery and Formulation* 8(2): 126–143.

Tran, Phuong Ha-Lien, Tran Thao Truong-Dinh, Zong Zhu Piao, Toi Van Vo, Jun Bom Park, Jisung Lim, Kyung Teak Oh, Yun-Seok Rhee, and Beom-Jin Lee. 2013. "Physical properties and in vivo bioavailability in human volunteers of isradipine using controlled release matrix tablet containing self-emulsifying solid dispersion." *International Journal of Pharmaceutics* 450(1–2): 79–86.

Truong, Duy Hieu, Tuan Hiep Tran, Thiruganesh Ramasamy, Ju Yeon Choi, Hee Hyun Lee, Cheol Moon, Han-Gon Choi, Chul Soon Yong, and Jong Oh Kim. 2016. "Development of solid self-emulsifying formulation for improving the oral bioavailability of erlotinib." *Aaps Pharmscitech* 17(2): 466–473.

Vanitasagar, S, and N Subhashini. 2013. "Novel self-nanoemulsion drug delivery system of fenofibrate with improved bio-availability." *International Journal of Pharmacy and Biological Sciences* 4(2): 511–521.

Wang, Lijuan, Jinfeng Dong, Jing Chen, Julian Eastoe, and Xuefeng Li. 2009. "Design and optimization of a new self-nanoemulsifying drug delivery system." *Journal of Colloid and Interface Science* 330(2): 443–448.

Weerapol, Yotsanan, Sontaya Limmatvapirat, Jurairat Nunthanid, and Pornsak Sriamornsak. 2014. "Self-nanoemulsifying drug delivery system of nifedipine: Impact of hydrophilic–lipophilic balance and molecular structure of mixed surfactants." *AAPS PharmSciTech* 15(2): 456–464.

Yao, Jing, Yun Lu, and Jian Ping Zhou. 2008. "Preparation of nobiletin in self-microemulsifying systems and its intestinal permeability in rats." *Journal of Pharmacy and Pharmaceutical Sciences* 11(3): 22–29.

Zhang, BE, WB Lu, and WW Chen. 2008. "Study on self-microemulsifying drug delivery system of Jiaotai Pill active components." *Zhong Yao Cai = Zhongyaocai = Journal of Chinese Medicinal Materials* 31(7): 1068–1071.

Zhang, Ping, Ying Liu, Nianping Feng, and Jie Xu. 2008. "Preparation and evaluation of self-microemulsifying drug delivery system of oridonin." *International Journal of Pharmaceutics* 355(1–2): 269–276.

Zidan, Ahmed S, Bader M Aljaeid, Mahmoud Mokhtar, and Tamer M Shehata. 2015. "Taste-masked orodispersible tablets of cyclosporine self-nanoemulsion lyophilized with dry silica." *Pharmaceutical Development and Technology* 20(6): 652–661.

12 Phytonanotechnology for Sustainable Agriculture

Reshmi S Nair, Shriya Iyer, Athira J,
and Asha Anish Madhavan
Amity University, Dubai, United Arab Emirates

CONTENTS

12.1 INTRODUCTION

Agriculture is one of the major sources of revenue for many countries across the globe. The majority of the world population depends on farming for their source of revenue. In recent years, there has been a rise in environmental issues and weather conditions causing a dent in the agriculture sector. Due to these factors, developing countries are facing an increasing lack of food and crop productivity. At the same time in developed countries, there is an excess of food supplies because of technological advancement. Moreover, both developed and developing countries are striving to improve the agriculture industry. Developing countries seek to increase the crop yield, help crops fight against disease, and be less prone to the effects of climate change. Developed countries seek to address the increasing demand for a nutritious and safe food supply for the people. In addition, the agriculture sector faces many challenges regarding the reduction of quality harvest: an increased food demand with rising populations, a decrease in productivity, a rise in illness outbreaks among crops, a lack of areas suitable for farming, depreciating soil quality, and short shelf life. This raises questions about the survival of the agriculture sector in the face of high demands and drastic weather changes, while providing quality products.

The efficiency of agriculture has been improved by the use of chemicals via fertilizers and pesticides. The use of chemicals has decreased disease and pest attacks, and remedied the problem of unfavorable soil with the required minerals. On the other hand, fertilizers pose a threat to the environment. The majority of fertilizer does not remain intact due to rain or humidity, causing soil and water pollution. Chemicals from both fertilizers and pesticides contaminate water, as well as leaving residue on crops, which in turn leads to adverse effects on human and aquatic life. Precise delivery of required nutrients and controlling the rate of supply of chemicals could help decrease the level of contamination. Hence there is an urge for new methods to supply the required nutrient and chemical demands for growing a healthy and increasing yield of crops. Thus, to supply the rising demand for food products and simultaneously alleviate numerous problems in the agriculture industry, scientists are using nanotechnology.

Nanotechnology is an interdisciplinary field that includes applications in the chemical, mechanical, electronics, and medical industries. It is an engineering department that deals with the manipulation of atoms and molecules pertaining to materials, systems, or functions that come under the range of 100 nm. It is considered an emerging field of science that is set to play a major role in the field of agriculture and food science in upcoming years. Nanotechnology provides "smart" alternatives by using nanomaterials to revolutionize the agriculture industry in order to decrease – or even eradicate – the effects of conventional practices on the environment, hence increasing the efficiency of products. They can be used to help alter the release of chemicals into the plant system, thus resulting in a decreased need for a frequent supply. Nanotechnology can be employed to develop devices such as nanosensors, nanocapsules, and nanoparticle delivery systems. Nanotechnology advances the detection of pest attack or lack of nutrients, has a quick response to infected sites, adjusts the absorption rate of chemicals by crops, assists in the removal of contaminants from the soil and targets the delivery of nutrients to the specific site (Prasad,

Bhattacharyya, and Nguyen 2017). Nanodevices improve both the speed and the power of recognition of variations in the crop.

The Green Revolution increased the use of chemical fertilizer but the crop does not absorb the vast majority of fertilizer chemicals, since they are washed off into the environment. The washed off chemicals are drained into the soil or water, causing chemical pollution. These chemicals have a huge effect on living beings. However, nanotechnology reduces the amount of fertilizers to be used, hence reducing the price. Nanotechnology advances agriculture by providing: a targeted delivery, more control over the rate of supply, an increase in the absorbance, and lengthening the life of crops. These factors make the industry significantly more effective and decrease the environmental pollution caused by the chemicals.

Nanotechnology produces pesticides and fertilizers by developing nanoparticles, nanocapsules, nanocrystals, and nanosensors. Nanoparticles provide a safer and ulti-mately better way to allow nutrients to be absorbed into the crops by not harming the environment, thus decreasing pollution. Nanocapsules are small systems that are able to manipulate their outer shells in order to regulate the delivery and concentra-tion of the supply of chemicals and consequently reduce runoffs (Nuruzzaman et al. 2016). Nanocrystals aid in developing fertilizers and pesticides that have higher effectiveness with much lower doses. Nanosensors are engineered to monitor the variations in soil humidity, chemical imbalances, and soil quality in order to ensure the quality and well-being of the crops. The field of nanotechnology enables us to miniaturize biosensors. Sensors of a sufficiently small size help to detect the cause of an illness before it is visible to the naked eye. It aids in quickly recognizing myco-toxins found in crops (Kranz et al. 2007).

Nanofertilizers and nanopesticides are specific methods to develop chemicals required by the crop to increase yield and quality while keeping the environment in mind. They help to decrease the contamination of soil and water and lower the chance of disease in humans and aquatic animals due to these chemicals (Solanki et al. 2015). Nano pesticides provide security against attack from bacterial or microbial diseases, as well as insect attack, by delivering specific chemicals to eliminate them. Their small size enhances the ability of the plant to efficiently absorb the nutrients.

Nanotechnology is a method of decreasing the amount of chemical wasted, reduc-ing the loss of nutrients, and improving the yield by controlling such pest and disease attacks. The use of nanotechnology assists in lowering the cost of buying fertilizer and pesticide chemicals. Nanomaterial advances current agriculture technology by adding smart features, such as target delivery and controlled discharge. This makes the agriculture technology more proficient as well as reducing the adverse effects on the ecosystem. Such nanomaterials can increase the quantity of quality products obtained from harvest while being sustainable.

12.2 SUSTAINABLE DEVELOPMENT

Sustainability is an important emerging need of the hour, a concept that emphasizes the application of new efficient methods to meet the growing demands of the pres-ent generation without compromising the needs of the future generation. In the world of agriculture, the major challenges faced involve unpredictable climate change, rapid industrialization, and the accumulation or runoff of fertilizers and pesticides which cause major environmental concerns such as soil and water pollution, as well as harming human health. By 2050, the population is estimated to increase to up to 6–9 billion, which intensifies existing agricultural problems. As a result, modern-day technological advancements could be employed to tackle this issue (Chen and Yada 2011). Nanotechnology, a recently developed field offers various benefits to the field of agriculture by aiding in soil-waste management and precision farming practices which are aimed at not only reducing the cost of production but also maximizing output. Discussed below are the beneficial applications of nanotechnology in the world of agriculture.

12.2.1 Role of Nanotechnology in Agriculture

The ability to control matter at nanoscale gives rise to the emergence of nanomaterials with new and innovative properties capable of addressing numerous environmental and societal issues. Nanotechnology can play a major role in agriculture by influencing the following factors (Shang et al. 2019):

   i) *Crop growth*: Nanotechnology, with the development of nanofertilizers, can accelerate plant growth, improve crop yield, and speed up the germination process.
  ii) *Crop improvement*: Nanotechnology, with the involvement of biotechnology, can help in the making of transgenic plants by using the genetic material from the target species (with the desired trait) carried to the host plant by means of nano-scale carriers and enter the nucleus of the host plant. Inside the nucleus, the desired trait is multiplied and the transgenic plant is formed.
 iii) *Crop protection*: The latest research in the field of nanotechnology involves the development of nanopesticides to protect crops from pathogens and pests. Researchers have been able to develop "micro-fabricated vessels" to enable ease of studying of various plant pathogens and device effective counter-attacks against them.
  iv) *Soil enhancement*: Places that suffer from soil deficiency due to the lack of specific ions not being present in the soil hamper the growth of the plant. Nanomaterials can be used to encapsulate the deficient ion and deliver it to the plant near the roots or in the surrounding soil by means of "nanoscale carriers" and "clay nanotubes".
   v) *Environment safety*: Due to the application of pesticides, chemical compounds, and contaminants, they accumulate in the soil, which ultimately harms the crop in the long run, hence the importance of reducing chemical runoff and soil degradation. The application of "nano-oligodynamic metallic particles" and "photo-catalytical processes" can help: they work as absorbents, and break down the toxic components respectively.

The development of nanoscale carriers, microfabricated xylem vessels, photo-catalytical process, clay nanotubes, and nano-oligodynamic metallic particles are further elaborated below.

• *Nano-scale carriers*: They are encapsulations or entrapments of target (herbicides, pesticides, fertilizers, or plant hormones) that can be anchored to the plant roots, surrounding soil structures or organic matter which carry, deliver, and release them into the plant body in a controlled manner. There are different types of nano-vehicles/nano-scale carriers that are developed to perform this function, namely: micelle, dendrimers, nanocapsules, nanospheres, liposomes, graphene, and nanotubes. The diagrammatic representation of each type is shown in Figure 12.1 (Da Silva-Candal et al. 2017). Different types of nano-scale carriers are shown in Figure 12.1.

A schematic representation of application is shown in Figure 12.2. As one of the major challenges mentioned earlier involves the runoff of chemicals causing environmental degradation, nano-scale carriers can help in reducing the chemical runoff by ensuring no overuse of pesticides and effective uptake of fertilizers by the plant (Ditta 2012).

Micro-fabricated xylem vessels: Agricultural crops are always under threat of being attacked by pests or by plant disease, thus it is important to know the causal organisms of diseases and study their interaction with the plant. The use of traditional means is a cumbersome process. But using micro-fabrication to develop artificial xylem vessels with nano-sized features means that it is now possible to

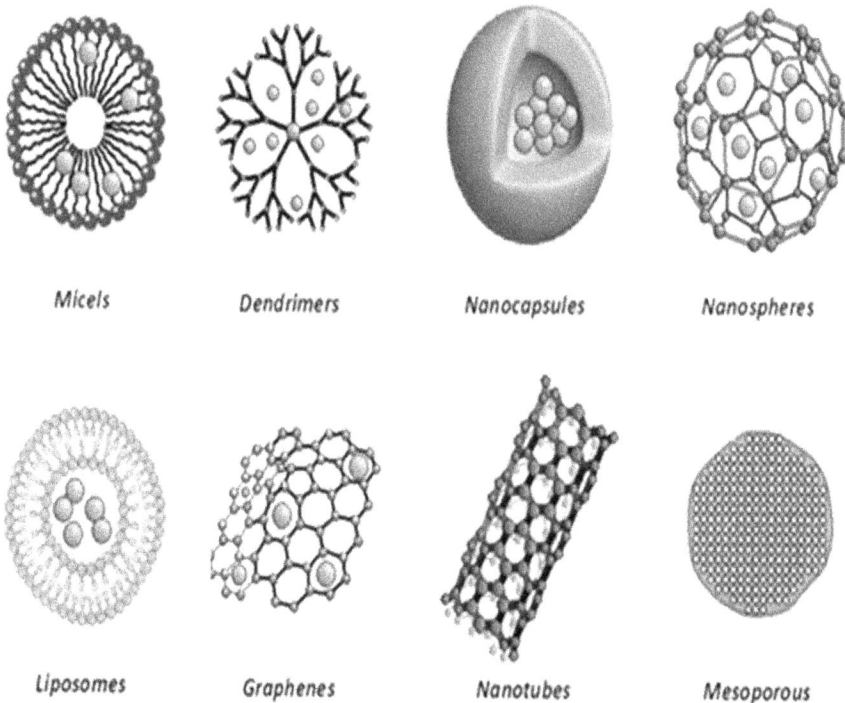

| | | | |
| Micels | Dendrimers | Nanocapsules | Nanospheres |

| Liposomes | Graphenes | Nanotubes | Mesoporous |

FIGURE 12.1   Different types of nano-scale carriers to deliver fertilizers, pink spherical structure representing the target molecule (Da Silva-Candal et al. 2017).

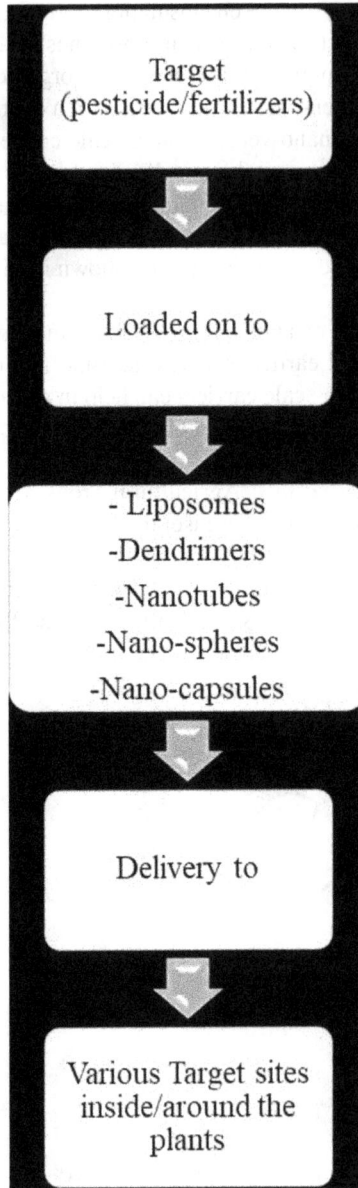

FIGURE 12.2 A schematic representation of the application of nano-scale carriers.

understand the mechanism involved and so help in developing treatment strategies to best counter the problem (Cursino et al. 2009). Figure 12.3 shows the application of micro-fabricated xylem vessels.

The artificial xylem vessels are constructed using polydimethylsiloxane (PDMS) which forms microfluid channels (employing lithographical techniques). Different types of xylem- inhabiting species are introduced to this micro-fabricated system.

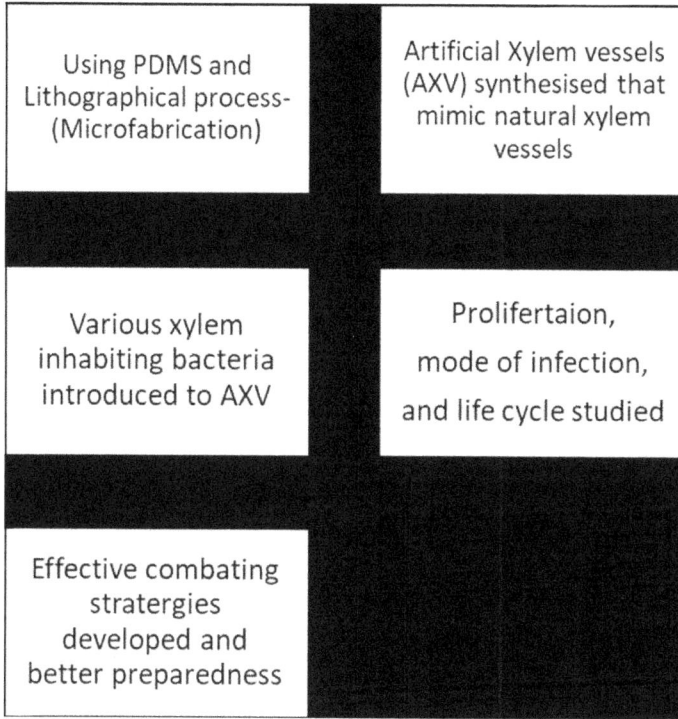

FIGURE 12.3 The schematic diagram above shows the application of micro-fabricated xylem vessels.

As this mimics the naturally occurring xylem vessel, we can now keep track of different bacterial species/strain proliferation and life cycle, and understand the mode of infection that occurs in natural plants. Therefore, artificially developed micro-fabricated vessels enable the study of plant pathogens in greater detail, helping agriculturists combat the pathogen with more preparation and efficiency. By studying how the bacterial species affect the artificial xylem, it aids in developing early warning systems and gives an idea about the symptoms to watch out for (Zaini et al. 2009).

- *Photo-catalytical process*: Pressing concerns for environment safety are due in part to the accumulation of pesticides and fertilizers that affect not only the water but also the soil that the plant grows in. In order to tackle this, the chemical compounds present in the pesticides must be degraded/decomposed. To achieve this, chemical compounds are exposed to ultraviolet (UV) light in the presence of a catalyst (a nanoparticle) (Chen and Yada 2011). Due to the exposure of UV, the outermost electron is energized and moves to a higher energy level, resulting in electron hole pairs. These serve as oxidizing agents and help in the degradation process, shown in the diagram below (Colmenares and Luque 2014).

    The photo-catalytic process is shown in Figure 12.4.

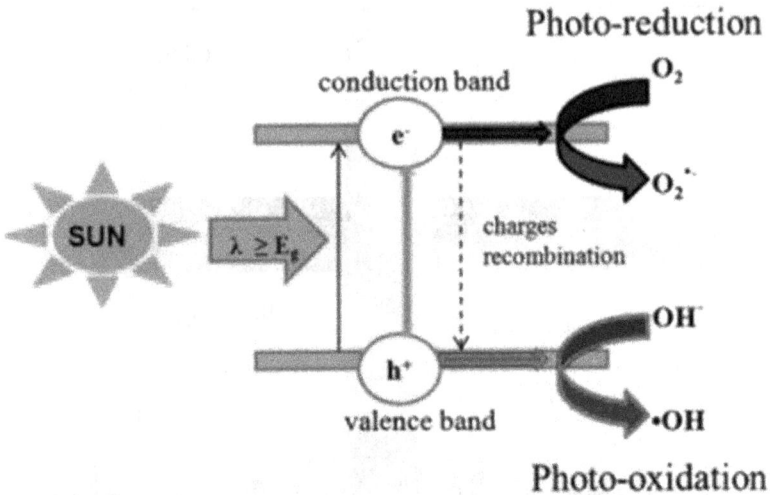

FIGURE 12.4 Photo-catalytic process resulting in degradation (reduction and oxidation products) (Colmenares and Luque 2014).

- *Clay nanotubes*: Halloysite (clay nanotube) has been manufactured to control the release of pesticide, as well as delivering nutrient-specific ions to nutrient-deficient soils, resulting in optimum usage of pesticides as well as enhancement of soil respectively. Clay nanotube can be used to load and deliver hydrophilic molecules/ions. At times when the target is highly soluble, it is important to control the release rate, which is usually done by mixing a viscous polymeric compound which is compatible with the target (Lvov et al. 2008).

  As precision farming aims to reduce production costs, using better carriers for pesticides in the form of nanotubes, as well as soil enhancement strategies, are effective solutions ("Nanotechnology – The Secret of the Fifth Industrial Revolution and the Future of the Next Generation" 2015). Figure 12.5 displays the mode of operation of halloysite in agriculture.

- *Nano-oligodynamic metallic and magnetic particles*: Water quality is always a concern for agriculture as irrigation plays a vital role in the growth of crops. It is therefore of paramount importance that water is free from radioactive contaminants, endocrine disrupting compounds, and various other coliform bacteria. To improve the water quality, nano-based oligodynamic metallic particles were developed. The anti-bacterial property of the silver particle has already been proven and is found to produce ROS-reactive oxygen species that can break DNA strands and remove chemicals, as well as biological contaminants (Street 2014). Furthermore, radioactive contaminants can be removed from water by the employment of magnetic nanoparticles (MNP). In order to attract radioactive contaminants, chelator extractants are covalently bonded on the MNP surface which results in the formation of a MNP-chelator complex. This complex provides excellent

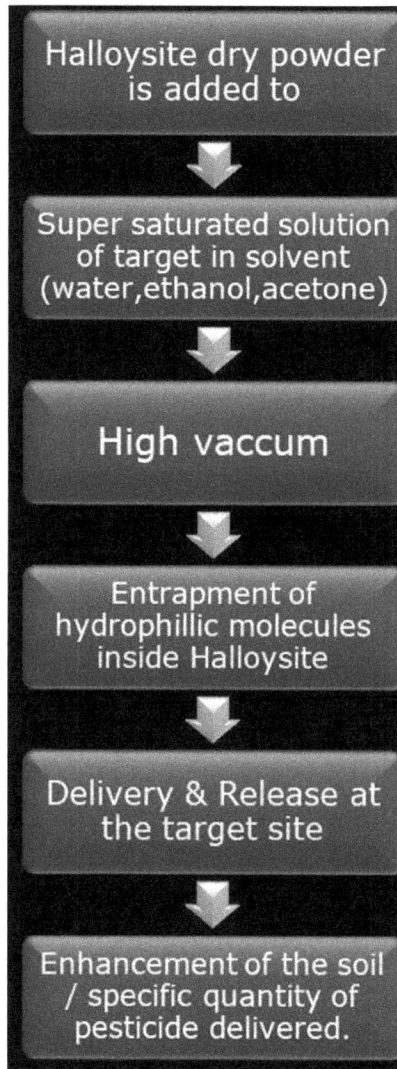

FIGURE 12.5   Flowchart displays the mode of operation of halloysite in agriculture.

benefits when it comes to the removal of contaminants attributed to the high surface-volume ratio, resulting in better kinetics for adsorption (Dubey et al. 2017). Figure 12.6 shows the application of MNP.

12.2.2   SUSTAINABLE INTENSIFICATION

A sustainable approach in the utilization of natural resources and its products has been a priority in recent years, thereby reducing the environmental impact. Sustainable intensification can be understood to be a way of augmenting the social outcomes of

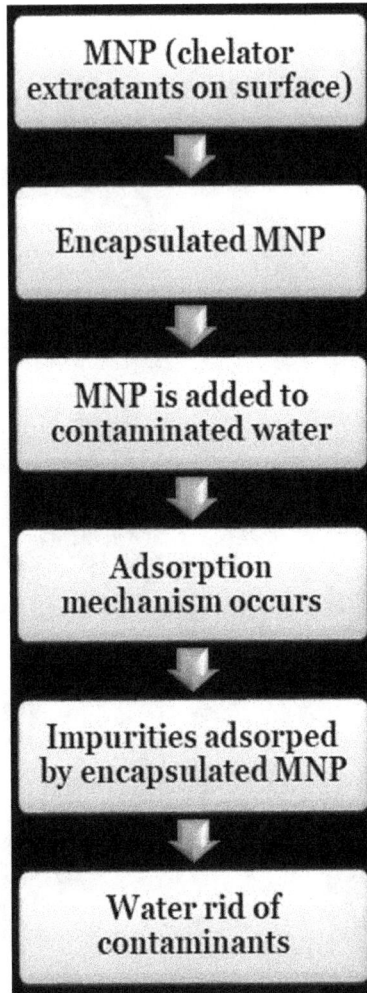

FIGURE. 12.6 The flowchart summarizes the effective application of MNP in contaminated water.

recent agricultural revolutions. Sustainable Intensification (SI) is defined as a method to enhance the agricultural yield focusing on environmental conservation without the conversion of additional non-agricultural land (Pretty 2018). SI includes various agricultural systems progressing towards the substantial enhancement of environmental outcomes with the aid of nanotechnology, precision farming, robotics, and various delivery systems which aim to ensure the targeted input of agrochemicals (Pretty 2018, Garbach et al. 2016, Smith 2013). Moreover, concepts of land-sharing and land-sparing have been improved to deliver the intensification of agricultural resources, which reduces non-agricultural land expansion and highly biodiverse habitats (Phalan et al. 2011). During recent years, a targeted methodology towards agriculture, along with sustainability, has categorized this as 'green thinking'. According to research statistics, SI has been predominant in 100 nations with a total of 163 million farms under cultivation by the year 2018. This approximates to 29% of farms worldwide, covering 453 million hectares of land (Pretty 2018, Pretty et al. 2018). Considering human population growth, food security, and over-exploitation of agricultural land, sustainable intensive farming methods are inevitable for an increased crop yield, thereby maintaining the ecosystem. Also, the pest management process has reduced the need for insecticides by more than 70% with the promotion of ecological services of biological control, using nectar plants and the like (Gurr et al. 2016).

12.2.3 Soil and Waste Management

12.2.3.1 Soil Management

Soil harbors many living organisms and is the reservoir of nutrients which is of great significance, especially for the growth of plants in agriculture. Therefore managing this resource is vital; traditional ways of doing this can be enhanced by the application of nanotechnology (Duhan et al. 2017). Ideal soil management practices firstly include monitoring the nutrient level in soil which can be effectively performed using nanosensors. Nanosensors are devices that can precisely detect and quantify the components of soil in real-time. Studying the components and chemical composition of soil helps us understand the additional nutrients the soil requires accurately, thereby reducing excess supply and waste of chemical fertilizers.

Secondly, ensuring no toxic chemical compounds enter the food chain is made possible with the use of nanofertilizers and nanopesticides. They work mainly on the mechanism of entrapment or encapsulation in which the needed dosage of the compound is encapsulated (loaded), delivered to the target site, and released in a controlled manner (Ditta 2012). This helps in making sure the chemical compounds are not used excessively, thus protecting the integrity of the soil.

Thirdly, it addresses the deficiency of mineral ions in the soil. Soil quality, as mentioned in the previous section, can be improved by means of nanoparticles. For instance, zinc ions being made bioavailable in areas that contain zinc-deficient soil can ameliorate the situation (White et al. 2012).

12.2.3.2 Waste Management

Agriculture produces a large quantity of waste including weed, fertilizers, damaged crops, animal waste, and more. Waste produce, and its negligent disposal, is a major concern to the industry. It causes unpleasant odors and chemical releases that have an adverse impact on the environment. This waste could be utilized as organic fertilizers for improving the quality of the land. It could also be a source of electrical energy in the form of natural gas. Nanotechnology-developed catalysts contribute to the bioconversion of waste into a renewable source of electricity production. Producing energy from bio waste using nanotechnology is helpful to eradicate high power systems, as well as helping the development of rural towns (Ditta 2012).

Nanomaterials possess high reactivity and are developed to be selective in nature, which makes it possible for them to act successfully on persistent waste contaminants with more success than traditional chemical methods. The effectiveness of the nano-material majorly depends on the selection of nanoparticles for the specific waste product and its conditions. Metal and carbon-based nanomaterials are the most common and most successful products currently (Solanki et al. 2015).

Nanotechnology also plays a role in the reduction of produced waste from agriculture. Precision farming is a method of using technology to increase output while decreasing input. The utilization of nanosensors to detect the precise nutrients required by the crop, and nanomaterials for the targeted delivery of fertilizer pesticides helps in decreasing the amount of fertilizers used, while increasing the number of healthy crops during harvest. Thus, monitoring the conditions on crop development using nanotechnology reduces agricultural waste and decreases pollution.

12.3 NANOFERTILIZERS

Ultimately in agriculture, higher crop productivity is the desired outcome, one-third of crop productivity, however, depends on the influence of fertilizers. Due to this, nutrient use efficiency (NUE) of crops is always a major yardstick to determine how well the plant absorbs nutrients from the fertilizers. Low NUE has become a prevalent concern in developing countries, because of low organic matter in soil and imbalanced fertilization processes, leading to a condition termed "multi-nutrient deficiency". Nanofertilizers have been developed to enhance the NUE of crops. Studies conducted by Subramanian et al. (2015) have shown that the monitored release of fertilizers by nanoparticles is possible. Motakef Kazemi and Salimi (2019) recently worked on employing a biomaterial to carry, deliver and release fertilizers rich in NPK (nitrogen, phosphorus, and potassium) and proved that nanoparticles are excellent contenders for increasing crop productivity.

12.3.1 APPLICATION OF NANOFERTILIZERS

12.3.1.1 Chitosan Nanoparticle for NPK Loading and Release

(Motakef Kazemi and Salimi 2019) worked on developing nanoparticles of chitosan (natural polysaccharide used in the food, medical and agricultural industries) that can load and release NPK fertilizers. For loading, urea, calcium phosphate, and potassium chloride were used as sources for nitrogen, phosphorus, and potassium respectively. These precursors were added to the solution of chitosan nanoparticles, and by means of adsorption mechanisms, the loading takes place. For the release of nutrients, the chitosan nanoparticle was first suspended in water after which nitrogen titration, UV-Vis spectroscopy, and atomic absorption spectroscopy waswere used to calculate the amount of release of nitrogen, phosphorus, and potassium respectively.

On analyzing the release rates, it was found that chitosan nanoparticle loaded with P was the fastest released, followed by K and N, attributed to the presence of more anionic species in the precursor of P and K than N (Motakef Kazemi and Salimi 2019).

12.3.1.2 Nano-Clays and Zeolites for Cationic Nutrient Ions

These materials naturally possess a honeycomb-like structure which helps in the increase of fertilizer efficiency. Cationic nutrients ions (NH_4^+, K^+, Ca^{2+}, Mg^{2+}) are captured, stored, and can be released with control. This is exemplified by studies carried out by (Leggo 2000). In the case of using a zeolite charged with ammonium, the zeolite helps break down or solubilize phosphate into forms that allow for plant uptake, ultimately improving the yield of the crop.

12.3.1.3 Surface-Modified Zeolites Using HDTA for Anionic Nutrient Ions

Normal zeolites cannot trap anion-based nutrient ions, which is why there is a need to perform surface modification on the zeolite to enable it to trap anion-based ions. This is done by adding a surfactant named hexadecyltrimethylammonium (HDTA). This produces a fertilizer carrier that is capable of slowly releasing anionic-nutrient ions (NO_3^-, PO_4^-, and SO_4^-). This approach was widely worked on by Li et al. 2003 (Li 2003).

12.3.1.4 Zinc Oxide Nanoparticles

Using carbon nanotubes or zeolites, it is now possible to deliver zinc oxide nanoparticles to specific locations near the plant/crop, thus enriching the zinc-deficient soil with active zinc ions that can result in plant uptake, hence increasing its NUE. When this mechanism was put into practice for barley plants, it was found to increase productivity by 91% compared to traditional fertilizers that contain zinc in the form of zinc sulfate, which on application only increased productivity by 31% (Sabir, Arshad, and Chaudhari 2014).

12.3.2 ADVANTAGES OF NANOFERTILIZERS

1) Nanoparticles prevent the precipitation of freely available nutrient ions as the nutrient ions are adsorbed on to the surface of the nanoparticle.
2) Reduction in the overuse of fertilizers, as fertilizer particles coated with nanomembranes, allows controlled release of the nutrients into the soil/plant root hairs.
3) Carbon nanotube sponges and carbon nanotube powders are known for their soaking properties which remove water contaminants such as pesticides, fertilizers, and wastes from pharmaceuticals protecting the environment and the plants.

12.3.3 LIMITATIONS OF NANOFERTILIZERS

1) Synthesis of nanofertilizers can be achieved either by physical or chemical methods. However, when physical methods like ball milling are employed, particles are often agglomerated, hence stabilizers are added in the form of polymer surfactants which may potentially harm the environment. Moreover, if zeolites are used as carriers, then the nutrients that can be loaded are limited to cationic forms ($K+$ and $NH4+$). If an anionic species needs to be loaded, then the zeolite has to be modified.

2) Fernández and Eichert (2009) have mentioned in their work that if nanoparticles are accumulated on the photosynthetic surfaces (which can occur when the nanofertilizer is designed to enter through the pores of leaves), it may lead to stomatal obstruction and hamper the gaseous exchange vital for the survival of the plant.

12.4 NANOPESTICIDES

Insects, pests, and pathogens are very evident threats to crops and plants. Traditional pesticides are chemical compounds that block the life cycle of an insect/pests or pathogens, thus killing them and protecting the plant from consequent attacks. However, a major challenge to be faced is that pests/insects are becoming immune to these pesticides and hence plants still get attacked. This leads to agriculturists using excess pesticides, which ultimately accumulate in the soil, harming both the plant and the environment. With the employment of nanopesticides, there is a means of control over the specificity and the quantity released, which enables more target-specific attacks and reduces excess usage. Prasad, Bhattacharyya, and Nguyen (2017) have developed a nano-encapsulated pesticide which has been found to be more beneficial than the traditional version, in terms of stability, permeability, and specificity. Application, limitation, and advantages of nanopesticides will be discussed in the sections below.

12.4.1 APPLICATION OF NANOPESTICIDES

12.4.1.1 Silver-Based Nanopesticides

The green synthesis of silver nanoparticles was accomplished with the use of bacterium Serratia sp. BHU-24, which served as a reducing agent due to the enzyme it possesses; silver was chosen due to its well-known antibacterial and fungal properties. This silver nanoparticle was specifically developed targeting the causative organism of spot blotch disease among wheat plants. The silver nanoparticles were successfully encapsulated after being identified as target-specific, and thence released into the target (the fungus *Bipolaris sorokiniana*). The process was successful and the target organism was killed, and validated by the studies of Lamsal et al. (2011).

12.4.1.2 Copper-Based Nanopesticides

(Li 2003); (Prasad, Bhattacharyya, and Nguyen 2017) both these research works explored the application of copper nanoparticles as pest-controlling agents. Copper nanoparticles synthesized were tested on grape trees and many fruit trees. On dissolving the copper nanoparticles in water and exposing it to the plants/trees, it was found that all fungal species affecting the growth of the plants were destroyed, which suggested that the synthesized copper-based nanopesticides showed anti-fungal/fungicidal properties.

12.4.2 ADVANTAGES OF NANOPESTICIDES

1) Supervision of insect pests via the manufacture of nanomaterial-based insecticides, as opposed to traditional pesticides that offer no opportunity to supervise the pest, making it harder to combat.

2) Reduced toxicity of compounds with the use of nanopesticides (controlled release into the environment).
3) Usage of bio-based material to make pesticides eco-friendly and target-specific, contrary to chemically synthesized traditional pesticides.
4) Self-regulating capability of the nanopesticide to perform biocidal activities only when inside the target organism (Nuruzzaman et al. 2016).

12.4.3 Limitations of Nanopesticides

1) Toxicological studies regarding these new forms of pesticides are still not completely understood, hence they are not necessarily one hundred percent toxicant-free.
2) The effect of nanomaterials used for pesticide application on human consumption is of concern.

12.5 NANOMATERIAL-BASED DEVICES

Among more recent research in agriculture is that of work on the detection mechanism, that precisely specifies the stages of crop growth, as well as attack from disease. By comparing the variations in the rate of release of protein during a healthy state and infected state, it is possible to detect the stages of the disease cycle. Nanomaterial devices such as nanocapsules, nanosensors are used for the easy detection of plant health degradation well before the illness is visible to the farmer. It helps in increasing fertilizer efficiency, enhancing the rate of absorption of nutrients, and delivering required nutrients directly at the specific site.

Nanodevices provide a more efficient and faster method of accurate readings using minimal amounts of sample and power. Nanomaterials are used to make agriculture devices smarter; they are smart in that they recognize the presence of pathogens and can respond accurately at the right time or else signal farmers take action. Hence nanodevices can be considered as a preventive mechanism as well as an early warning method that can be utilized for delivering chemicals in a controlled manner.

12.5.1 Precision Farming (SSCM)

Precision agriculture is a farming management concept of measuring and responding to inter- and intra-field variation in crops and its environment to form a decision support system for whole farm management in order to acquire the maximum output from the available resources. Nowadays, nanotechnology is extensively used in modern agriculture to make true the concept of precision agriculture. The Green Revolution resulted in an increase in the use of fertilizers and pesticides that led to the loss of the ecosystem, and resistance against pesticides and severe pollution. Precision farming or site-specific crop management (SSCM) is a precise farming method that involves studying and measuring the variation in environmental conditions of the field or the crop, in order to maximize the output using minimum resources with the help of technology. It aims to increase the crop yields substantially while using the least amount of fertilizers by monitoring the growth of the crop and providing targeted action. Precision farming utilizes information technology, GPS, and remote sensors to measure the variations to determine the exact source of concern. Precision farming also aids in decreasing the agricultural waste, hence controlling the environmental pollution significantly.

Nanomaterial-based devices for the targeted delivery of nutrients to crops and sensors for precision farming are made available by using nanotechnology. Precision farming can be enhanced with the use of nanotechnology-based devices. Modern nanodevices and methods have shown the potential to be a solution to several issues with typical agricultural practices, and can thus revolutionize the field. A large variety of elements can be used to prepare nanomaterials, such as metal oxides, magnetotactic elements, quantum dots, polymeric units, lipid, emulsion, and more (Zheng et al. 2005). The result and efficiency of nanomaterials in the stages of growth and functions may change depending on the crop. Nano-encapsulated fertilizers aid in the gradual delivery of chemicals and nutrients, helping in accurate dosage. Nano-encapsulation protects the required nutrient entrapped within from early and fast degradation, as well as increasing the capacity of chemicals to regulate the attack of pests for extended periods of time. Nano-encapsulated pesticide hence decreases both the quantity of chemicals required and their release into the environment, making it an eco-friendly alternative.

Carrier systems enhanced by nanotechnology enable control or the delayed release of chemicals, increasing pest-control-efficiency duration and targeted delivery straight to the disease site. This enables nano-encapsulated pesticides to reduce pollution of land and water by leaching, and hence decrease the adverse effect on non-target plants and animal life (Nuruzzaman et al. 2016). Nanoparticles aid in decreasing leaching and the release of harmful chemicals.

Atrazine is a chemical used to control the growth of unwanted plants and weeds in sugarcane farms. Atrazine binds to a specific site during the photosynthesis process, and in doing so, hinders photosynthesis from proceeding. Such obstruction of the photosynthetic route can cause an increase in the quantity of reactive oxygen species (ROS) to be released. This increase in the ROS release causes a development in oxidative stress by exerting pressure on the antioxidant and photoprotective mechanisms. This leads to the mutation to cells of protein. Atrazine also causes leaf chlorosis, and hinders cell growth that results in the demise of the leaf or plant. Excessive use of atrazine can cause land or water pollution as a result of its degradation under environmental factors. Atrazine has adverse effects on the metabolism and growth of flora and fauna as well as marine life. Human beings are also affected by the use of atrazine. Exposure to atrazine can lead to damage in the reproductive system and even cause tumor growth.

The employment of nanoparticles as a carrier system for atrazine has had promising results. It has subsequently decreased the pollution of the environment. Polymeric nanocapsules were developed with poly(epsilon-caprolactone) (PCL). It is a decomposable aliphatic polyester that is non-hazardous to flora and fauna. Studies show increased stability, less mobility of atrazine on land, and higher efficiency of encapsulation of atrazine in the PCL nanocapsules (Oliveira et al. 2015). The surface functionalization of nanoparticles is shown in Figure 12.7.

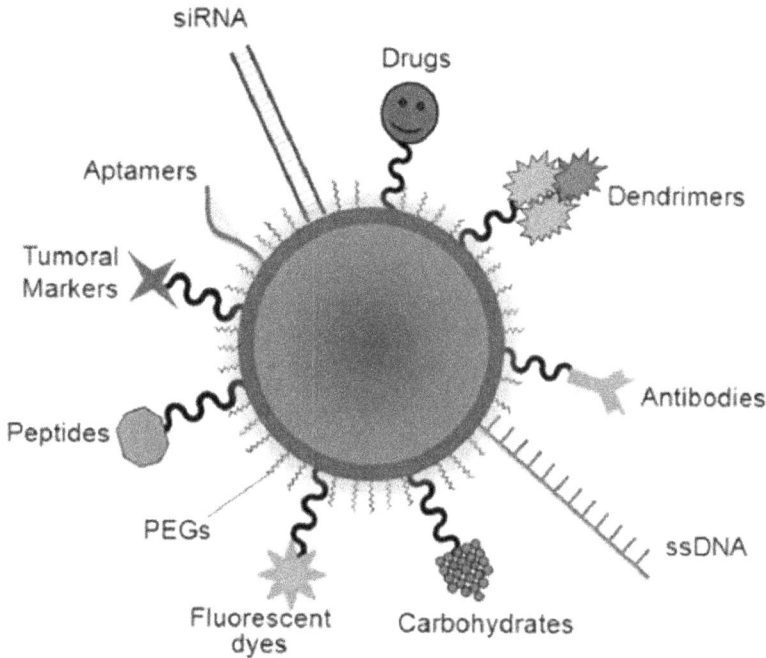

FIGURE 12.7 Surface functionalization of nanoparticles. Nanoparticle with surface functionalization allowing the attachment of a variety of targeting molecules including drugs, antibodies, and nucleic acids (Conde et al. 2014).

Nanosensors that are made up of carbon nanotubes (CNTs) or nanocantilevers are in nanoscale, i.e. of minuscule size, enabling them to capture and study single protein or chemical structures. Nanoparticle surfaces can be developed in a specific way to activate a signal in the detection of variations in the chemical or electronic level of the crop due to the activity of pests (Conde et al. 2014). They also use the properties of dendrimers. Dendrimers are small spheres that have a core and many branches outward; the end of these branches recognize the target chemical molecule or proteins and bind to it. This causes a chain of enzymatic reactions against the pathogen.

12.5.2 NANOSENSORS

To study the approximate quantity of nutrients found in the soil and its quick estimation is of greater importance in order to understand the conditions required for the efficiency of the soil. Quantifying the mineral concentration in the soil will ameliorate the richness as well as decrease the strain of surplus components. So the emergence of nanosensors for determining the nutrient concentration has proven to play a significant role in the improvement of agriculture.

Nanosensors have the ability to monitor and acquire information about the crop, field, and fertilizer conditions at all times. They can provide instantaneous results regarding the rate at which the crop is growing, illness, or attack of insects on the crop, and even about a lack or excess of specific fertilizers (Chen and Yada 2011). Real-time updates have decreased the need for excessive amounts of pesticides and fertilizers used, thereby helping in reducing the cost of production and also reducing the effects of pesticide/fertilizer chemicals in land and water pollution. Nanosensors are helpful for farmers as they give precise data on the fertilizers' necessities, which reduces the costs of buying excess fertilizers. Farmers are thus able to save the unused fertilizer. Conventional ways to analyze the soil composition, humidity of soil, and pest attack are time-consuming or require manual labor. Utilizing nano-sensors, we can obtain immediate and precise data, meaning that the crops can be protected more quickly and efficiently. As compared to conventional sensors used for detection, nanosensors have higher sensitivity as a result of greater value of surface-to-volume ratio, instant results, higher reliability, and accuracy in data. Furthermore, they only require samples in nanounits.

12.5.2.1   Nanobiosensors to Identify Soil Nutrients

The utilization of fertilizers is rapidly growing in the agriculture industry all over the world. Fertilizers are used to increase the rate of plant growth and effective soil minerals. However, fertilizers containing heavy metals such as Pb and Ar have ill effects on the environment by polluting soil and water that leads to biomagnification (Delgadillo-Vargas, Garcia-Ruiz, and Forero-Álvarez 2016). An effective way of using fertilizer is through its addition in appropriate weather conditions, during the suitable phase of growth, and in a specific amount. This can be done with accurate analysis of the components of the soil. Normal methods of studying soil content are costly and give slow results, whereas biosensors provide an ecological alternative that gives precise instant results on fertilizers in soil and water.

A study was carried out by (Mura et al. 2014) to identify the presence of nitrate in water. This was achieved with the use of a calorimetric assay that was developed using Au nanoparticles. The gold nanoparticles were stabilized with the help of modified cysteamine. Modified cysteamine has an attraction towards nitrite; it binds with the nitrate molecules and indicates a change in color.

12.5.2.2   Nanobiosensors to Monitor Plant Diseases

The crop yield is affected by insects, pests, and unwanted plant growth, causing a dent in the income of farmers. Therefore it is important to protect the crop using appropriate methods. Nanobiosensors allow a smart take on farming by keeping a check on soil humidity and components. They detect the pathogens that attack the crop well before any macro changes take place that farmers can see. Nanomaterials such as quantum dots are frequently used. Table 12.1 shows the tested fungicide with FRAC MoA group and nanocarriers.

Nanosensors made of quantum dots are used to identify the occurrence of virus Polymyxa betae. It is a yellow vein virus responsible for rhizomania disease in sugar beet, after plant root sap samples pre-treatment for virus extraction (Algar and Krull 2007). Nanopores in nanosensors can also be used to recognize the attack of diseases in crops. Zinc oxide nanoparticles have been used to build a nanobiosensor along with Au nanoparticles to detect the presence of the fungi Trichoderma harzianum.

TABLE 12.1
The Tested Fungicide with the FRAC MoA Group and Nanocarrier Are Listed. Experiments Conducted on Crop Plants against Targeted Pests or Exploring Environmental Factors Such as Off-Target Toxicity and Soil Leaching Are Specified (Worrall et al. 2018)

Fungicide	Nanoparticle	Crop	Target Pest	Toxicity	Reference
Ferbam [44]	Gold	Tea leaves	-	-	(Hou et al. 2016)
Pyraclastrobin	Chitosan/MSN	-	P. asparagi	-	(Cao et al. 2016)
Carbendazim	Chitosan/Pectin	Cucumber Maize Tomato	F. oxyporum and A. parasiticus	E. coli and S. aureus	(Sandhya et al. 2016)
Myclobutanil Metalaxyl	Magnetic nanocomposites	-	-	-	(Wang et al. 2018)
Pyrimethanil	MSN	Cucumber	-	-	(Zhao et al. 2017)
Clove essential oil	Chitosan	-	A. niger	-	(Hasheminejad, Khodaiyan, and Safari 2019)

MSN = mesoporous silica nanoparticles

12.5.2.3 Nano-Electronic Sensors

Figure 12.8 shows the plant-monitoring nanosensor network with nanosensors. Electronic nanosensors are made up of nanomaterials that show specific optical and electronic properties that help in detecting signaling molecules in crops. Nanosensors are used to convert the chemical release of a crop into electronic messages to monitor the crop health electronically. It converts the chemical signals that are a result of attack or lack of minerals into electronic and optical signals. Essentially, nano-electronic sensors function by detecting the charge in the system. Molecular binding of chemicals alters the charge density of the nanodevice, causing it to sense the generated signal. Nano-electronic sensors like chemiresistors and transistor-based sensors work by studying the variation in charge density. When charge is exchanged from the adsorbed elements and the nanodevice, there exists a variation in its surface charge density. This changes the Fermi energy and conductance of the sensor which is used to detect the molecule-sensor activity (Kulkarni et al. 2014).

12.5.3 TARGETED DELIVERY APPROACH

Sustainable agriculture involves significant methods of delivering genes or nucleotides into the crop molecules, as well as chemicals and other organic elements ("Performance Functional Foods – 1st Edition" 2020). While using fertilizers, roughly 0.1% enter the crop systems; the majority is lost into the environment. During typical procedures of spraying, crop-protecting agents with a very small amount of chemicals hit the target sites of the crop that causes issues in the effective progress of the crop, due to a paucity of nutrients reaching the crop system. The drains of these chemicals are largely caused by it being washed of, or chemical processes, such as reaction with sunlight, hydration, as well as microbial degradation (Sekhon 2014). Techniques have been created for the delivery of chemicals directly into the seeds of the crops, such as electroporation and ultrasound. There still exists a gap in practices within agriculture. There is a need to increase the efficiency of delivery and growth of the crop. This gap promises to be filled with the use of nanomaterials because of their properties which allow for the targeting of precise cells. Nanomaterials have a significant surface area and attachment which shows their ability to proficiently deliver chemicals (Sankaran et al. 2010). Nanodevices study the presence of disease, lack of nutrients, or other illnesses, much earlier than the visible symptoms show. Smart nanosystems can monitor the delivery of supplements and treat plant illness.

Nanomaterials are used to directly deliver specifically to the area where the crop has been attacked by insects or pests. The nanodevices are mechanized to supply chemicals specific to the amount required by the crop into the cell. The recent advancement of nanotechnology-based synthesis of slow or controlled release fertilizers, pesticides, and herbicides has, therefore, received extra attention in agriculture farming. Nanotechnology is a method to alter the properties of the nano-encapsulation such that there is a slow release of the chemical required. It helps in controlling the rate of the discharge of nutrition to be delivered to the target site. Nanomaterials

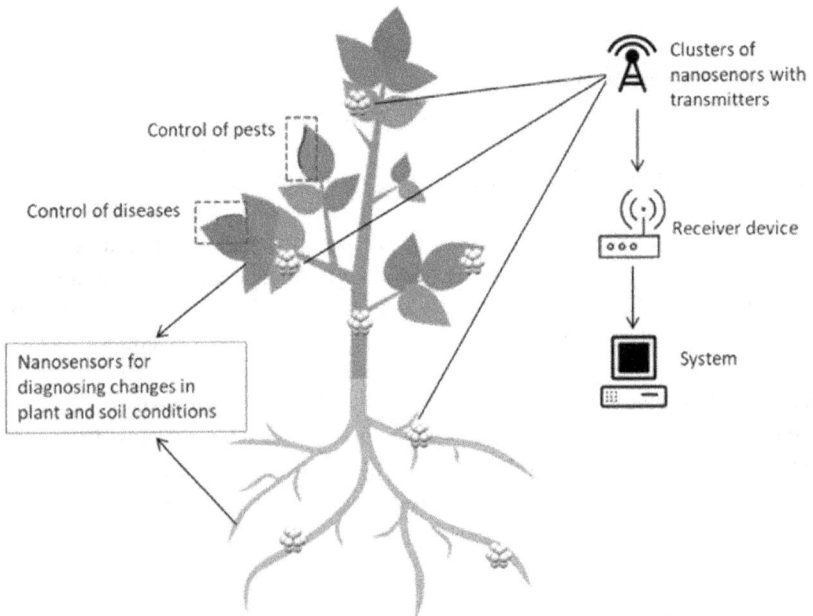

FIGURE 12.8 Plant monitoring nanosensor network with nanosensors, transmitter, and detector.

are able to recognize and precisely react to the external triggers such as change in humidity, temperature, and lack of nutrients. They are able to release medicine in a controlled manner accurately, as a response to these external factors. Target delivery intends to control the release of precise quantities of nutrients that are required by the crop regularly for specific periods of time in order to increase the efficiency while also reducing the runoff and pollution of the chemicals (Subramanian and Lee 2012). Nanomaterials require a substantially small amount of pesticide in micro- or nano-scale, thereby reducing the need for chemicals, and so increasing efficiency.

Nanocapsules consist of tiny cavities that store the chemical. These chemicals are absorbed and dispersed into the cavities that are sheltered from degradation thus increasing its efficiency of delivery to the crop cells. At the same time, for the delivery systems to be useful, the capsule chemical should be directed at the specific sites in the required concentration with minimum degradation. Compared to conventional encapsulation, such methods are less precise. Nanospheres show properties that show increased efficiency in the delivery and precision due to their small size. Nanoscales help them enter the cell system much more easily (Kranz et al. 2007).

Capping KNO_3 with graphene oxide films increases the time taken for the nitrate to be released into the crop. Capping is cheap and can be used for the large-scale production of fertilizers. Nanoparticles that are encapsulated using DNA fragments are delivered to the gene of the crop by using gene gun technology in order to obtain a favorable outcome in the crops. Nanoparticles help in increasing the stability of chemicals, simultaneously decreasing the degradation and its discharge into the environment. These increase the effectiveness of using chemicals in agriculture as well as help reduce the amount of chemicals used on crops (Wäckers, Romeis, and van Rijn 2007).

12.6   CONCLUSIONS

With the onset of the Green Revolution in efforts to combat a rising demand for food, the agricultural industry began the excessive use of chemicals for enhanced production and improved quality. However, not all the fertilizers applied were utilized by the plant, in turn resulting in the accumulation of excessive chemicals, thus affecting the flora and fauna. Nonetheless, the advancement in technology has paved a way for a sustainable and effective solution which incorporates the use of nanomaterials to tackle these problems. Nanoscale carriers have been employed for the targeted delivery of specific doses of pesticides and fertilizers in a regulated manner. Additionally, the high surface-to-volume ratio of nanomaterials due to its smaller size leads to an increase in the absorption rate of nutrients by the crops. Early prediction and diagnosis of diseases or pest attacks are achieved with quick precision using nanotechnology, thereby decreasing the overuse of pesticides. The application of nanotechnology makes agriculture more proficient and reduces the adverse effects on the environment.

REFERENCES

Afsharinejad, Armita, Alan Davy, Brendan Jennings, and Conor Brennan. 2016. "Performance Analysis of Plant Monitoring Nanosensor Networks at THz Frequencies." *IEEE Internet of Things Journal* 3(1): 59–69. doi:10.1109/jiot.2015.2463685.

Algar, W. Russ, and Ulrich J. Krull. 2007. "Quantum Dots as Donors in Fluorescence Resonance Energy Transfer for the Bioanalysis of Nucleic Acids, Proteins, and Other Biological Molecules." *Analytical and Bioanalytical Chemistry* 391(5): 1609–18. doi:10.1007/s00216-007-1703-3.

Cao, Lidong, Huirong Zhang, Chong Cao, Jiakun Zhang, Fengmin Li, and Qiliang Huang. 2016. "Quaternized Chitosan-Capped Mesoporous Silica Nanoparticles as Nanocarriers for Controlled Pesticide Release." *Nanomaterials* 6(7): 126. doi:10.3390/nano6070126.

Chen, Hongda, and Rickey Yada. 2011. "Nanotechnologies in Agriculture: New Tools for Sustainable Development." *Trends in Food Science and Technology* 22(11): 585–94. doi:10.1016/j.tifs.2011.09.004.

Colmenares, Juan Carlos, and Rafael Luque. 2014. "Heterogeneous Photocatalytic Nanomaterials: Prospects and Challenges in Selective Transformations of Biomass-Derived Compounds." *Chemical Society Reviews* 43(3): 765–78. doi:10.1039/C3CS60262A.

Conde, João, Jorge T. Dias, Valeria Grazão, Maria Moros, Pedro V. Baptista, and Jesus M. de la Fuente. 2014. "Revisiting 30 Years of Biofunctionalization and Surface Chemistry of Inorganic Nanoparticles for Nanomedicine." *Frontiers in Chemistry* 2(July). doi:10.3389/fchem.2014.00048.

Cursino, Luciana, Yaxin Li, Paulo A. Zaini, Leonardo De La Fuente, Harvey C. Hoch, and Thomas J. Burr. 2009. "Twitching Motility and Biofilm Formation Are Associated WithtonB1inXylella Fastidiosa." *FEMS Microbiology Letters* 299(2): 193–99. doi:10.1111/j.1574-6968.2009.01747.x.

Da Silva-Candal, Andrés, Bárbara Argibay, Ramón Iglesias-Rey, Zulema Vargas, Alba Vieites-Prado, Esteban López-Arias, Emilio Rodríguez-Castro, et al. 2017. "Vectorized Nanodelivery Systems for Ischemic Stroke: A Concept and a Need." *Journal of Nanobiotechnology* 15(1). doi:10.1186/s12951-017-0264-7.

Delgadillo-Vargas, Olga, Roberto Garcia-Ruiz, and Jaime Forero-Álvarez. 2016. "Fertilising Techniques and Nutrient Balances in the Agriculture Industrialization Transition: The Case of Sugarcane in the Cauca River Valley (Colombia), 1943–2010." *Agriculture, Ecosystems and Environment* 218(February): 150–62. doi:10.1016/j.agee.2015.11.003.

Ditta, Allah. 2012. "How Helpful Is Nanotechnology in Agriculture?" *Advances in Natural Sciences: Nanoscience and Nanotechnology* 3(3): 033002. doi:10.1088/2043-6262/3/3/033002.

Dubey, Shikha, Sushmita Banerjee, Siddh Nath Upadhyay, and Yogesh Chandra Sharma. 2017. "Application of Common Nano-Materials for Removal of Selected Metallic Species from Water and Wastewaters: A Critical Review." *Journal of Molecular Liquids* 240(August): 656–77. doi:10.1016/j.molliq.2017.05.107.

Duhan, Joginder Singh, Ravinder Kumar, Naresh Kumar, Pawan Kaur, Kiran Nehra, and Surekha Duhan. 2017. "Nanotechnology: The New Perspective in Precision Agriculture." *Biotechnology Reports* 15(September): 11–23. doi:10.1016/j.btre.2017.03.002.

Fernández, V., and T. Eichert. 2009. "Uptake of Hydrophilic Solutes Through Plant Leaves: Current State of Knowledge and Perspectives of Foliar Fertilization." *Critical Reviews in Plant Sciences* 28(1–2): 36–68. doi:10.1080/07352680902743069.

Garbach, Kelly, Jeffrey C. Milder, Fabrice A. J. DeClerck, Maywa Montenegro de Wit, Laura Driscoll, and Barbara Gemmill-Herren. 2016. "Examining Multi-Functionality for Crop Yield and Ecosystem Services in Five Systems of Agroecological Intensification." *International Journal of Agricultural Sustainability* 15(1): 11–28. doi:10.1080/147359 03.2016.1174810.

Gurr, Geoff M., Zhongxian Lu, Xusong Zheng, Hongxing Xu, Pingyang Zhu, Guihua Chen, Xiaoming Yao, et al. 2016. "Multi-Country Evidence That Crop Diversification Promotes Ecological Intensification of Agriculture." *Nature Plants* 2(3). doi:10.1038/nplants.2016.14.

Hasheminejad, Nayeresadat, Faramarz Khodaiyan, and Mohammad Safari. 2019. "Improving the Antifungal Activity of Clove Essential Oil Encapsulated by Chitosan Nanoparticles." *Food Chemistry* 275(March): 113–22. doi:10.1016/j.foodchem.2018.09.085.

Hou, Ruyan, Zhiyun Zhang, Shintaro Pang, Tianxi Yang, John M. Clark, and Lili He. 2016. "Alteration of the Nonsystemic Behavior of the Pesticide Ferbam on Tea Leaves by Engineered Gold Nanoparticles." *Environmental Science and Technology* 50(12): 6216–23. doi:10.1021/acs.est.6b01336.

Kranz, Heiko, Erol Yilmaz, Gayle A. Brazeau, and Roland Bodmeier. 2007. "In Vitro and In Vivo Drug Release from a Novel In Situ Forming Drug Delivery System." *Pharmaceutical Research* 25(6): 1347–54. doi:10.1007/s11095-007-9478-y.

Kulkarni, Girish S., Zhaohui Zhong Karthik Reddy, and Xudong Fan. 2014. "Graphene Nanoelectronic Heterodyne Sensor for Rapid and Sensitive Vapour Detection." *Nature Communications* 5(1): 1–7. doi:10.1038/ncomms5376.

Lamsal, Kabir, Sang Woo Kim, Jin Hee Jung, Yun Seok Kim, Kyong Su Kim, and Youn Su Lee. 2011. "Application of Silver Nanoparticles for the Control OfColletotrichumSpeciesIn Vitroand Pepper Anthracnose Disease in Field." *Mycobiology* 39(3): 194–99. doi:10.5941/myco.2011.39.3.194.

Leggo, Peter J. 2000. *Plant and Soil* 219(1/2): 135–46. doi:10.1023/a:1004744612234.

Li, Z. 2003. "Use of Surfactant-Modified Zeolite as Fertilizer Carriers to Control Nitrate Release." *Microporous and Mesoporous Materials* 61(1–3): 181–88. doi:10.1016/s1387-1811(03)00366-4.

Lvov, Yuri M., Dmitry G. Shchukin, Helmuth Möhwald, and Ronald R. Price. 2008. "Halloysite Clay Nanotubes for Controlled Release of Protective Agents." *ACS Nano* 2(5): 814–20. doi:10.1021/nn800259q.

Motakef Kazemi, Negar, and Ali Asghar Salimi. 2019. "Chitosan Nanoparticle for Loading and Release of Nitrogen, Potassium, and Phosphorus Nutrients." *Iranian Journal of Science and Technology, Transactions A: Science* 43(6): 2781–86. doi:10.1007/s40995-019-00755-9.

Mura, S., G. Greppi, P. P. Roggero, E. Musu, D. Pittalis, A. Carletti, G. Ghiglieri, and J. Irudayaraj. 2014. "Functionalized Gold Nanoparticles for the Detection of Nitrates in Water." *International Journal of Environmental Science and Technology* 12(3): 1021–28. doi:10.1007/s13762-013-0494-7.

"Nanotechnology Application in Agriculture: A Review RL Manjunatha, Dhananjay Naik and KV Usharani." 2020. Accessed March 17. http://www.phytojournal.com/archives/2019/vol8issue3/PartT/8-2-109-768.pdf.

"Nanotechnology - The Secret of Fifth Industrial Revolution and the Future of Next Generation." 2015. *Nusantara Bioscience* 7(2). doi:10.13057/nusbiosci/n070201.

Nuruzzaman, Md., Mohammad Mahmudur Rahman, Yanju Liu, and Ravi Naidu. 2016. "Nanoencapsulation, Nano-Guard for Pesticides: A New Window for Safe Application." *Journal of Agricultural and Food Chemistry* 64(7): 1447–83. doi:10.1021/acs.jafc.5b05214.

Oliveira, Halley Caixeta, Renata Stolf-Moreira, Cláudia Bueno Reis Martinez, Renato Grillo, Marcelo Bispo de Jesus, and Leonardo Fernandes Fraceto. 2015. "Nanoencapsulation Enhances the Post-Emergence Herbicidal Activity of Atrazine against Mustard Plants." *PLoS One* 10(7). doi:10.1371/journal.pone.0132971.

"Performance Functional Foods –1st Edition." 2020. www.elsevier.com. Accessed March 15. https://www.elsevier.com/books/performance-functional-foods/watson/978-1-85573-671-9.

Phalan, B., M. Onial, A. Balmford, and R. E. Green. 2011. "Reconciling Food Production and Biodiversity Conservation: Land Sharing and Land Sparing Compared." *Science* 333(6047): 1289–91. doi:10.1126/science.1208742.

Prasad, Ram, Atanu Bhattacharyya, and Quang D. Nguyen. 2017. "Nanotechnology in Sustainable Agriculture: Recent Developments, Challenges, and Perspectives." *Frontiers in Microbiology* 8(June). doi:10.3389/fmicb.2017.01014.

Pretty, Jules. 2018. "Intensification for Redesigned and Sustainable Agricultural Systems." *Science* 362(6417): eaav0294. doi:10.1126/science.aav0294.

Pretty, Jules, Tim G. Benton, Zareen Pervez Bharucha, Lynn V. Dicks, Cornelia Butler Flora, H. Charles, J. Godfray, et al. 2018. "Global Assessment of Agricultural System Redesign for Sustainable Intensification." *Nature Sustainability* 1(8): 441–46. doi:10.1038/s41893-018-0114-0.

Pretty, Jules, and Zareen Pervez Bharucha. 2014. "Sustainable Intensification in Agricultural Systems." *Annals of Botany* 114(8): 1571–96. doi:10.1093/aob/mcu205.

Sabir, Sidra, Muhammad Arshad, and Sunbal Khalil Chaudhari. 2014. "Zinc Oxide Nanoparticles for Revolutionizing Agriculture: Synthesis and Applications." *The Scientific World Journal* 2014: 1–8. doi:10.1155/2014/925494.

Sandhya, Sandeep Kumar, Dinesh Kumar, and Neeraj Dilbaghi. 2016. "Preparation, Characterization, and Bio-Efficacy Evaluation of Controlled Release Carbendazim-Loaded Polymeric Nanoparticles." *Environmental Science and Pollution Research* 24(1): 926–37. doi:10.1007/s11356-016-7774-y.

Sankaran, Sindhuja, Ashish Mishra, Reza Ehsani, and Cristina Davis. 2010. "A Review of Advanced Techniques for Detecting Plant Diseases." *Computers and Electronics in Agriculture* 72(1): 1–13. doi:10.1016/j.compag.2010.02.007.

Sekhon, Bhupinder. 2014. "Nanotechnology in Agri-Food Production: An Overview." *Nanotechnology, Science and Applications*, May, 31. doi:10.2147/nsa.s39406.

Shang, Yifen, Md. Kamrul Hasan, Golam Jalal Ahammed, Mengqi Li, Hanqin Yin, and Jie Zhou. 2019. "Applications of Nanotechnology in Plant Growth and Crop Protection: A Review." *Molecules* 24(14): 2558. doi:10.3390/molecules24142558.

Smith, Pete. 2013. "Delivering Food Security without Increasing Pressure on Land." *Global Food Security* 2(1): 18–23. doi:10.1016/j.gfs.2012.11.008.

Solanki, Priyanka, Arpit Bhargava, Hemraj Chhipa, Navin Jain, and Jitendra Panwar. 2015. "Nano-Fertilizers and Their Smart Delivery System." *Nanotechnologies in Food and Agriculture*: 81–101. doi:10.1007/978-3-319-14024-7_4.

Street, Anita. 2014. *Nanotechnology Applications for Clean Water : Solutions for Improving Water Quality*. Amsterdam: Elsevier.

Subramanian, Vivek, and Takhee Lee. 2012. "Nanotechnology-Based Flexible Electronics." *Nanotechnology* 23(34): 340201–340201. doi:10.1088/0957-4484/23/34/340201.

Subramanian, Kizhaeral S., Angamuthu Manikandan, Muthiah Thirunavukkarasu, and Sharmila Rahale. 2015. *Nanotechnologies in Food and Agriculture*. Edited by Mahendra Rai, Caue Ribeiro, Luiz Mattoso, and Nelson Duran. Cham: Springer International Publishing. https://doi.org/10.1007/978-3-319-14024-7.

Wäckers, Felix L., Jörg Romeis, and Paul van Rijn. 2007. "Nectar and Pollen Feeding by Insect Herbivores and Implications for Multitrophic Interactions." *Annual Review of Entomology* 52(1): 301–23. doi:10.1146/annurev.ento.52.110405.091352.

Wang, Xuemei, Xiaomin Ma, Pengfei Huang, Juan Wang, Tongtong Du, Xinzhen Du, and Xiaoquan Lu. 2018. "Magnetic Cu-MOFs Embedded within Graphene Oxide Nanocomposites for Enhanced Preconcentration of Benzenoid-Containing Insecticides." *Talanta* 181(May): 112–17. doi:10.1016/j.talanta.2018.01.004.

White, Philip J., John W. Crawford, María Cruz Díaz Álvarez, and Rosario García Moreno. 2012. "Soil Management for Sustainable Agriculture." *Applied and Environmental Soil Science* 2012: 1–3. doi:10.1155/2012/850739.

Worrall, Elizabeth, Aflaq Hamid, Karishma Mody, Neena Mitter, and Hanu Pappu. 2018. "Nanotechnology for Plant Disease Management." *Agronomy* 8(12): 285. doi:10.3390/agronomy8120285.

Zaini, Paulo A., Leonardo De La Fuente, Harvey C. Hoch, and Thomas J. Burr. 2009. "Grapevine Xylem Sap Enhances Biofilm Development ByXylella Fastidiosa." *FEMS Microbiology Letters* 295(1): 129–34. doi:10.1111/j.1574-6968.2009.01597.x.

Zhao, Pengyue, Lidong Cao, Dukang Ma, Zhaolu Zhou, Qiliang Huang, and Canping Pan. 2017. "Synthesis of Pyrimethanil-Loaded Mesoporous Silica Nanoparticles and Its Distribution and Dissipation in Cucumber Plants." *Molecules* 22(5): 817. doi:10.3390/molecules22050817.

Zheng, Lei, Fashui Hong, Shipeng Lu, and Chao Liu. 2005. "Effect of Nano-TiO(2) on Strength of Naturally Aged Seeds and Growth of Spinach." *Biological Trace Element Research* 104(1): 83–92. United States. doi:10.1385/BTER:104:1:083.

13 Nanotechnology and Its Potential Applications in the Field of Biotechnology

Nimmy Srivastava
Amity University Jharkhand, Ranchi, India

CONTENTS

13.1 INTRODUCTION

In early 1985, chemists studying the behavior pattern of carbon discovered that pure carbon could be manipulated to form symmetrical 60-atom spheres resembling a dome now known as "buckyballs". Subsequently, researchers realized that this new technology could play a major role in the transformation of existing industries (Thomas and Narvaez, 2005).

Nanotechnology is a novel scientific approach of engineering techniques involved in the design and synthesis of materials in sub-100 nm range with the objective of designing structures and devices having novel functions (Figure 13.1). The prefix

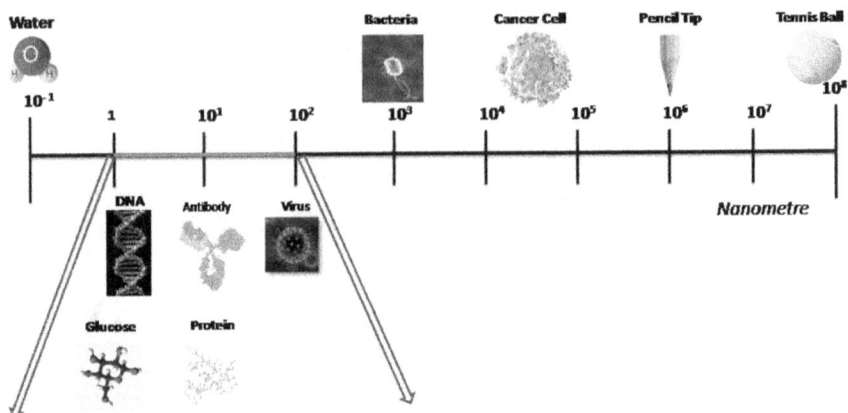

FIGURE 13.1 Integration of biomolecules in nanoscale range.

"nano" is derived from the Greek word meaning "dwarf." One nanometer (nm) is about the width of 6 carbon atoms or 10 water molecules (De Morais et al., 2014). At this microscale scientists can easily manipulate atoms to generate efficient materials with customized properties (Jain, 2005).

13.1.1 HISTORY

The origins of nanotechnology commenced with the historic lecture given by Richard Feynman in 1959 at the California Institute of Technology titled "There's Plenty of Room at the Bottom", where he bestowed the concept of building objects from the bottom up. However, at that time the idea did not gain much attention (De Morais et al., 2014). Later, in 1986 Eric Drexler published "Engines of Creation", highlighting the use of nanotechnology. He anticipated a molecular nanotechnology field that would allow the manufacture of fabricated products from the bottom up with decisive molecular control (Morrow et al., 2007).

Since then, nanotechnology has grown as a distinctive field, ranging from conventional device physics to a completely new approach with the objective of incorporating its application in various fields of science like organic chemistry, molecular biology, semiconductor physics, microfabrication, etc. (Fakruddin et al., 2012).

13.1.2 NANOBIOTECHNOLOGY

A confluence of the two fields – nanotechnology and biotechnology – has led to the evolution of nanobiotechnology. It is believed that the competent association of these two fields has led the way to a new class of multifunctional system for biological application (Figure 13.2) with better precision and a higher recognition rate.

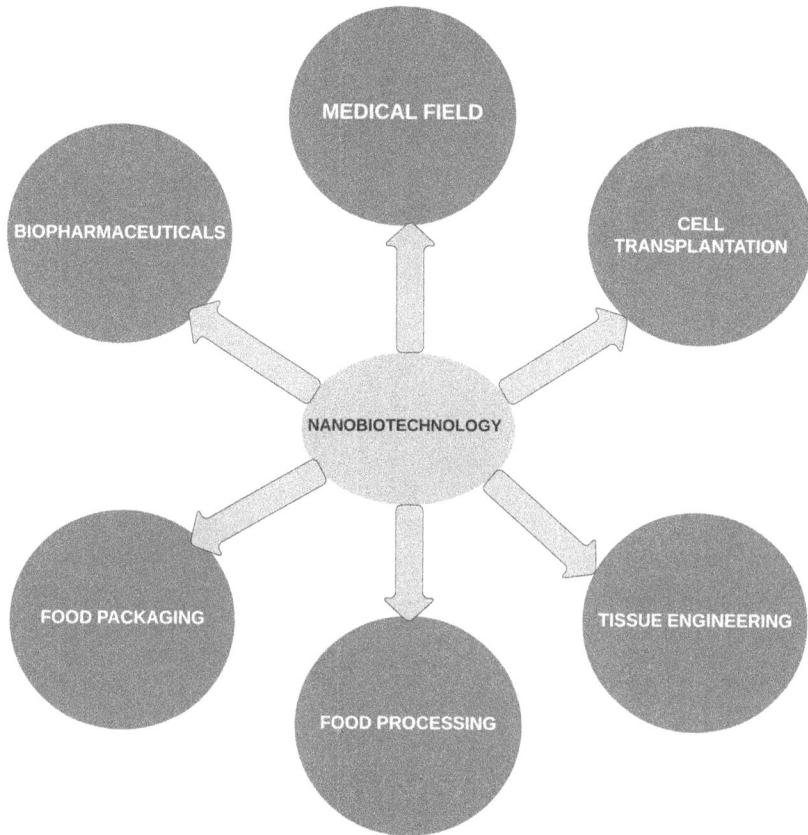

FIGURE 13.2   Various applications of nanobiotechnology.

13.1.3   ADVANTAGES OF NANOBIOTECHNOLOGY

The physiological condition arising by virtue of pathogen infection and the anatomical transformation of these tissues conceivably broadens the scope for the production of detection materials of a nanoscale level. Nanomaterials can be designed and built at the same size as cell components, making them an ideal candidate for interaction with cells (Thomas and Narvaez, 2006). Nanomaterial-based products have various other advantages:

- Every tissue infected by some disease shows a peculiar physiological condition and this can be utilized for designing drug delivery systems and drug-targeting molecules (Vasir and Labhasetwar, 2005).
- Nanoparticles have the capability to be assembled in higher concentrations at a specific location, compared to conventional drugs (Vasir et al., 2005).

- There is an increase in vascular permeability as well as impaired activity of lymphatic drainage in tumors, which improves the effect of a nanoparticle-based system in tumor tissue, through better transportation and retention rate of drugs (Maeda et al., 2000, Matsumra and Maeda, 1986).
- A nanoparticle-based system has the capacity to scrupulously target disease or inflamed tissues (Allen and Cullis, 2004).

13.2 APPLICATIONS OF NANOBIOTECHNOLOGY

13.2.1 MEDICAL APPLICATIONS

13.2.1.1 Diagnosis

In the medical field, currently available disease diagnostic methods mostly rely upon the manifestation of apparent symptoms of the disease. After recognizing the symptoms, a medical professional concludes the diagnosis with the specific ailment of the patient and its severity range. With time, discernible symptoms appearing in a patient's treatment show a drop-off in performance. One of the prime objectives of nanobiotechnology is designing new methodologies to diagnose various diseases at an early stage (Shaffer, 2005). Several researchers around the world have successfully demonstrated the use of metal- and semiconductor-based nanoparticles in biomedical applications.

13.2.1.2 Detection

Our contemporary clinical analysis shows the presence of a pathogen or a molecule by detecting its binding with a specific antibody. Such tests are accomplished by conjugating antibodies with various types of dyes and then observation is done under fluorescence or electron microscopy. The use of dyes has few limitations, as they are not specific for a particular pathogen and its detection. Semiconductor nanocrystals, popularly known as quantum dots (QDs), can be used as they can endure repeated cycles of excitation and light emission (Drexler, 1992), and hence can also be used in the diagnosis of diseases such as cancer (Moghimi et al., 2005).

13.2.1.3 Bioimaging

In recent years, advances have been made in intracellular imaging with the labeling of target molecules with quantum dots that facilitate the investigation of intracellular signals with the use of confocal fluorescence microscopy (Lin and Datar, 2006, Guccione et al., 2004), or correlation imaging.

13.2.2 THERAPEUTIC APPLICATIONS

A new formulation of drugs can be developed with the help of nanotechnology with fewer side effects, and can be used in many applications.

13.2.2.1 Drug Delivery

Therapeutic nanoparticles can be delivered to the specified site of infection with the assistance of radio or magnetic signals, where even standard drugs are unable to reach. These nanodrugs can be designed for the release of its formulation, especially in the presence of certain molecules or any external stimulus like heat. Encapsulating drugs in nanosized material e.g. dendrimers, a hollow capsule made of polymer, and nanoshell which can be used for controlled drug delivery. Electrospun nanofibers serve as a promising drug delivery vehicle as they resemble an extracellular matrix (Cunha et al., 2011). Co-axial electrospinning is also utilized successfully for this purpose as it has the capacity to produce nanotubes or nanofibers embedded with drugs (Chakraborty et al., 2009). Polymers used in the fabrication of nanoparticles mostly include poly (DL-lactic-coglycoloc acid), polyvinyl alcohol, poly(ethylene-co-vinyl-acetate) (Kompella et al., 2013). A co-delivery technique can be used for the administration of different drugs, and consequently incompatible drugs can be dispensed inside the body of the patient. Nanodrugs have also been used in theranostics for the diagnosis and treatment of disease at the same time (Shi et al., 2010).

13.2.2.2 Gene Therapy

In most cases it has been observed that the development of a disease is the result of some defective gene or its products. For the purpose of correcting these genetic disorders, there has arisen recently a new method. Gene therapy has been introduced based on the technique for the delivery of repaired genes or the replacement of a defective one (Ariga, 2006; Sahoo et al., 2007). Normal genes could be inserted within the genome which replace the defective genes. The swapping of normal and defective genes may be achieved through reverse mutation, which reinstates the genes, so they may return to their normal function (Hanakawa et al., 2005).

Human cells have a diameter of a few microns and cell organelles are found within the nanoscale range. Nanoparticle-based devices have an advantage as they can enter cells easily and can interact with the genome in a better way, as compared to larger devices (Kompella et al., 2013). Nonmaterial could be applied to replace the conventional viral vectors for gene delivery. There are several other major advantages of using nanoparticles in gene therapy:

- The structure of nanoparticles protects genes of interest, carried by them from degradation by nucleases present in the cell.
- Nanoparticles minimize any side effects, as they direct the nucleic acid to the specific site of action only.
- The sustainability of nanoparticles is longer, as compared to other gene delivery vehicles (Kompella et al., 2013).

13.2.3 Cell Transplantation and Tissue Engineering

One extensive research area in the medical biotechnology field is cell transplantation and engineering, and scientists all over the world are focused on exploring new biological substitutes to restore the normal function of organs and damaged tissues (Atala, 2005; Langer, 1999)

The predominant objective of tissue engineering is to reconstruct the damaged tissues by the use of growth factors, biopolymers which may be injected, and bio-materials which provide support to the developing cells (Venugopal et al., 2012) and direct interaction between cells and their microenvironment by creating signal molecules at a nanoscale level. Cells receive signals and respond to the information presented by the surrounding environment and this is important for control of behav-ior of cells (Wheeldon et al., 2011).

The electrospinning technique is the most widely used method to construct a bio-logical system compatible with biomaterial, and has shown promising results. This method transforms polymers solution in a filamentous form with diameters falling in a range from few micrometers to nanometers. The electrospinning method works on the electrostatic principle (De Morais et al., 2014) whereby help from a syringe polymer solution is provided to the system, which is subjected to different electrical voltages. The end result is a solid fiber (Greiner and Wendorff, 2007). An advanced version of the electrospinning method is the compound spinneret method, where two components can be fed by co-axial capillary channels and are integrated into a composite fiber (Chakraborty et al., 2009).

The biological applications of nanofibers are:

Skin grafting – Skin is the external organ of the human body which serves as a defensive barricade from the external environment (Blanpain, 2010; Martin, 1997), and any damage caused to it should be quickly repaired. The conventional treatments available are insufficient in the prevention of scar formation, nor do they promote fast healing in the patient (Middelkoop et al., 2004). The main focus of tissue engi-neering is the regeneration of skin, and to improve the healing process (Fujmori et al., 2006; Zuijlen et al., 2000). Various studies have been conducted to design ideal skin substitutes, and mostly polymeric matrices associated with stem cells and growth factors are used (Jin et al., 2013; Jin et al., 2011; Mohamed and Xing, 2012 and Steffens et al., 2013). It has been reported that fibroblast and keratinocytes are also used (Jin et al., 2013; Chandrasekaram et al., 2011).

Bone transplant – Injuries of the bone result in severe pain and can also lead to the disability of millions of people worldwide (Giannoudis et al., 2005). For the replacement of bone, the designing of scaffolds is based on the physical properties of bones, i.e. mainly strength, hardness, and porosity (Vasita and Katti, 2006). In the medical field, the most widely used synthetic bioactive bone substitute material is based on calcium phosphate (Wepener et al., 2012). Considerable studies have been conducted on Hydroxyapatite (HAp) which is chemically similar to inorganic com-ponent present in bone matrix (Wepener et al., 2012) for the development of nanofi-bers incorporated with HAp (Venugopal et al.,2012; Wepener et al., 2012; Sasmazel, 2011, Song et al., 2012) and it has shown satisfactory results.

Bladder transplant – To efficaciously replace a bladder in a patient damaged by cancer, infection, or inflammation, it is necessary to design some alternative bio-active substitute. According to some studies performed on various biodegradable and biocompatible polymers, it has been observed that PLGA and polyether ure-thane (PU) showed topography of scaffold, and influences cell response (Thapa et al., 2003). It can be seen that nanometer topography like bladder tissue shows better tissue integration.

13.2.4 FOOD PROCESSING AND PACKAGING

Since the prehistoric era, we humans have been trying to devise and improve food preservation techniques. With the help of nanobiotechnology, food spoilage can be decreased manifold, by improving the shelf life of food and food packaging techniques. The main focus of the food industry is on safe food packaging. Bio-nanocomposite and its potential application for the packaging of food items and edible and biodegradable nanocomposite films have gained much attention in recent years.

The use of nanomaterials as antimicrobial agents is also under research. Among available metal nanoparticles, predominantly silver, along with zinc and sulfur, have shown antimicrobial activity. Utilizing this property of metal nanoparticles, biocompatible polymeric silver nanocomposites (Sharma et al., 2009; Costa et al., 2011) have been prepared, making use of various naturally available polymers like starch, gelatin, sodium alginate, chitosan, carboxymethyl cellulose (Sharma et al., 2009; Vimala et al., 2011).

These natural biopolymers can also be used to prepare nanocomposite films applicable in the food industry for food packaging and to keep food fresh for a longer duration. Immunosensors in the form of nanowires are also under research, and are used for the detection of microbes present in packed food items (Sastry et al., 2011; Goncalves et al., 2012). Another area where nanoparticles can be applied is the enhancement of soil health, which in turn increases the productivity of the crop plant (Sastry et al., 2011).

With the advancement in nanomaterials and techniques, an effective pathogen detection method can be developed to monitor food-borne pathogens, which are one of the major causes of the outbreak of food-borne disease. The traditional microbe-detection method like colony count is time-consuming and laborious. Sometimes pathogen is present in low numbers, making detection somewhat difficult. All these factors pose difficulties for the quality control of semi-perishable food items.

Nanomaterials allow for the binding of different biomolecules like unicellular bacteria, toxins molecules, protein, and nucleic acids (Lahiff et al., 2010), due to both the nanoscale size and large surface area. Other important properties of nanomaterials like excellent optical and electronic property allow for the use of nanomaterials for pathogen detection and the development of biosensors (Figure 13.3) with improved response times (Gilmartin and Kennedy, 2012). The biosensor and its components are shown in Figure 13.3.

13.2.5 BIOPHARMACEUTICALS

One of the blooming industries in recent years is pharmaceutical, with the main aim of drug development for various diseases. Nanoscale techniques for the development of drugs may prove to be a major boon for the industry, which every year invests millions of dollars on R&D activities. With the use of nanotechnology, we

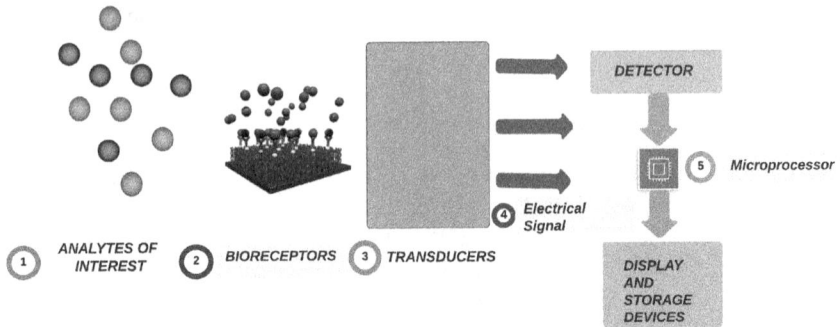

FIGURE 13.3 Biosensor: an analytical device and its components. (1) Analytes of interest. (2) Bioreceptors receive the analytes (different types of bioreceptors: (i) affinity binding, e.g. antibodies, peptides, nucleic acid; (ii) biocatalytic, e.g. enzymes, microorganisms, organelles; (iii) biomimetic, e.g aptamers). (3) Interaction between analytes and bioreceptors results in change of solution properties present around transducers, and transducers translate this change into a physical change of energy and then to an electrical signal. (4) Electrical signals are passed to microprocessors and after amplification and analysis they are converted to concentration units. (5) Concentration units are transferred to display and storage devices.

can easily engineer target molecules and atoms by lashing them to biomembranes and controlling their time and place of reaction (Fakruddin et al., 2012). It is a fast process with a minimal requirement of a few easily available reagents. This technique can potentially reduce drug discovery costs along with the advantage of specific drug development (Table 13.1). The list of FDA-approved biopharmaceuticals is presented in Table 13.1. Potential Pharmaceuticals, a pharma company based in Kentucky, has been working in this area of biopharmaceutical products from the beginning.

13.3 CONCLUSION

In spite of the concern and debate encompassing nanobiotechnology and nanoparticle-based products, worries about its noxious effect on human health and deleterious impact on the environment, the fact that nanotechnology has massive potential and hope for the future cannot be ignored. With more advances and research, it may lead to some major innovations in the fields of drug delivery, gene therapy, biosensors, clinical diagnosis, and bioimaging, all of which are already emerging. The most recent quantum dot technology has already been given the license. In coming years, we will witness more R&D activity in laboratories concerning nanoproducts and creeping into commercialization. American Pharmaceutical Partners is already testing the encapsulation of the cancer drug Taxol in nanopolymer Paclitaxel. Advances in the nanomedicine field have already opened opportunities for its application in various medical areas which are currently under investigation.

TABLE 13.1
List of FDA-Approved Biopharmaceuticals

Drug Product/Brand Name	Delivery Route	Manufacturer	Indication(s)	FDA Approval Date
1) Nanoparticle Drug Component: Liposomes				
AmBisome®	Intravenous	Gilead Sciences	Systemic fungal infections	FDA 1997
DaunoXome®	Intravenous	Gilead Sciences	HIntravenous-related Kaposi's Sarcoma	FDA 1996
DepoCyt®	Intravenous	Sigma-Tau	Lymphomatous malignant meningitis	FDA 1999/2007
DepoDur®	epidural space	Pacira Pharmaceuticals	For treatment of chronic pain in patients requiring long-term around-the-clock opioid analgesic on a daily basis	FDA 2004
Doxil® (Caelyx, outside US)	Intravenous	OrthoBiotech Schering-Plough	AIDS-related Kaposi's Sarcoma, multiple myeloma, ovarian cancer	FDA 1995
Inflexal® V	Cellular	Berna Biotech Ltd.	Influenza vaccine	Switzerland 1997
Marqibo®	Intravenous	Onco TCS	Acute lymphoid leukemia, Philadelphia chromosome-negatIntravenous, relapsed or progressed	FDA 2012
Mepact™	Intravenous	Takeda France SAS	Non-metastasizing resectable osteosarcoma	Europe 2009
Myocet®	Intravenous	Perrigo	Metastatic breast cancer	Europe 2000
Visudyne®	Intravenous	Bausch and Lomb	Photodynamic therapy of age-related macular degeneration, pathological myopia, ocular histoplasmosis syndrome	FDA 2000
2) Nanoparticle Drug Component: Lipid-Based (Non-Liposomal) Formulations				
Abelcet® Marketed outside USA as Amphocil®	Intravenous	Enzon	Systemic fungal infections	FDA 1995 and 1996
Amphotec®	subcutaneous	Sequus	invasIntravenous aspergillosis, patients who are intolerant of amphotericin B	FDA 1996

(Continued)

TABLE 13.1 (CONTINUED)
List of FDA-Approved Biopharmaceuticals

Drug Product/Brand Name	Delivery Route	Manufacturer	Indication(s)	FDA Approval Date
3) Nanoparticle Drug Component: PEGylated Proteins, Polypeptides, Aptamers				
Adagen®	Intravenous	Enzon	Adenosine deaminase deficiency – severe combined immunodeficiency disease	FDA 1990
Cimzia®	subcutaneous	UCB	Crohn's disease, rheumatoid arthritis	FDA 2008
Neulasta®	subcutaneous	Amgen	Febrile neutropenia, In patients with nonmyeloid malignancies; prophylaxis	FDA 2002
Oncaspar®	subcutaneous	Amgen	Acute lymphoblastic leukemia	FDA 1994
Pegasys®	subcutaneous	Nektar Hoffmann-La Roche	Hepatitis B and C	FDA 2002
PegIntron®	subcutaneous	Enzon Schering-Plough	Hepatitis C	FDA 2001
Somavert®	subcutaneous	Nektar Pfizer	Acromegaly, second-line therapy	FDA 2003
Macugen®	intravitreal	OSI Pharmaceuticals Pfizer	Neovascular age-related macular degeneration	FDA 2004
Mircera®		Hoffman-La Roche	Anemia associated with chronic renal failure in adults	FDA 2007
4) Nanoparticle Drug Component: Nanocrystals				
Emend®	oral capsule	Merck, Elan	nausea in chemotherapy patients	FDA 2003
Megace ES®	oral solution	Endo Pharmaceuticals	Anorexia, cachexia	FDA 2005
Rapamune®	oral solution oral tablet	Wyeth, Elan	Immunosuppressant for kidney transplant	FDA 2002
Tricor®	oral tablet	Abbot	Hypercholesterolemia, hypertriglyceridemia	FDA 2004
Triglide®	oral tablet	SkyePharma First Horizon	Lipid disorders	FDA 2005
5) Nanoparticle Drug Component: Polymer-Based Nanoformulations				
Copaxone®	subcutaneous	TEVA	Multiple sclerosis	FDA 1996/2014
Eligard®	subcutaneous	Tolmar Pharmaceuticals	Advanced prostate cancer	FDA 2002

(Continued)

TABLE 13.1 (CONTINUED)
List of FDA-Approved Biopharmaceuticals

Drug Product/Brand Name	Delivery Route	Manufacturer	Indication(s)	FDA Approval Date
Genexol®	Intravenous	Lupin Ltd.	Metastatic breast cancer, pancreatic cancer	South Korea 2001
Opaxio®	Oral	Cell Therapeutics	Glioblastoma	FDA 2012
Renagel®	oral tablet	Genzyme	Hyperphosphatemia	FDA 2000
6) Nanoparticle Drug Component: Protein–Drug Conjugates				
Abraxane®	Intravenous	Abraxis BioScience Astra Zeneca	Metastatic breast cancer, non-small-cell lung cancer	FDA 2005
Kadcyla®	Intravenous	Genentech	Metastatic breast cancer	FDA 2013
Ontak®	Intravenous	Eisai Inc	Primary cutaneous T-cell lymphoma, CD25-positIntravenouse, persistent or recurrent disease	FDA 1994/2006
7) Nanoparticle Drug Component: Surfactant-Based Nanoformulations				
Fungizone® (also referred to as "conventional AMB")	Intravenous	Abbott	Systemic fungal infections	FDA 1966
DiprIntravenousan®	Intravenous	Zeneca Pharmaceuticals	anesthetic	FDA 1989
Estrasorb™	transdermal	Novavax	Hormone replacement therapy during menopause	FDA 2003
8) Nanoparticle Drug Component: Metal-Based Nanoformulations				
Feridex®	Intravenous	AMAG pharmaceuticals	Intravenouser/spleen lesion MRI	FDA 1996
Feraheme™ (Ferumoxytol)	Intravenous	AMAG pharmaceuticals	Treatment of iron deficiency anemia in adults with chronic kidney disease	FDA 2009
Elestrin ®	transdermal	BioSanté	Treatment of hot flashes in menopausal women	FDA 2006

(Continued)

TABLE 13.1 (CONTINUED)
List of FDA-Approved Biopharmaceuticals

Drug Product/Brand Name	Delivery Route	Manufacturer	Indication(s)	FDA Approval Date
Ferrlecit®	Intravenous	Sanofi Avertis	Iron deficiency in chronic kidney disease	FDA 1999
INFeD®	Intravenous	Sanofi Avertis	Iron deficiency in chronic kidney disease	FDA 1957
DexIron®/Dexferrum®	Intravenous	Sanofi Avertis	Iron deficiency in chronic kidney disease	FDA 1958
9) Nanoparticle Drug Component: Virosomes				
Gendicine®	Intratumor	Shenzhen SiBione	Head and neck squamous cell carcinoma	People's Republic of China 2003
Rexin-G®	Intravenous	Epeius Biotechnologies	For all solid type tumors	Philippines 2007

Source: Reisner et al., 2012; Caster et al., 2017; Anselmo and Mitragotri, 2016; Grumezescu, 2017; Ventola, 2017; D'Mello et al., 2017

Abbreviations: FDA – US Food and Drug Administration; MRI – magnetic resonance imaging; AIDS – acquired immunodeficiency syndrome.

REFERENCES

Allen, T. M. and P. R. Cullis. 2004. Drug delivery systems: Entering the mainstream. *Science* 303(5665):1818–1822.

Anselmo, A. C. and S. Mitragotri. 2016. Nanoparticles in the clinic. *Bioeng. Transl. Med.* 1(1):10–29.

Ariga, T. 2006. Gene therapy for primary immunodeficiency diseases: Recent progress and misgivings. *Curr. Pharm. Des.* 12(5):549–556.

Atala, A. 2005. Technology insight: Applications of tissue engineering and biological substitutes in urology. *Nat. Clin. Pract. Urol.* 2(3):143–149.

Blanpain, C. 2010. Stem cells: Skin regeneration and repair. *Nature* 464(7289):686–687.

Caster, J. M., A. N. Patel, T. Zhang and A. Wang. 2017. Investigational nanomedicines in 2016: A review of nanotherapeutics currently undergoing clinical trials. *Wiley Interdiscip. Rev.* 9(1):e1416.

Chakraborty, S., I. C. Liao, A. Adler and K. W. Leong. 2009. Electrohydrodynamics: A facile technique to fabricate drug delivery systems. *Adv. Drug Deliv. Rev.* 61(12):1043–1054.

Chandrasekaran, A., R. Venugopal, J. Sundarrajan. and S. Ramakrishna. 2011. Fabrication of a nanofibrous scaffold with improved bioactivity for culture of human dermal fibroblasts for skin regeneration. *Biomed. Mater.* 6(1): 015001.

Cunha, C., S. Panseri and S. Antonini. 2011. Emerging nanotechnology approaches in tissue engineering for peripheral nerve regeneration. *Nanomedicine* 7(1):50–59.

Costa, C., A. Conte, G. G. Buonocore and M. A. Del Nobile. 2011. Antimicrobial silver-montmorillonite nanoparticles to prolong the shelf life of fresh fruit salad. *Int. J. Food Microbiol.* 148(3):164–167.

De Morais, M. G., V. G. Martins, D. Steffens and, P. Pranke. 2014. Biological applications of nanobiotechnology. *J. Nanosci. Nanotechnol.* 14(1):1007–1017.

D'Mello, S. R., C. N. Cruz, M. L. Chen, M. Kapoor, S. L. Lee and K. M. Tyner. 2017. The evolving landscape of drug products containing nanomaterials in the United States. *Nat. Nanotechnol.* 12(6):523.

Drexler, E. K. 1992. *Nanosystems: Molecular Machinery, Manufacturing and Computation.* New York: John Wiley & Sons.

Drug approvals and databases. https://www.fda.gov/Drugs/InformationOnDrugs/default.htm. Accessed 16 Jan 2020.

Fakruddin, Md., Z. Hossain and H. Afroz. 2012. Prospects and applications of nanobiotechnology: A medical perspective. *J. Nanobiotechnology* 10(1):31.

Fujimori, Y., K. Ueda, H. Fumimoto, K. Kubo and Y. Kuroyanagi. 2006. Skin regeneration for children with burn scar contracture using autologous cultured dermal substitutes and superthin auto-skin grafts: Preliminary clinical study. *Ann. Plast. Surg.* 57(4):408–414.

Giannoudis, P. V., H. Dinopoulos and E. Tsiridis. 2005. Tsiridis bone substitutes: bone substitutes: An update. *Injury* 36:S20–27.

Gilmartin, N. and R. O' Kennedy. 2012. Nanobiotechnologies for the detection and reduction of pathogens. *Enzyme Microb. Technol.* 50(2):87–95.

Gonçalves, C., P. Pereira, P. Schellenberg, P. J. Coutinho and F. M. Gama. 2012. Multifaceted development and application of biopolymers for biology. *J. Biomater. Nanobiotechnol.* 3(2):178.

Greiner, A. and J. H. Wendorff. 2007. Electrospinning: A fascinating method for the preparation of ultrathin fibers. *Angew. Chem. Int. Ed. Engl.* 46(30):5670–5703.

Grumezescu, A. M. 2017. *Nanoscale Fabrication, Optimization, Scale-Up and Biological Aspects of Pharmaceutical Nanotechnology.* New York: William Andrew.

Guccione, S., K. C. Li and M. D. Bednarski. 2004. Vascular-targeted nanoparticles for molecular imaging and therapy. *Methods Enzymol.* 386:219–236.

Hanakawa, Y., Y. Shirakata, H. Nagai, Y. Yahata, S. Tokumaru, K. Yamasaki, M. Tohyama, K. Sayama and K. Hashimoto. 2005. Cre-loxP adenovirus-mediated foreign gene expression in skin-equivalent keratinocytes. *Br. J. Dermatol.* 152(6):1391–1392.

Jain, K. K. 2005. The role of nanobiotechnology in drug discovery. *Drug Discov. Today* 10(21):1435–1442.

Jin, G., M. P. Prabhakaran and S. Ramakrishna. 2011. Stem cell differentiation to epidermal lineages on electrospun nanofibrous substrates for skin tissue engineering. *Acta Biomater.* 7(8):3113–3122.

Jin, G. M. P., D. Prabhakaran, D. Kai, S. K. Annamalai, K. D. Arunachalam and S. Ramakrishna. 2013. Tissue engineered plant extracts as nanofibrous wound dressing. *Biomaterials* 34(3):724–734.

Kompella, U. B., A. C. Amrite, R. P. Ravi and S. A. Durazo. 2013. Nanomedicines for back of the eye drug delivery, gene delivery, and imaging. *Prog. Retin. Eye Res.* 36:172–198.

Lahiff, E., C. Lynam, N. Gilmartin, R. O' Kennedy and D. Diamond. 2010. The increasing importance of carbon nanotubes and nanostructured conducting polymers in biosensors. *Anal. Bioanal. Chem.* 398(4):1575–1589.

Langer, R. 1999. Selected advances in drug delivery and tissue engineering. *J. Control. Release* 62(1–2):7–11.

Lin, H. and R. H. Datar. 2006. Medical applications of nanotechnology. *Natl Med. J. India* 19(1):27–32.

Maeda, H., J. Wu, T. Sawa, Y. Matsumura and K. Hori. 2000. Tumor vascular permeability and the EPR effect in macromolecular therapeutics: A review. *J. Control. Release* 65(1–2):271–284.

Martin, P. 1997. Wound healing--Aiming for perfect skin regeneration. *Science* 276(5309):75–81.

Matsumura, Y. and H. Maeda. 1986. A new concept for macromolecular therapeutics in cancer chemotherapy: Mechanism of tumoritropic accumulation of proteins and the antitumor agent smancs. *Cancer Res.* 46(12 Pt 1):6387–6392.

Moghimi, S. M., A. C. Hunter and J. C. Murray. 2005. Nanomedicine: Current status and future prospects. *FASEB J.* 19(3):311–330.

Middelkoop, E., A. J. van den Bogaerdt, E. N. Lamme, M. J. Hoekstra, K. Brandsma and M. M. Ulrich. 2004. Porcine wound models for skin substitution and burn treatment. *Biomaterials* 25(9):1559–1567.

Mohamed, A. and M. M. Xing. 2012. Nanomaterials and nanotechnology for skin tissue engineering. *Int. J. Burns Trauma* 2(1):29–41.

Morrow, K. J. Jr., R. Bawa and C. Wei. 2007. Recent advances in basic and clinical nanomedicine. *Med. Clin. North Am.* 91(5):805–843.

Reisner, D. E., R. Bawa, S. Brauer, J. Alvelo, W. Zheng, C. Vulpe, R. Bawa, J. Alvelo and M. Gericke. 2012. Bionanotechnology. In: *Handbook of Research on Biomedical Engineering Education and Advanced Bioengineering Learning: Interdisciplinary Concepts*, ed. Ziad, O.A-F., vol. 1, pp. 436–489. IGI Global.

Sahoo, S. K., S. Parveen and J. J. Panda. 2007. The present and future of nanotechnology in human health care. *Nanomedicine* 3(1):20–31.

Sasmazel, H. T. 2011. Novel hybrid scaffolds for the cultivation of osteoblast cells. *Int. J. Biol. Macromol.* 49(4):838–846.

Sastry, R. K., H. B. Rashmi and N. H. Rao. 2011. Nanotechnology for enhancing food security in India. *Food Policy* 36(3):391–400.

Shaffer, C. 2005. Nanomedicine transforms drug delivery. *Drug Discov. Today* 10(23–24):1581–1582.

Sharma, V. K., R. A. Yongard and Y. Lin. 2009. Silver nanoparticles: Green synthesis and their antimicrobial activities. *Adv. Colloid Interface Sci.* 145(1–2):83.

Shi, J., A. R. Votruba, O. C. Farokhzad and R. Langer. 2010. Nanotechnology in drug delivery and tissue engineering: From discovery to applications. *Nano Lett.* 10(9):3223–3230.

Song, W., D. C. Markel, S. Wang, T. Shi, G. Mao and W. Ren. 2012. Electrospun polyvinyl alcohol-collagen-hydroxyapatite nanofibers: A biomimetic extracellular matrix for osteoblastic cells. *Nanotechnology* 23(11):115101.

Steffens, D., M. Lersch, A. Rosa, C. Scher, T. Crestani, M. G. Morais, J. A. Costa and P. Pranke. 2013. A new biomaterial of nanofibers with the microalga Spirulina as scaffolds to cultivate with stem cells for use in tissue engineering. *J. Biomed. Nanotechnol.* 9(4):710–718.

Thapa, A., D. C. Miller, T. J. Webster and K. M. Haberstroh. 2003. Nano-structured polymers enhance bladder smooth muscle cell function. *Biomaterials* 24(17):2915–2926.

Thomas, T.C. and R.A. Narvaez. 2006. The convergence of biotechnology and nanotechnology: Why here, why now? *Journal of Commercial Biotechnology* 12(2):105–110.

van Zuijlen, P. P., A. J. van Trier, J. F. Vloemans, F. Groenevelt, R. W. Kreis and E. Middelkoop. 2000. Graft survival and effectiveness of dermal substitution in burns and reconstructive surgery in a one-stage grafting model. *Plast. Reconstr. Surg.* 106(3):615–623.

Vasir, J. K. and V. Labhasetwar. 2005. Targeted drug delivery in cancer therapy. *Technol. Cancer Res. Treat.* 4(4):363–374.

Vasir, J. K., M. K. Reddy and V. Labhasetwar. 2005. Nanosystems in drug targeting: Opportunities and challenges. *Curr. Nanosci.* 1(1):47–64.

Vasita, R. and D. S. Katti. 2006. Nanofibers and their applications in tissue engineering. *Int. J. Nanomed.* 1(1):15–30.

Ventola, C. L. 2017. Progress in nanomedicine: Approved and investigational nanodrugs. *Pharm. Ther.* 42(12):742.

Venugopal, J. R., M. P. Prabhakaran, R. Ravichandran, K. Dan S. Ramakrishna. 2012. Biomaterial strategies for alleviation of myocardial infarction. *J. R. Soc. Interface* 9(66):1–19.

Vimala, K., Y. M. Mohan, K. Varaprasad, N. Redd, N. S. Ravindra, N. S. Naidu and K. M. Raju. 2011. Fabrication of curcumin encapsulated chitosan-PVA silver nanocomposite films for improved antimicrobial activity. *J. Biomater. Nanobiotechnol.* 2:55–64.

Wepener, I., W. Richter, D. van Papendorp and A. M. Joubert. 2012. In vitro osteoclast-like and osteoblast cells' response to electrospun calcium phosphate biphasic candidate scaffolds for bone tissue engineering. *Mater. Sci. Mater. Med.* 23(12):3029–3040.

Wheeldon, I., A. Farhadi, A.G. Bick, E. Jabbari and A. Khademhosseini. 2011. Nanoscale tissue engineering: spatial control over cell-materials interactions. *Nanotechnology* 22(21):212001.

14 Crop Improvement and Applied Nanobiotechnology

Rajani Sharma and Shilpa Prasad
Amity University Jharkhand, Ranchi, India

CONTENTS

14.1 INTRODUCTION

Nanotechnology is a revolution in the field of science. Its nanosize (between 100–2500 nanometers) has expanded its area of application, as it minimizes the use of resources, and in turn, decreases the toxic level with maximum input. It has gained a significant position in the mechanical, electrical, and physical worlds. Despite being new to the biological and scientific world, it has a broad range of applications. In the agricultural field, nanotechnology offers a solution to satisfy the alarming rate of population growth. Researchers have implemented nanoparticles as immune elicitors, for the reclamation of abandoned land, nano-based biosensors to detect the presence of pesticides in agro-products, and the improvement of quality and yield of crops (Tripathi et al., 2018). It is preferred in energy production, as it increases efficiency, using the photovoltaic technique and plasmonic enhancement. This property has also been implemented in agriculture to increase the efficiency of photosynthesis, particularly to overcome abiotic stress. The best part of the technology is its nanosize which makes it compatible for absorption by plant cells.

Nanotechnology has a better mechanism of delivery with boosted efficiencies leading to a reduction in the use of harmful chemicals. The integration of nanotechnology with organic farming is the prime focus of nanobiotechnology. Nanotechnology possesses the ability to increase the immunogenic response and can also increase the quality of food with the process of biofortification. This chapter basically focuses on the role of nanotechnology in the field of sustainable agriculture. Nanotechnology can lead the mission for a sustainable agriculture with limited resources with striking developments in the field of science (Mishra et al., 2016). It is an emerging technique to solve the current demand and challenges of elevating the world population.

Nanoparticles can be created through various physical, chemical, as well as biological methods. Fungi, bacteria, and plant viruses are reported for synthesis of nanoparticles in microbial systems to gain control over shape, size, and other desired properties of the synthesized nanomaterials (Singh et al., 2015). Some of the commonly adopted methods for the conception of nanoparticles have been mentioned in Table 14.1.

Nanoparticles are used even when plants are availing of favorable conditions to further enhance the production. During adverse conditions, it supports the plant to adapt and maintain the production level. Mishra et. al. in 2016 stated that nanoparticles synthesized from metal-based, metal oxide, and carbon have been seen to improve the yield of many crop plants by increasing the rate of germination, number of flowers, root elongation, and plant biomass (Mishra et al., 2016).

Crops are generally stored for a longer time before consumption. With aging, the biochemical properties of seed have led to the loss of the germination property. Silver nanoparticles have proven instrumental in the germination of even aged seeds of rice by enhancing α-amylase activity, which is the primary requirement for germination (Mahakham et al., 2017). This chapter will deal with all the fields in agriculture where nanotechnology is emerging for the improvement of seed quality, soil physiology, and the immune system of plants and nanosensors.

14.2 CHALLENGES IN THE FIELD OF AGRICULTURE

The alarming increase in population is exploiting nature in many ways. It also exerts effects on the environment that lead to climate change. The increasing rate of population utilizes land in many non-agricultural requirements. This is directly increasing the demand for agricultural production in limited land under an adverse environment. This chapter focuses on the concerns and problems related to sustainable agriculture, and their remedies with the use of nanobiotechnology. Through this chapter we will discuss the basic requirements for sustainable agriculture, the traditional methods used, and the current technology implemented to satisfy increasing hunger and surrounding demands. The sustainability of agriculture depends on the availability of farming land, soil quality, irrigation, postharvest problems, and care during storage and transportation.

TABLE 14.1
Different Physical, Chemical and Biological Techniques to Synthesize Nanoparticles

Methods	Process
Physical Methods	
High energy ball milling (HEBM)	The method was developed by John Benjamin in 1970. In this method the chemical bonding between the milled material is ruptured by the kinetic energy of moving balls. Such nanoparticles can withstand high pressure and temperature.
Inert gas condensation (IGC)	It is based on the principle of condensation of atomic vapor under high pressure. This is believed to be the most primitive method and most suited for platinum and silver NPs (Maicu, 2014).
Physical vapor deposition (PVD)	In this method thin films and NPs are synthesized by aggregation of vapors from the solid source (Okuyama et al., 2006).
Laser pyrolysis	This is the method to create NPs from metal and non-metal oxides at the interface of the laser beam and the molecular flow of gaseous/vapor phase reactants (D'Amato et al., 2013).
Flame spray pyrolysis (FSP)	In this method the precursors are evaporated or decomposed to form metal vapors which then nucleates due to supersaturation and finally aggregate either by physical or chemical interactions to form NPs. This technique is quite complex and produces more functional NPs.
Electrospraying technique	This is the method to create NPs with flexible and controlled parameters. In this method the desired polymers and solvent is taken by the syringe in an electrochemical device. Under high voltage the solvent is vaporized in the form of charged droplets leaving behind NPs or fibers as end product (Bhushani and Anandharamakrishnan, 2014).
Melt mixing	In this technique the polymer is mixed with modified nanofillers by the process of kneading, injecting, or molding (Karak, 2009). Such a method forms polymer consisting of NPs as the filters to achieve desired material characteristics (Lin et al., 2006).
Chemical Methods	
Sol–gel method	In this method the nanoparticles are created by the process of hydrolysis and condensation. Hydrolysis creates a get surface on water over which solid particles are condensed (Brinker and Scherer, 1990).
Microemulsion technique	In this monodispersed spherical droplet are synthesized with water-in-oil or oil-in-water surfactant (Solanki and Murthy, 2011).
Hydrothermal synthesis	This is an inexpensive method to synthesize NPs metal oxides and lithium iron phosphate with optimized properties (Abedini et al., 2013) under controlled pressure and temperature either by batch hydrothermal or continuous hydrothermal process.
Polyol synthesis	It is used to synthesize metal, metal oxide, and magnetic NPs with ethylene glycol as a reaction medium. Ethylene glycols act as reducing and stabilizing agent (Rahaman and Green, 2009).

(Continued)

TABLE 14.1 (CONTINUED)
Different Physical, Chemical and Biological Techniques to Synthesize Nanoparticles

Methods	Process
Chemical vapor deposition (CVD)	Under high temperature and pressure, the solid films are synthesized from vapor phase.
Plasma enhanced chemical vapor deposition (PECVD)	It functions similar to CVD but requires gaseous precursor under vaccum conditions with power supply.
Biological Methods	
Biogenic synthesis using microorganisms	Microorganisms target metal ions and convert them to element metals by enzymatic process either inside the cell or outside the cell.
Biomolecules as templates to design nanoparticles	The biomolecules like amino acids, lipids, and nucleic acid contain charges which act as an excellent template to bind transition metals. For example during the formation of AuNps, DNA hydrogel crosslink before incorporating transition metal ions like gold, Au(III) metal ions to Au NPs due to reduction of Au(III) (Zinchenko et al., 2014).
Plant extracts for nanoparticle synthesis	This is the most preferred technique to synthesize NPs due to its eco-friendly and non-toxic method. This method is preferred to synthesize NPs of bi-metallic alloys, metal oxides, noble metals, etc. (Iravani, 2011).

14.2.1 SMALL AND FRAGMENTED LAND-HOLDINGS

Agriculture has always been a major sector of the economy in most countries, including India. It maintains a noticeable share of the Gross Domestic Product (GDP) (Madhusudhan, 2015). At the present time, increasing population and the demand for food has led to extra concerns in this field. This has created a major problem in agriculture known as land fragmentation, which means the owner has small pieces of land at a dispersed distance (Alemu et al., 2017). Fragmentation of land creates irregularity in the shape and size of land. Analysts state that such lands are not suited for farming, which further results in the wastage of land along the border due to fencing. However, fragmentation has benefits such as a variety of soils, a different microclimate, and potential for crop rotation. However, extra investment for traveling and time management is difficult for monitoring in those lands.

14.2.2 AVAILABILITY OF SEED

In agriculture, seed is a basic requirement, along with fertilized land. The model that uses seed, pulls us 10,000 years back when the propagative nature of seed was utilized in the flooded river plains of North Africa and the Middle East. Viability and vigor of seed control the cultivation quality or productivity. A vigor seed provides proper emergence of seedling and plant variety, which result in an improvement of production rate and quality. Hectoliter Mass (HLM), Protein Content (PC), and Falling Number (FN) determine the quality of crop production. A recent study says that India is on the way to a second green revolution with regards to seed production. India has increased to the Compound Growth Rate (CAGR) of 8.4 % in terms of volume from the FY (Financial Year) 2009 to FY 2015. It is capable of satisfying 3.5 million tons of hunger, and from FY 2014 to FY 2019 the CAGR reached 14.1%. At this stage, even with these figures, there is still a concern to increase the production (ICRA, 2015; Technavio, 2015) due to an alarming increase in the population that will need 130 million tons of rice by 2025.

14.2.3 PROPER IRRIGATION FACILITY

A fertile land with good seed quality also requires a good irrigation facility. It is always advisable to reuse the water, but the source of water also determines the quality of food. Running water from industry or black water may contain some harmful chemicals which can affect both human and soil health. The increasing demand for water in industries and domestic purposes highlights special concerns on the need to increase the water-holding capacity of soil. Scientists are continuously seeking crop improvement with less water demand.

14.2.4 POSTHARVEST LOSSES (PHL)

A major challenge in the need to fulfill the demand for hunger and increasing crop production is the postharvest losses due to insufficient storage systems and inappropriate transportation systems. Such mismanagement is termed "Food Loss" (Bourne, 1977). This means that available food is not consumed due to its destruction. PHL is a major challenge for developing countries. Loss occurs at every stage from production till consumption. This loss may be due to natural causes like hailstorm or flood or from the effects of manual labor. In developing countries, including India, harvesting is done manually. The process is very slow which delays the harvesting time and increases the loss. A survey in Punjab, India, states that there is an increase of 67% of loss in wheat and 10.3% of loss in paddy harvesting (Grover, Davinder and Singh 2013). During the process of drying, the presence of rodents and birds damages large amounts of food. Harvesting is then followed by the steps of threshing and cleaning. A delay in threshing, as well as inadequate cleaning invites the growth of molds, which is a major factor for PHL. Crops are also at risk from mold infection when the moisture content is more than 13 15% (Khan, 2010). The physical conditions of the place where harvested crops are stored is also an important factor, resulting in PHL. Inappropriate temperatures and infection during storage may also lead to heavy loss. Finally, the transportation facility also contributes to loss due to poor handling, packaging, and road conditions (Alavi et al., 2012).

14.3 TECHNOLOGY TO OVERCOME AGRICULTURAL CHALLENGES

Farmers have very few techniques to support farming. They plow their land using animals, use organic farming, and apply the handpicking technique to get rid of weeds. They used to have animal wastage as a source of fertilizer. To control the pH of land and nutrients, crop rotation used to be common practice (Degu et al., 2019). Agroforestry systems were very common at that time, in which nitrogen is fixed to the soil by the tree itself and their biomass used as fertilizer (Brown et al., 2018). Such ground level practices were never questioned previously, due to less population and hence lower demand. But with a changing situation, the need to modernize all techniques and equipment is required. This need is at the root of such techniques as the development of hydroponic plants. Nevertheless, every system has its own lacuna, which gives momentum to the search for more advances. The increase in demand has led to many wrong practices at an alarming rate. The use of chemical fertilizer has undoubtedly given the solution to many problems but in turn, also raised other problems. It affects not only the soil health but also the animal and plant kingdom (Tomar et al., 2015). Therefore, to satisfy the increasing demands, scientists have introduced many techniques for sustainable agriculture. Organic farming is always preferred and is still a very common practice. It is based on indigenous knowledge and is high in demand. It favors the ecosystem and satisfies the economic demand. The basic mechanism of organic farming is to increase the deficient natural ecology and nourish the edaphic factors. Organic farming preserves soil organic matter, enhances water quality, and conserves biodiversity (Kesavan and Swaminathan, 2007). The microbiota present in soil improves the porosity and permeability of the soil while feeding on the organic matter. Microbiomes solubilize phosphorous that is bound to iron and aluminum and thus make it available for plants. Microbes obtain the carbon source from plants in exchange for phosphorous, nitrogen, sulfur, and micronutrients from the soil. Researchers have manipulated the gene of such microbes using the gene-silencing technique. This technique also has major problems, as the newly introduced organism has to face the challenge of acclimatizing to the new environment (Qiua et al., 2019). Additionally, these microbes may get rejected by the plant microenvironment, which depends on the host immune system and indigenous colonies. So, the introduced microbes may cause dysbiosis and thus ecological integrity (Cook, 1993). Considering as a whole all the challenges and advantages of the different techniques in agriculture, nanobiotechnology has advanced at a flying rate to improve food production. A brief summary of the above challenges and the role of nanotechnology in overcoming them is presented in Figure 14.1.

14.4 NANOTECHNOLOGY TO ENHANCE THE MECHANISM OF PHOTOSYNTHESIS

In 1970, dwarfism of wheat (*Triticum aestivum*) contributed to the Green Revolution. Dwarfism decreased the lodging and prevented the loss of energy in non-yield biomass (Swaminathan, 1993). The other major crops were maize (*Zea mays*), rice (*Oryza* sativa), and soya bean (*Glycine max*) in this row. One of the usual methods of increasing the production of these plants is to prolong their growing season, as they do not get proper ecology to grow in different locations (Vass, 2012). Their growth is largely dependent on the exploitation of solar energy during the process of photosynthesis, to increase its efficiency to absorb the required amount of light with maximum conversion of light energy to chemical energy. In the process of

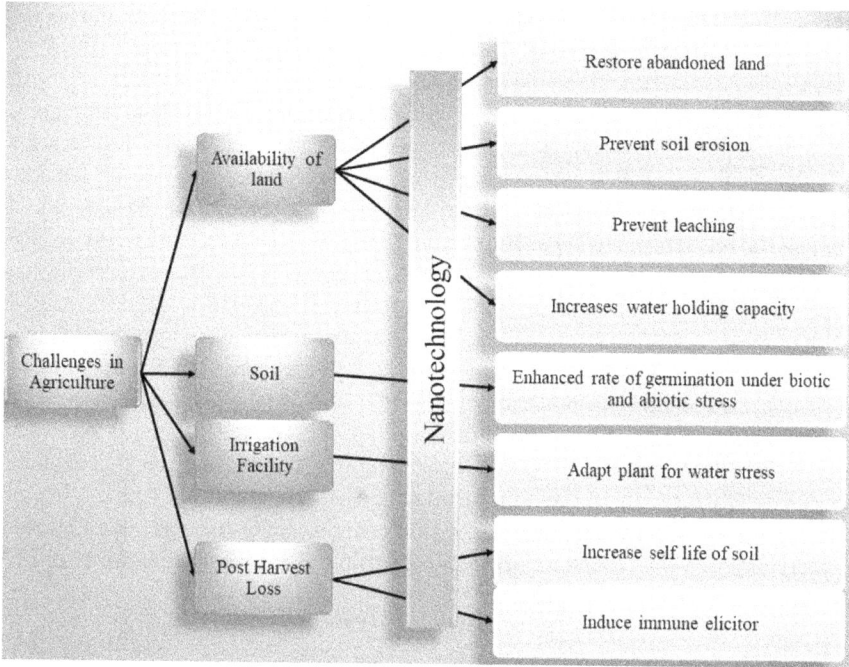

FIGURE 14.1 Major challenges in the field of agriculture and the role of nanotechnology to override these challenges.

photosynthesis, light energy is held by the light harvesting complex to synthesize starch, thus releasing oxygen. We essentially classify plants as C3 and C4, depending upon the requirement of the light source. There are many plants which are short-day, which means they need a smaller amount of light during photoperiodism. The increased intensity of light in such plants may cause photoinhibition due to the release of the reactive oxygen species (Vass, 2012). Despite this, the plant has a mechanism to fight against such deleterious effects by dissipating the spare energy in the form of heat. Researchers are looking to overcome this problem using different light-intensity controlling techniques. Quantum dots (QDs) has always been a choice in this respect because of the adjustable wavelength property (Li et al., 2020). The emerging importance of nanotechnology has enabled us to branch out with the use organic-inorganic nanoparticles, metal nanoparticles, and semiconducting polymer nanoparticles, along with QDs. Increasing the efficiency of photosynthesis is a promising factor for sustainable agriculture when fundamentally under abiotic stress conditions. There are various artificially designed light-harvesting systems which can increase the light absorption by the plants during abiotic conditions. In 1972, Fujishima and Honda succeeded in the photolysis of water with semiconductor electrodes which mimic photosynthesis (Fujishima and Honda, 1972). This technique was largely implemented, and enhanced the efficiency of photosynthesis. Semiconductor-based nanoparticles are more efficient than semiconductor alone in the photolysis of water. The common semiconductor-based nanoparticles are TiO_2-Pt, TiO_2-CdS,

TiO2-CdS-Pt, CdS-Pt, CdS-graphene-Pt, CdSe-MoS$_2$, and MoS$_2$-graphene/TiO$_2$ (Tang et al., 2008; Girginer et al., 2009; Zhou et al., 2012; Maitra et al., 2014).

A single nanoparticle like gold can even behave as antenna molecules when illuminated close to their plasmon resonance. However, the major drawback with gold is that it has lower fluorescence enhancement due to its spherical shape. But changing its geometry to rod or shell can reduce the optical absorption and can lead to surface-enhanced fluorescence (Kuhn et al., 2006). Silver nanoparticles have the capability to increase the mechanism of Light Harvesting Complex (LHC) up to 41% when loaded inside the inverse opal TiO$_2$ film (Li et al., 2012), due to the increased efficiency of power conversion at Schottky junctions in Au:TiO$_2$ (Dong et al., 2015). Silver nanoparticles have comparatively higher fluorescence enhancement due to lower optical absorption. The fluorescence of gold nanoparticles can further be increased by placing two or more metal structures together with small gaps between them (Le Ru et al., 2006). The LHC is based on the mechanism of FRET (Forster Resonance Energy Transfer); it was investigated that maximum fluorescence enhancement can be seen at a distance of 12 nm with metal nanoparticles (Bujak et al., 2012; Olejnik et al., 2012). Using these techniques, the silver nanoparticles have proved efficient in light collection, which led to a 5–20-fold increase in fluorescence activity of the PSI in cyanobacteria (Kim et al., 2011).

14.4.1 Nanotechnology Enhancing Germination Percentage

In sustainable agriculture, the germination percentage (GP), germination index (GI), germination rate (GR), and seedling vigor index (SVI) are matters of concern. Germination is the process in which seeds absorb water by the process of imbibition, and embryonic axis elongates in the form of roots followed by shooting. The processing time and rate of germination highly determine the quality production. There are various metal-based nanoparticles which have accelerated the germination process even under stress conditions. Such metals are Au, Cu, Fe, FeS2, TiO2, Zn, and ZnO. Even carbon-based nanoparticles have shown good results in the process of seed germination (Gopinath et al., 2014). Priming seeds with these metals have not only proved good for seed generation but also in the lengthening of seeds and roots. ZnO nanoparticles have been proven to give better results for promoting seed germination with respect to Ag, CuO, and TiO$_2$ in rice. This may be because Zn is the major constituent of Indole Acetic Acid (IAA) (Pandey et al., 2010), which increases the rate of sprouting or rooting. In an experiment priming rice seedling with nanoparticles, an increase was seen in the vigor index in the sixth day of germination without showing any toxic effect. The enhanced result is due to better imbibition of primed seed in the first 4 hours and later 24 hours of the process. Further, α-amylase activity of the seed was also seen to increase three-fold (Mhakham et al., 2017). After germination, the role of the root begins with the hydrogenase activity that plays a major role in the absorption of nutrition and water. The priming of seeds with metal-based nanoparticles also proved good in this aspect. Nanoparticles not only influence the phenotype, but also affect the molecular level. They increase the expression of genes PIP1 and PIP2. These are the aquaporin genes which are responsible for the increased water uptake by the seed: this is better for the imbibition process (Mhakham et al, 2017).

The role of nanoparticles is not only limited to the enhanced germination rate but also allows for an increase in viability and vigor during storage periods, even under stress conditions.

14.4.2 Nanotechnology Can Restore the Abandoned Land

Nanotechnology has also been shown to play an important role in restoring the quality of soil which has deteriorated due to mining, the deposition of industrial and hospital waste, and erosion. The main challenge to the renewal of such an area is that there are very high degrees of deformity, which include: imbalances in pH, poor water-holding capability, less content of moisture, low Soil Organic Carbon (SOC) content, high level of toxicity, and may also contain radioactive compound.

The restoration of such soil will bring about ecological balance. Zero-valent iron nanoparticles, iron oxides nanoparticles, phosphate-based nanoparticles, iron sulfide nanoparticles, and C nanotubes are some of the major nanoparticles that are useful in solving such issues. Nanocarbons are the most preferred among these. The use of nanocarbon is favorable for the environment; also, it is the nanosize carbon which comes from coconut shells. It has been seen to increase the water-holding capacity of the soil. The oxidized form of nanocarbon has the ability to hold Cu^{2+} and Cd^{2+} that help to restore the nutrition for plants (Nair et al., 2010).

Another approach of nanotechnology regarding reclamation of soil is the use of zeolite crystals. Zeolites are the crystalline cage-like structure having the capability of exchanging ions and selectively holding water molecules. They are the hydrated form of alkali or alkali earth cations of aluminosilicates (Mumpton, 1985). These associated properties of zeolite make it suitable for soil remediation like the removal of toxins, increasing the clay-silt fractions, improving nutrient levels and the water-holding capacity. It has a high cation exchange capacity which makes it suitable to maintain the pH of the soil. The abandoned mine soils may contain certain harmful gases like H_2S and SO_2 and the good absorbing capability of zeolite can remove these gases, favoring the growth of vegetation. Mine soils lack the macronutrient element which is required for the growth of vegetation. These include nitrogen, phosphorous, and potassium. To replace nitrogen in the soil, nitrogen-fixing bacteria can play a major role, but the inappropriate condition of the soil may prevent their growth. Tagging zeolite with NH^{4+} can help to replenish N to the soil. Further NH^{4+} form will prevent nitrogen from volatilization (Lai et al., 1986).

Apatite (e.g., Ca10(PO4)6(OH)2), a rock phosphate, is one major source of phosphorous, a macronutrient for plant growth. Abandoned lands have comparatively high pH, which prevents the dissolution of phosphorous in the soil and the addition of zeolite to apatite with the aim to create a sink for Ca^{2+} maintaining the pH, and leading to bioavaibility of phosphorous to the plants (Liu, 2011; Knox et al., 2003).

A major basic concern, along with managing nutrition and pH of the mining soil, is to prevent them from erosion due to the action of rainfall and wind. Nanoparticles have proven their efficiency with regard to solving this concern. In acidic soil, the addition of Ca-tagged zeolite increases the stability of the soil in wet conditions due to aggregation into large particles. In sodic soil, the calcium type of zeolite exchanges its Ca with the Na present in the soil, leading to aggregation of the soil, which increases soil hydraulic conductivity and reduces clay dispersion (Baalousha, 2009). Another technique to stimulate soil aggregation, in order to prevent its erosion, is to apply polyelectrolyte and nanoparticles which enhance the flocculation rate (Ovenden and Xiao, 2002). Hence, nanoparticles are quite effective in restoring the soil quality, even under adverse conditions.

So, nanotechnology can be applied to improve the quality of seed and soil which are the basic requirements for sustainable agriculture (Figure 14.2 and Table 14.2).

14.4.3 NANOPARTICLES AND SECONDARY METABOLITES IN PLANTS

Secondary metabolites in plants are the organic substances that are not directly involved in normal plant growth and development, but they play vital roles in various signaling cascades like the defense mechanism against microorganisms, etc. Secondary plant products are considered for their vital role in the survival of the plant in its ecosystem, to maintain circadian cycle, protection against plant pathogens, mechanical injury, and other types of biotic and abiotic stress, thereby leading to an increase in plant tolerance level. Experimental evidence has also shown the enhancement of secondary metabolites through treatment using nanoparticles under *in vivo* and *in vitro* conditions (Mishra et al., 2016). Various strategies have been reported about nanoparticles to improve the yield of secondary metabolites in plants under *in vitro* conditions. Their implementation under such *in vivo* conditions is still in progress (Nair et al., 2010; Krishnaraj et al., 2012). Nanoparticles possess the potential to be used as novel effective elicitors in plant biotechnology for the elicitation of secondary metabolite production (Fakruddin et al., 2012). "Elicitors" can be

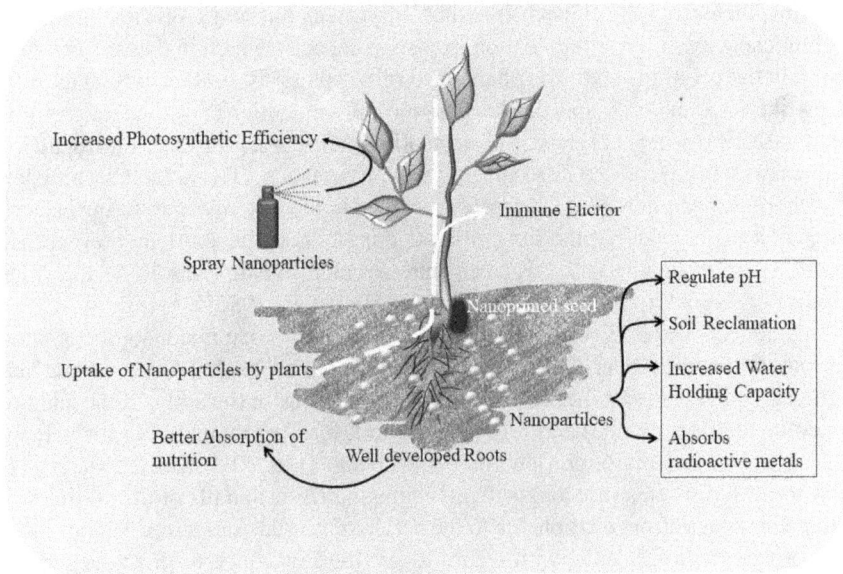

FIGURE 14.2 Application of nanobiotechnology to improve soil quality as well as seed germination.

TABLE 14.2
Different Nanoparticles Used to Improve Plant Variety

Nanoparticles	Plant	Activity	Reference
Nanocarbon	Tomato (*Lycopersicon esculentum*)	Improve seed germination by enhancing water uptake.	Khodakovskaya et al., 2009
Nanocarbon with nitrogen fertilizer	Rice (*Oryza sativa*)	More flowers and fruit. Increased production.	Kammann et al., 2011 Fan et al., 2012
		Preventing leaching by slow release of fertilizers.	Wu, 2013
Nanocarbon-polymer	Ryegrass (*Lolium perenne*)	Removing metal ions from contaminated groundwater and soil.	Khaydarov et al., 2012
Clinoptilolite with NH4+	Sweet corn (*Zea mays*)	Increased efficiency to use nitrogen.	Perrin et al., 1998
	Radish (*Raphanus sativus*)	Slow-release N-fertilizer. Prevent injury by urea.	Lewis et al., 1984
Zinc oxide nanoparticles	Tomato (*Lycopersicon esculentum*)	Increases the efficiency of nutrient delivery system.	Derosa et al., 2010
Zeolite	Maize (*Zea mays*) and barley (*Hordeum vulgare*).	Enhance the growth and yields. Control Zn concentration in plant tissues.	Chlopecka and Adriano, 1996
	Maize (*Zea mays*) and oat (*Avena sativa*)	Enhanced the growth of maize and oat. Decreased the accumulation of Cd, Pb, and Zn.	Knox et al., 2003
	Soybean (*Glycine max*)	Abated the Pb toxicity.	Mahmoodabadi, 2010
Nzvi (Nanoscale Zero-Valent Iron Particles)	*Oryza Sativa*	Reduced the Cd accumulations in rice.	Watanabe et al., 2009
	Changbai Larch (*Larix olgensis*)	Increase in root length.	Bao-Shan et al., 2004
Ceo2	*Coriandrum sativum*	Elongation of root. Effective ROS Scavenger.	Hernandez-Viezcas et al., 2013
	Tomato	Reduced Fusarium wilt Increased chlorophyll contents, catalase, and total fruit weight.	Adisa et al., 2018
Carbon nanotubes	*Arabidopsis*	Increased electron transport rate.	Folbeth et al., 2016
	Cicer arietinum	Enhanced rate of both root and shoot.	Tripati et al., 2011
Ag NPs	Chickpea (*Cicer atietinum*)	Reduce chickpea Fusarium wilt.	Fernández et al., 2016

defined as chemicals or bioagent from various sources which initiate or progress the biosynthesis of specific compounds responsible for physiological and morphological changes in the target living organism, when provided in very low concentrations to a living cell system. Studies have noted the role of nanoparticles as elicitors (Nayak et al., 2010; Zhang et al., 2013; Ghasemi et al., 2015; Yarizade and Ramin, 2015) for enhancing the expression level of genes related to the production of secondary metabolite (Ghasemi et al., 2015). Nanoparticles have successfully offered a new strategy in enhancing the secondary metabolite production. The response and dose of NPs are species-specific. Hence, the selection of an appropriate concentration of nanoparticle is essential for recognizing higher benefits for a target agro-economic trait.

14.4.4 Nanotechnology to Fight against Plant Disease

Technology is not only limited to the external environment of the soil, but also acts as an immune stimulator in plants. Chitosan, a biopolymer, is significantly preferred to synthesize nanocapsule for these purposes because of its biodegradable and non-toxic character. Two of such chitosan-based nanoparticles (CNPs) are Cu-CNPs and Zn-CNPs, that have shown better results than Bavistin fungicide against *Curvularia* leaf spot (CLS) disease of maize (Choudhary et al., 2017, 2019), mycelial growth and blast disease and *Pyricularia grisea* infection in finger millet (Sathiyabama and Manikandan, 2018). Along with this, Cu-CNPs have boosted the growth of maize, resulting in an 11.6% increase in grain weight, whereas Zn-CNPs elevated the defense enzyme activities (Choudhary et al., 2019).

The defense mechanism of plants is regulated by elicitors, a foreign molecule associated with plant pathogens. Encapsulating such particles in chitosan improve their stability and continuous release which boosts the immunity of plant naturally. This technique was successful in controlling the action of *Rhizoctonia solani* against tomato (Qian et al., 2011). *Pseudomonas syringae pv. Syringae* releases an elicitor harpin against *R. solani*, which on encapsulation in the fibrous network of chitosan develops immunity in tomato against the same. The encapsulation of harpin in chitosan (H-CNPs) increases its bioavaibility and has a slow release rate which allows it to last for a longer time, resulting in the boosting of innate immunity of plants against *R. solani* (Nadendla et al., 2018). Biosynthesized silver nanoparticles led to an increase in phenolic group secretion which brings about natural resistance against bacterial infection in tomato. Gold nanoparticles also have the ability to enhance the photosynthetic efficiency which makes tomato plant energetically stable against *Alternaria solani* infection (Quiterio-Gutierrez et al., 2019). Magnesium oxide-associated nanoparticles (MgO-NPs) are also very effective against *A. solani*. The nanoparticles of MgO stimulate the formation of reactive oxygen species (ROS), ethylene (ET), Jasmonic acid (JA), and sialic acid, signaling a pathway along with the accumulation of tyloses and β-1,3-glucanase, making tomato resistant to early blight. Another successful disease resistance inducing nanoparticle is Cerium oxide nanoparticles (CeO$_2$NPs). The application of CeO$_2$NPs to soil or leaves results in an increase in catalase, fruit weight, and chlorophyll contents, even under the severe infection of Fusarium wilt (Adisa et al., 2018).

Pests are another major concern hindering sustainable agriculture. There are many pesticides available but the dose that is required to treat pests sometimes proved toxic to plants, as well as other life. The majority do not even reach the target site because of degradation by microbes, oxygen, and light. The introduction of nanoparticles to the delivery system of pesticides in plants has solved most of these problems. Tagging these pesticides with nanoparticles improves their delivery to the plant system and lengthens their presence for action against pests and even protects them from degradation. Researchers have essentially encapsulated the pesticides to enhance the activeness of pesticides. The entrapment of metalaxyl in mesoporous silica nanoparticles increases its release rate into the soil 7-fold, as compared to free metalaxyl within a tenure of 30 days (Wanyika H, 2013). Leaching is one of the problems which reduces the efficiency of pesticides. A positive result against this was seen by encapsulating tebuconazole and carbendazim in solid lipid nanoparticles (SLNs), making it suitable for commercial utilization (Sarlak et al., 2014). Validamycin is a very promising antibody against *Rhizoctonia solani*. It is proven to be more efficient with better germicidal efficacy when loaded to nano-calcium carbonate. *F. oxysporum* and *Aspergillus parasiticus* are the saprophytic fungi which remain inside the plants as endophytes. To fight against such fungus, the chemical needs to be delivered as far as their residence (Sathiyabama et al., 2019). Most of the fungicides lose their activity on the way because of the degradation of their active components by light, microbes, and oxygen. Carbendazime is one of the fungicides used to prevent the growth of such fungus in the plants. Nanocapsules formed of carbendazim-loaded chitosan protect their active components from degradation, and increase the activity against *F. oxysporum* and *Aspergillus parasiticus* by 20% and 16% respectively (Sandhya et al., 2017). Many of the essential oils have an antimicrobial activity but their volatile nature reduces the efficiency. Capping such essential oils with nanoparticles prevents volatilization, slow release, and controls the time for their release.

14.4.4.1 Nanobiosensors

The era prior to that of nanotechnology was dominated by the use of chemical fertilizers and pesticides, which led to heavy accumulation in plants. The Pesticide Data Program (PDA) annually analyzes fruits, vegetables, and crops to evaluate the presence of pesticides. In the Caribbean area, it has been observed that an increase in the demand of banana has elevated the application of pesticides, generating potential risk factors for health (Llorent-Martinez et al., 2011). Such pesticides can develop a range of health issues, from a minor allergy to more severe consequences like cancer, sterility, and even death ((Nicolopoulou-Stamati et al., 2016). This severity has nailed the safety measures to use pesticides. There are various techniques that enable the detection of pesticide concentration in agro-products. Gas chromatography (GC) and liquid chromatography (LC) are the most preferable classic methods to detect the concentration of pesticides, specifically organophosphorous pesticides (OOPs) (Cabrera et al., 2009; Sharma et al., 2010). But the low detection level of (OOPs) has preferred GC-MS and ultra-high-performance liquid chromatography-tandem mass spectrometry (UHPLC-MS/MS) over GC (Jenkin et al., 2001; Corazza et al., 2019). With advances in technology, new pesticides have also evolved which are thermolabile and polar in nature, make it unmanageable to be detected by GC-MS. These traditional methods need highly skilled handling, and are time-consuming and expensive (Songa and Okonkwo, 2016). Issues with time and health risks have switched the focus to techniques towards those involving a biosensor. This consists of a receiver, a transducer, and a biological compound like antibody and enzyme. The biological compounds have the efficiency to detect a key molecule present in the pesticides (Ali et al., 2017). The integration of nanomaterials with the biosensors and sensors has increased the efficiency of the detection of pesticides and their analytes (Mulchandani et al., 2001). There are various sensors based on nanotechnology which have been developed and proved successful; Some are mentioned in Table 14.3.

These techniques are cost-effective, more precise, and can detect even lower concentrations of pesticides, making it suitable for use as a sensor. Nanosensors have also shown their efficiency in detecting viral and bacterial infection in association with microarray techniques (Rathna et al., 2018).

14.4.5 NANOBIOTECHNOLOGY IN GENE MANIPULATION

Nanotechnology is nowadays considered an important technology that improves traditional agricultural practices and thereby offers sustainable development by reducing the waste of agricultural inputs (Jampilek and Kralova, 2015; Dubey and Mailapalli, 2016). Genetic engineering has always played a pioneering role in every biological field. Genetic manipulation is an important aspect for sustainable agriculture (Joga et al., 2016).

Nanosize has proved more efficient in controlled delivery systems with minimal side effects (Shojaei et al., 2019). The manipulated size of nanoparticles changes their physical and chemical properties, along with an increase in their surface area, which in turn increases the reactivity (Ghormade et al., 2011). It is quite easy to create DNA-tagged nanoparticles through different processes such as absorption, weak bond interaction, capsulation, and entrapment (Nuruzzaman et al., 2016; Pandey, 2018). Nanomaterials also improve the stability of agrochemicals and protect them from degradation and subsequent release into the environment, which increases the effectiveness and thus reduces the quantities of agrochemicals required. Nanobiotechnology also provides opportunities for molecular transporters to modify genes and even produce new organisms (Lyons, 2010). Nanobiotechnology techniques implicate nanoparticles, nanocapsules, and nanofibers to carry foreign DNA and the chemicals that facilitate to modify the target genes. Viral gene delivery vectors face numerous challenges that include limited host range, limited size of inserted genetic material, transportation across the cell membrane, and also trafficking the problem of the nucleus (Ghormade et al., 2011). Recent breakthroughs in nanobiotechnology provide important measures to overcome this challenge by completely replacing the genetic material of one species by another (Torney et al., 2007). It has been reported that silicon dioxide nanoparticles have been devised to deliver DNA fragments/sequences to the target species, such as tobacco and corn plants without any undesirable side effects (Cheng et al., 2016).

TABLE 14.3

Sensors Based on Nanoparticles which Can Analyze the Presence of Pesticides in Agro-Products

Nano-Based Sensor	Pesticides	Minimum Concentration Detected	Sample Detected	Method Used	References
Carbon nanosphere (CB) based screen printed sensor	Carbonfuran	10 µmol/L	Wheat and maize	Cyclic voltammetry	Della Pelle et al., 2018
AuNPs based electrodes reduced with graphene oxide	Carbonfuran	2×10^{-8} mol/L	Vegetables	Molecular polymer printing (MIP)	Tan et. al., 2015
AuNPs based electrode coated with (3-mercaptopropyl)-trimethoxysilane (MPS)	Carbofuran	1.0nM	Fruits	Molecular polymer printing (MIP)	Song et al., 2016
Tin oxide with fluorine (FTO)-AuNPs	Dimethyldithiocar-bamate	0.02mg/Kg	Edible parts of plants	Square wave voltammetry	Bucur et al., 2018
	Diuron	2.6×10^{-7} mol/L	Water	Square wave voltammetry	de Araujo et al., 2018
Guar agarose gum bioconjugated urease with AuNPs	Glyphosphate (GF)	0.5 ppm	Water	Amperometric and potentiometric	Vaghela et al., 2018
CuNPs based genosonsors	Dithiocarbamates	8.4–489.3 ng/ml	Tomato and mango	Spectrophotometry	Ghoto et al., 2019
Citrate trapped-AuNPs	Organophosphate pesticides	118ng/ml	Rice	Pattern recognition techniques, including hierarchical cluster analysis (HCA) and linear discriminant analysis (LDA)	Fahimi-Kashani and Hormozi-Nezhad, 2016
Pt-Au nanoparticles electrodeposited on multi-walled carbon nanoparticles (MWNTs)	Organophosphate pesticides	0.08 nmol/L	Cabbage	Enzymatic Reaction	Miao et al., 2016

A nanoparticle (NP)-assisted delivery system is used to create novel crop through genetic manipulation can be achieved through gene-gun technology by coating the DNA with NPs (Lyons, 2010). The use of chitosan nanoparticle-entrapped SiRNA delivery vehicles has also provided scope for crop improvement (Zhang et al., 2010). Similarly, advances in the nanomaterial-based specific delivery of CRISPR/Cas9 single guide RNA (sgRNA) is very promising. It is comprised of CRISPR repeat-spacer arrays and Cas proteins: An RNA-directed defense system in prokaryotes. It has successfully been used for genome editing in plants (Miller et al., 2017). Still, there remain many challenges in this field, including delivery efficiency (Shang et al., 2019).

14.5 POTENTIAL RISKS OF NANOBIOTECHNOLOGY

The development of novel nanotechnological tools has improved the quality of land, availed the plants of proper absorption of minerals and water, manipulated many genetic systems in plants, upregulated the immunological response, and minimized the use of chemical fertilizers, herbicides, and pesticides. The smaller size of nanoparticles has exposed more of its surface area, thus increasing its reactivity. Nanotechnology is also considered a green technology, as it is synthesized from renewable resources. It emits less greenhouse gases, is cost-effective, reduces waste, and is energy efficient (Prasad et al., 2014, 2016). In animals, nanoparticles are generally considered foreign particles, and can be eliminated from the immune system. But sometimes, due to its nanosize, it can easily surpass from interacting with immune cells. In the embryo it can cause lethal effects (Exbrayat et al., 2015). The charges on nanoparticles can disrupt the cell membrane (Hondroulish et al., 2014). Sometimes, NPs can absorb certain particles which are harmful to organisms like CNT nanosponges which contain iron and sulfur. They are efficient in absorbing water contaminants like fertilizers, oil, and pesticides which can penetrate human cells, thereby endangering life (Porter et al., 2007). Various metal-oxide-based NPs are used in plants as antimicrobial agents. The release of such NPs from plant waste to the environment can also target other useful microbes and organisms like fish, nematodes, and other cell cultures (Bondarenko et al., 2013). A major concern is the need to pay for the edible parts of plants, as the use of high concentration nanosilver on such sites can generate free radicals, which lead to the damage of DNA. TiO_2 based nanoparticles are mostly used by the researcher to increase the efficiency of photosynthesis, and increase the bioavailability of nutrition to the plants. It does not cause any harmful effect on plants but increase in its use may contaminate the soil. TiO_2 has the ability to generate free radicals from water in the presence of sunlight, provoking DNA damage and apoptosis in earthworms (Lapied et al., 2011). EFSA (European Food Safety Authority) proposed the assessment of the passage of nanoparticles in the food chain (EFSA Scientific Committee, 2011). Nanotechnology, like other techniques, also needs to create a balance with social, economic, and ethical challenges. There is still a need to know more about nanoparticles. Such lack of information means the adoption of maximum precautions to be taken. While engineering with nanoparticles, it is always advisable to use gloves, mask, and other safety equipment. Various research on mice suggested that nanostructured particles are more hazardous than the macroparticles (Tran et al., 1999, 2000; Barlow et al., 2005; Duffin et al., 2007).

Nanoparticles can also be applied to plants as spray, which creates the impression of its presence in the air. Through air, it can enter our lungs (ICRP, 1994) through breathing and from lungs to other organs via the bloodstream (Nemmar et al., 2002). A study of the impact of nanoparticles on pigs states that it can easily penetrate through the skin through passive diffusion, and be localized in the dermal as well as the epidermal layer (Ryman-Rasmussen et al., 2006).

The FDA (Food and Drug Administration) properly regulate the product, as well as their waste impact concerning both biotic and abiotic factors (FDA, 2014).

14.6 FUTURE PROSPECTS

Agriculture constitutes an important part of the employment sector in most countries, along with its impact on economic development. Recession never affects the field of agriculture, rather demand increases day by day. But the basic requirement for agriculture, that of land, is involved in industrial and residential purposes. The challenge of this era is to feed the increasing population with limited resources. Plant products are not only limited for food, but are also required for commercial and medicinal purposes. Plant tissue culture has multiplied the number of plants, but land problems persist. Nanotechnology has introduced a promising field, which is of maximum benefit with minimal utilization of resources. Nanotechnology enables the utilization of abandoned land. It has formulated the reclamation technique, which seeks to utilize land which is left vacant for agriculture. The technique can even make use of aged seeds which have lost their germination quality. Nanotechnology increases the shelf life of seeds preventing postharvest loss. The technique is not limited to the storage system, rather it can be used to boost the innate immunity of the plants, which will naturally increase the quality and yield. The technique has also introduced a sensor technique, leading to early detection of disease or contamination, preventing heavy loss of resources. In the case of drought-prone areas, nanotechnology can be used to increase the water-holding capacity of plants. The energy-quenching property of nanoparticles can be utilized to maximize the absorption of sun rays to increase the efficiency of photosynthesis. Such techniques can be used to grow long-day plants in land which receives limited sun energy. The best of nanotechnology is its size which can be used far below its toxic level with maximum proficiency. Its size also makes it compatible for entering plant cells easily, which can also pass to their daughter cell, leaving long term effects. The technique is proved best for the drug delivery drug in plants, which can be used to help them resist herbicides and pesticides.

In short, we can say that nanotechnology can meet the general requirements of agriculture which are to increase the productivity and yield, lessen the generation of waste products, create disease-resistant plants and enable whole year production. Nanotechnology may be evidence of the instigation of another green revolution. There are still unanswered facts about the role of nanotechnology for sustainable agriculture, making it a major area of interest for research. There is still the requirement to formulate different types of nanoparticles varying in size and shape, which is its uniqueness and a major factor to maximize its utilization. This is well proven by its demand, as well as its application in every field, including mechanical, quantum science, medical, electrical, biological, agriculture, and many more.

14.7 SUMMARY

Nanotechnology is a breakthrough that has the potential to improve the quality as well as quantity of sustainable agriculture. Recent advances in nanoscience have had a great impact on agricultural practices and the food industry. The afore-mentioned milestones can be achieved using various strategies that could help in the development of new-generation pesticides, minimize the use of fertilizers, control pests and diseases, help in the understanding of the mechanism of host-parasite interactions, remove contaminants from soil and water bodies, improve the self-life of vegetables and flowers, water management, regenerating soil fertility, reclamation of salt-affected soils, checking acidification of irrigated lands, and the stabilization of erosion-prone surfaces. This article has attempted to highlight some of the most promising and important applications of nanotechnology in agriculture. We have also focused on the strategies that are promising for the advancement of scientific and technological knowledge that are used presently. Nanotechnology provides new agrochemical agents as well as delivery mechanisms that could improve crop productivity, together with minimizing the use of pesticides. Nanotechnology can enhance agricultural production through different approaches such as improving agrochemicals, nanobiosensors, nanodevices that could genetically modify crops, nanocomposites, hydroponics, organic farming, health, and breeding of animal/poultry, and through postharvest management.

In the field of agriculture, nanotechnology has emerged as a boon to raise crop production with quality enrichment. The emergence of engineered nanomaterials and their actions within the frame of sustainable agriculture have revolutionized the world agriculture situation, elevating it to a better position. In sustainable agriculture, nanobiotechnology provides strong assurance of better management and conservation of crops for the future.

REFERENCES

Abedini, A., A. R. Daud, M. A. A. Hamid, N. K. Othman, and E. Saion. 2013. A review on radiation-induced nucleation and growth of colloidal metallic nanoparticles. *Nanoscale Research Letters* 8(474): 1–10. doi: 10.1186/1556-276X-8-474.

Adisa, I. O., V. L. R. Pullagurala, S. Rawat et al. 2018. Role of cerium compounds in fusarium wilt suppression and growth enhancement in tomato (Solanum Lycopersicum). *Journal of Agricultural and Food Chemistry* 66(24): 5959–5970. doi: 10.1021/acs.jafc.8b01345.

Alavi, H. R., R. L. Clarete, A. M. Htenas, R. J. Kopicki, and A. W. Shepherd. 2012. Trusting trade and the private Sector for Food Security in Southeast Asia. *World Bank Publications: Directions in Development; Trade Washington, DC, USA, (2012).* http://documents.worldbank.org/curated/en/741701468170978381/Trusting-trade-and-the-private-sector-for-food-security-in-Southeast-Asia.

Alemu, G. T., Z. B. Ayele, and A. A. Berhanu. 2017. Effects of land fragmentation on productivity in Northwestern Ethiopia. *Advances in Agriculture* 2017. Article ID 4509605.

Ali, J., J. Najeeb, M. A. Ali, M. F. Aslam, and A. Raza. 2017. Biosensors: Their fundamentals, designs, types and most recent impactful applications: A review. *Journal of Biosensors and Bioelectronics* 8: 1–9. doi: 10.4172/2155-6210.1000235.

Baalousha, M. 2009. Aggregation and disaggregation of iron oxide nanoparticles: Influence of particle concentration, pH and natural organic matter. *Science of the Total Environment* 407(6): 2093–2101. doi: 10.1016/j.scitotenv.2008.11.022.

Bao-Shan, L., L. Chun-hui, F. Li-jun, Q. Shu-chun, and Y. Min. 2004. Effect of TMS (nano-structured silicon dioxide) on growth of Changbai Larch seedlings. *Journal of Forestry Research* 15(2): 138–140.

Barlow, P. G., A. C. Clouter-Baker, K. Donaldson, J. MacCallum, and V. Stone. 2005. Carbon black nanoparticles induce type II epithelial cells to release chemotaxins for alveolar macrophages. *Particle and Fibre Toxicology* 2(11). doi: 10.1186/1743-8977-2-11.

Bhushani, A., and C. Anandharamakrishnan. 2014. Electrospinning and electrospraying techniques: Potential food based applications. *Trends in Food Science and Technology* 38(1): 21–33. doi: 10.1016/j.tifs.2014.03.004.

Bondarenko, O., K. Juganson, A. Ivask, K. Kasemets, M. Mortimer, and A. Kahru. 2013. Toxicity of Ag, CuO and ZnO nanoparticles to selected environmentally relevant test organisms and mammalian cells *in vitro*: A critical review. *Archives of Toxicology* 87(7): 1181–1200. doi: 10.1007/s00204-013-1079-4.

Bourne, M. 1977. *Post Harvest Food Losses—The Neglected Dimension in Increasing the World Food Supply*. Department of Food Science and Technology, Cornell University, Ithaca, NY.

Brinker, C. J., and G. W. Scherer. 1990. The physics and chemistry of sol-gel. In: *Sol Gel Science*. Processing, Brinker, C. J., Scherer, G. W., Ed., Academic Press.

Brown, S. E., D. C. Miller, P. J. Ordonez, and K. Baylis. 2018. Evidence for the impacts of agroforestry on agricultural productivity, ecosystem services, and human well-being in high-income countries: A systematic map protocol. *Environmental Evidence* 7(1): 24.

Bucur, B., F. D. Munteanu, J. L. Marty, and A. Vasilescu. 2018. Advances in enzyme-based biosensors for pesticide detection. *Biosensors* 8(2): 27. doi: 10.3390/bios8020027.

Bujak, L., T. Brotosudarmo, N. Czechowski et al. 2012. Spectral dependence of fluorescence enhancement in LH2-Au nanoparticle hybrid nanostructures. *Acta Physiologica Polonica* 122: 252.

Cabrera, J. C., E. M. Hernández-Suárez, V. G. Saúco, J. Hernández-Borges, and M. Rodríguez-Delgado. 2009. Ángel Analysis of pesticide residues in bananas harvested in the Canary Islands (Spain). *Food Chemistry* 113(1): 313–319. doi: 10.1016/j.foodchem.2008.07.042.

Cheng, H. N., K. T. Klasson, T. Asakura, and Q. Wu. 2016. Nanotechnology in agriculture. In: *Nanotechnology: Delivering on the Promise*, Cheng, H. N., Doemeny, L., Geraci, C. L., Schmidt, D. G., Eds., ACS, Washington, DC, Vol. 2, pp. 233–242.

Chlopecka, A., and D. C. Adriano. 1996. Mimicked in-situ stabilization of metals in a cropped soil: Bioavailability and chemical form of zinc. *Environmental Science and Technology* 30(11): 3294–3303. doi: 10.1021/es960072j.

Choudhary, R. C., R. V. Kumaraswamy, S. Kumari et al. 2017. Cu-chitosan nanoparticle boost defense responses and plant growth in maize (Zea mays L.). *Scientific Reports* 7(1): 9754. doi: 10.1038/s41598-017-08571-0.

Choudhary, R. C., R. V. Kumaraswamy, S. Kumari et al. 2019. Zinc encapsulated chitosan nanoparticle to promote maize crop yield. *International Journal of Biological Macromolecules* 127: 126–135. doi: 10.1016/j.ijbiomac.2018.12.274.

Cook, R. J. 1993. Making greater use of introduced microorganisms for biological control of plant pathogens. *Annual Review of Phytopathology* 31: 53–80. doi: 10.1146/annurev.py.31.090193.000413.

Corazza, G., J. Merib, S. D. Carmo, L. Mendes, and E. Carasek. 2019. Assessment of a fully optimized DPX-based procedure for the multiclass determination of pesticides in drinking water using high-performance liquid chromatography with diode array detection. *Journal of Brazilian Chemical Society* 30: 1211–1221. doi: 10.21577/0103-5053.20190016.

D'Amato, R., M. Falconieri, S. Gagliardi, E. Popovici, E. Serra, G. Terranova, and E. Borsella. 2013. Synthesis of ceramic nanoparticles by laser pyrolysis: From research to applications. *Journal of Analytical and Applied Pyrolysis* 104: 461–469.

de Araújo, G. M., and F. R. Simões. 2018. Self-assembled films based on polypyrrole and carbon nanotubes composites for the determination of diuron pesticide. *Journal of Solid State Electrochemistry* 22(5): 1439–1448. doi: 10.1007/s10008-017-3807-9.

Degu, M., A. Melese, and W. Tena. 2019. Effects of soil conservation practice and crop rotation on selected soil physicochemical properties: The case of Dembecha District, Northwestern Ethiopia. *Applied and Environmental Soil Science* 2019. Article ID 6910879.

Della, P. F., C. Angelini, M. Sergi, M. Del Carlo, A. Pepe, and D. Compagnone. 2018. Nano carbon black-based screen printed sensor for carbofuran, isoprocarb, carbaryl and fenobucarb detection: Application to grain samples. *Talanta* 186: 389–396. doi: 10.1016/j.talanta.2018.04.082.

Derosa, M. C., C. Monreal, M. Schnitzer, R. Walsh, and Y. Sultan. 2010. Nanotechnology in fertilizers. *Nature Nanotechnology* 5(2): 91. doi: 10.1038/nnano.2010.2.

Dong, H., Z. Wu, A. El-Shafei et al. 2015. Ag-encapsulated Au plasmonic nanorods for enhanced dye sensitized solar cell performance. *Journal of Materials Chemistry A* 3(8): 4659–4668.

Dubey, A., and D. R. Mailapalli. 2016. Nanofertilisers, nanopesticides, nanosensors of pest and nanotoxicity in agriculture. In: *Sustainable Agriculture Reviews*, Lichtfouse, E., Ed., Springer, Cham, Switzerland, vol. 19, pp. 307–330.

Duffin, R., L. Tran, D. Brown, V. Stone, and K. Donaldson. 2007. Proinflammogenic effects of low-toxicity and metal nanoparticles in vivo and in vitro: Highlighting the role of particle surface area and surface reactivity. *Inhalation Toxicology* 19(10): 849–856. doi: 10.1080/08958370701479323.

EFSA Scientific Committee. 2011. Scientific opinion on guidance on the risk assessment of the application of nanoscience and nanotechnologies in the food and feed chain. *EFSA Journal* 9(5): 2140.

Exbrayat, J. M., E. N. Moudilou, and E. Lapied. 2015. Harmful effect of nanoparticles on animals. *Advanced Materials and Nanotechnology for Sustainable Energy Development.* Article ID 861092. doi: 10.1155/2015/861092.

Fahimi-Kashani, N., and M. R. Hormozi-Nezhad. 2016. Gold nanoparticle-based colorimetric sensor array for discrimination of organophosphate pesticides. *Analytical Chemistry* 88(16): 8099–8106. doi: 10.1021/acs.analchem.6b01616.

Fakruddin, M. D., Z. Hossain, and H. Afroz. 2012. Prospects and applications of nanobiotechnology: A medical perspective. *Journal of Nanobiotechnology* 10: 1–8.

Fan, L., Y. Wang, X. Shao et al. 2012. Effects of combined nitrogen fertilizer and nano-carbon application on yield and nitrogen use of rice grown on saline-alkali soil. *Journal of Food, Agriculture and Environment* 10: 558–562.

FDA. 2014. Guidance for industry considering whether an FDA-regulated product involves the application of nanotechnology. https://www.fda.gov/downloads/RegulatoryInformation/Guidances/UCM401695.pdf.

Fernández, J. G., M. A. Fernández-Baldo, E. Berni, G. Camí, N. Durán, J. Raba, and M. I. Sanz. 2016. Production of silver nanoparticles using yeasts and evaluation of their antifungal activity against phytopathogenic fungi. *Process Biochemistry* 51(9): 1306–1313.

Folberth, C., R. Skalský, E. Moltchanova, J. Balkovič, L. B. Azevedo, M. Obersteiner, and M. van der Velde. 2016. Uncertainty in soil data can outweigh climate impact signals in global crop yield simulations. *Nature Communications* 7: 11872.

Fujishima, A., and K. Honda. 1972. Electrochemical photolysis of water at a semiconductor electrode. *Nature* 238(5358): 37–38.

Ghasemi, B., R. Hosseini, and F. D. Nayeri. 2015. Effects of cobalt nanoparticles on artemisinin production and gene expression in Artemisia annua. *Turkish Journal of Botany* 39: 769–777. doi: 10.3906/bot-1410-9.

Ghormade, V., M. V. Deshpande, and K. M. Paknikar. 2011. Perspectives for nano-biotechnology enabled protection and nutrition of plants. *Biotechnology Advances* 29(6):
792–803.

Ghoto, S. A., M. Y. Khuhawar, T. M. Jahangir, and J. U. D. Mangi. Applications of copper nanoparticles for colorimetric detection of dithiocarbamate pesticides. *Journal of
Nanostructure in Chemistry* 9(2): 77–93. doi: 10.1007/s40097-019-0299-4.

Girginer, B., G. Galli, E. Chiellini, and N. Bicak. 2009. Preparation of stable CdS nanoparticles in aqueous medium and their hydrogen generation efficiencies in photolysis of
water. *International Journey of Hydrogen Energy* 34(3): 1176–1184. doi: 10.1016/j.
ijhydene.2008.10.086.

Gopinath, K., S. Gowri, V. Karthika, and A. Arumugam. 2014. Green synthesis of gold
nanoparticles from fruit extract of *Terminalia arjuna*, for the enhanced seed germination activity of *Gloriosa superb*. *Journal of Nanostructure in Chemistry* 4(3): 115. doi:
10.1007/s40097-014-0115-0.

Grover, Davinder, and J. M. Singh. 2013. Post-harvest losses in wheat crop in Punjab: Past and
present. *Agricultural Economics Research Review* 26: 293–297.

Hernandez-Viezcas, J. A., H. Castillo-Michel, J. C. Andrews et al. 2013. In situ synchroton
X-ray fluorescence mapping and speciation of CeO_2 and ZnO nanoparticles in soil cultivation soybean (glycin max). *ACS Nano* 7(2): 1415–1423.

Hondroulis, E., J. Nelson, and L. Chen-Zhong. 2014. Biomarker analysis for nanotoxicology.
In: *Biomarkers in Toxicology*, Grupta, D., Ed., Elsevier, pp. 689–695.

ICRA. 2015. Double-Digit Growth for Indian Seed Industry: New Delhi August 17.

ICRP. 1994. *Human Respiratory Tract Model for Radiological Protection*. Pergamon,
Elsevier Science Ltd., Oxford, England. *International Commission on Radiological
Protection*, Publication No.:66.

Iravani, S. 2011. Green synthesis of metal nanoparticles using plants. *Green Chemistry*
13(10): 2638–2650. doi: 10.1039/C1GC15386B.

Jampilek, J., and K. Kral'ova. 2015. Application of nanotechnology in agriculture and food
industry, its prospects and risks. *Ecological Chemistry and Engineering S-Chemia i
Inzynieriaekologiczna S* 22: 321–361.

Jenkins, A. L., R. Yin, and J. L. Jensen. 2001. Molecularly imprinted polymer sensors for
pesticide and insecticide detection in water. *Analyst* 126(6): 798–802. doi: 10.1039/
b008853f.

Joga, M. R., M. J. Zotti, G. Smagghe, and O. Christiaens. 2016. RNAi efficiency, systemic
properties, and novel delivery methods for pest insect control: What we know so far.
Frontiers in Physiology 7: 553. doi: 10.3389/fphys.2016.00553.

Kammann, C., S. Linsel, J. W. Gobling, and H. W. Koyro. 2011. Influence of biochar on
drought tolerance of Chenopodium quinoa: Wild and on soil-plant relations. *Plant and
Soil* 9: 115–123. doi: 10.1007/s11104-011-0771-5.

Karak, N. 2009. *Fundamentals of Polymers: Raw Materials to Finish Products*. Prentice-
Hall of India Pvt Ltd, New Delhi, 1st ed.

Kesavan, P. C., and M. S. Swaminathan. 2007. Strategies and models for agricultural sustainability in developing Asian Countries. *Philosophical Transactions of the Royal Society*
363(1492): 877–891. doi: 10.1098/rstb.2007.2189.

Khan, M. A. 2010. *Post Harvest Losses of RICE*, Khan S. L., Ed., Trade Development
Authority of Pakistan, Karachi, Pakistan.

Khaydarov R. A., R. R. Khaydarov, O. Gapurova, and R. Malish. 2012 Remediation of
metal ion-contaminated groundwater and soil using nanocarbon-polymer composition. In: Quercia, F., Vidojevic, D. (eds), *Clean Soil and Safe Water*. NATO Science
for Peace and Security Series C: Environmental Security. Springer, Dordrecht, pp.
167–182.

Khodakovskaya, M., E. Dervishi, M. Mahmood, Y. Xu, Z. Li, F. Watanabe, and A. S. Biris. 2009. Carbon nanotubes are able to penetrate plant seed coat and dramatically affect seed germination and plant growth. *ACS Nano* 3(10): 3221–3227. doi: 10.1021/nn900887m.

Kim, I., S. Bender, J. Hranisavljevic, L. M. Utschig, L. Huang, G. P. Wiederrecht, and D. M. Tiede. 2011. Metal nanoparticle PlasmonEnhanced light-harvesting in a photosystem I thin film. *Nano Letters* 11(8): 3091–3098. doi: 10.1021/nl2010109.

Knox, A. S., D. I. Kaplan, D. C. Adriano, T. G. Hinton, and M. D. Wilson. 2003. Apatite and phillipsite as sequestering agents for metals and radionuclides. *Journal of Environmental Quality* 32(2): 515–525. doi: 10.2134/jeq2003.0515.

Krishnaraj, C., E. G. Jagan, R. Ramachandran, S. M. Abirami, N. Mohan, and P. T. Kalaichelvan. 2012. Effect of biologically synthesized silver nanoparticles on Bacopa monnieri (Linn.) Wettst. Plant growth metabolism. *Process Biochemistry* 47(4): 651–658.

Kühn, S., U. Håkanson, U. Rogobete, and U. Sandoghdar. 2006. Enhancement of single-molecule fluorescence using a gold nanoparticle as an optical nanoantenna. *Physical Review Letters* 97(1): 017402.

Lai, T. M., and D. D. Eberl. 1986. Controlled and renewable release of phosphorous in soils from mixtures of phosphate rock and NH4-exchanged clinoptilolite. *Zeolites* 6(2): 129–132.

Lapied, E., J. Y. Nahmani, E. Moudilou et al. 2011. Ecotoxicological effects of an aged TiO_2 nanocomposite measured as apoptosis in the anecic earthworm Lumbricusterrestris after exposure through water, food and soil. *Environment International* 37(6): 1105–1110. doi: 10.1016/j.envint.2011.01.009.

Le Ru, E., P. G. Etchegoin, and M. Meye. 2006. Enhancement factor distribution around a single surface-enhanced Raman scattering hot spot and its relation to single molecule. *The Journal of Chemical Physics* 125(20): 204701.

Lewis, M. D., I. F. D. Moore, and K. L. Goldsberry. 1984. Ammonium-exchanged clinoptilo-lite and granulated clinoptilolite with urea as nitrogen fertilizers. In: *Zeo-Agriculture: Use of Natural Zeolites in Agriculture and Aquaculture*, Pond, W. G., Mumpton, F. A., Eds., Westview Press, Boulder, CO, pp. 105–111.

Li Yanjuan, W. L., H. Zhang, Y. Liu, L. Ma, B. Lei, and B. Lei. 2020. Amplified light harvest-ing for enhancing Italian lettuce photosynthesis using water soluble silicon quantum dots as artificial antennas. *Nanoscale* 12(1): 155–166. doi: 10.1039/C9NR08187A.

Li, H., X. Hu, W. Hong et al. 2012. Photonic crystal coupled plasmonic nanoparticle array for resonant enhancement of light harvesting and power conversion. *Physical Chemistry Chemical Physics: PCCP* 14(41): 14334–14339. doi: 10.1039/C2CP42438J.

Lin, B., U. Sundararaj, and P. Potschke. 2006. Mater. Melt mixing of polycarbonate with multi-walled carbon nanotubes in miniature mixers. *Macromolecular Molecular and Engineering* 291: 227–238.

Liu, R. 2011. In-situ lead remediation in a shoot-range soil using stabilized apatite nanoparti-cles. In: *Proceedings of the 85th ACS Colloid and Surface Science Symposium*, McGill University, Montreal, Canada.

Llorent-Martínez, E. J., P. Ortega-Barrales, M. F. D. Córdova, A. Ruiz-Medina, and O. B. Pilar. 2011. Trends in flow-based analytical methods applied to pesticide detection: A review. *Analytica Chimica Acta* 684: 30–39. doi: 10.1016/j.aca.2010.10.036.

Lyons, K. 2010. Nanotechnology: Transforming food and the environment. *Food First Backgr* 16: 1–4.

Madhusudhan, L. Agriculture role on Indian economy. *Business and Economics Journal* 6(4): 176. doi: 10.4172/2151-6219.1000176.

Mahakham, W., A. K. Sarmah, S. Maensiri, and P. Theerakulpisut. 2017. Nanopriming tech-nology for enhancing germination and starch metabolism of aged rice seeds using phytosynthesized silver nanoparticles. *Scientific Reports* 7(1): 8263. doi: 10.1038/s41598-017-08669-5.

Mahmoodabadi, M. R. 2010. Experimental study on the effects of natural zeolite on lead toxicity, growth, nodulation, and chemical composition of soybean. *Communications in Soil Science and Plant Analysis* 41(16): 1896–1902. doi: 10.1080/00103624.2010.495801.

Maicu, M. 2014. Synthesis and deposition of metal nanoparticles by gas condensation process. *Journal of Vacuum Science and Technology. Part A* 32: 02B113. doi: 10.1116/1.4859260.

Maitra, U., S. R. Lingampalli, and C. N. R. Rao. 2014. Artificial photosynthesis and the splitting of water to generate hydrogen. *Current Science* 106: 518–527.

Miao, S. S., M. S. Wu, L. Y. Ma, X. J. He, and H. Yang. 2016. Electrochemiluminescence biosensor for determination of organophosphorouspesticidesbasedonbimetallicPt-Au/multi-walledcarbonnanotubesmodifiedelectrode. *Talanta* 158: 142–151. doi: 10.1016/j.talanta.2016.05.030.

Miller, J. B., S. Zhang, P. Kos et al. 2017. Non-viral CRISPR/Cas gene editing in vitro and in vivo enabled by synthetic nanoparticle co-delivery of Cas9 mRNA and sgRNA. *Angewandte Chemie International Edition* 56(4): 1059–1063.

Mishra, P., P. K. Shukla, K. Pramanik, S. Gautam, and C. Kole. 2016. Nanotechnology for crop improvement. In: *Plant Nanotechnology*, Kole, C., Kumar, D., Khodakovskaya, M., Ed., Springer, Cham, pp. 219–256. doi: 10.1007/978-3-319-42154-4 9.

Mulchandani, A., W. Chen, P. Mulchandani, J. Wang, and K. R. Rogers. 2001. Biosensors for direct determination of organophosphate pesticides. *Biosensors and Bioelectronics* 16(4–5): 225–230. doi: 10.1016/S0956-5663(01)00126-9.

Mumpton, F. A. 1985. Using zeolites in agriculture. In: *Innovative Biological Technologies for Lesser Developed Countries, Congress of the United States, Office of Technology Assessment*, Washington, DC. OTA-BP-F-29.

Nadendla, S. R., T. S. Rani, P. R. Vaikuntapu, R. R. Maddu, and A. R. Podile. 2018. Harpin$_{Pss}$ encapsulation in chitosan nanoparticles for improved bioavailability and disease resistance in tomato. *Carbohydrate Polymers* 199: 11–19. doi: 10.1016/j.carbpol.2018.06.094.

Nair, R., S. H. Varghese, B. G. Nair, T. Maekawa, Y. Yoshida, and D. S. Kumar. 2010. Nanoparticulate material delivery to plants. *Plant Science* 179(3): 154–163. doi: 10.1016/j.plantsci.2010.04.012.

Nayak, P., A. S. Patnakar, B. Madhusudan, M. R. S. Rayasa, and E. B. Souto. 2010. Artemether loaded lipid nanoparticles produced by modified thin film hydration: Pharmacokinetics, toxicological and in vivo antimalarial activity. *European Journal of Pharmaceutical Sciences* 40(5): 448–455. doi: 10.1016/j.ejps.2010.05.007.

Nemmar, A., P. H. M. Hoet, B. Vanquickenborne et al. 2002. Passage of inhaled particles into the blood circulation in humans. *Circulation* 105(4): 411–414. doi: 10.1161/hc0402.104118.

Nicolopoulou-Stamati, P., S. Maipas, C. Kotampasi, P. Stamatis, and L. Hens. 2016. Chemical pesticides and human health: The urgent need for a new concept in agriculture. *Frontiers in Public Health* 4: 231. doi: 10.3389/fpubh.2016.00148.

Nuruzzaman, M., M. M. Rahman, Y. J. Liu, and R. Naidu. 2016. Nanoencapsulation, nanoguard for pesticides: A new window for safe application. *Journal of Agriculture and Food Chemistry* 64(7): 1447–1483.

Okuyama, K., M. Abdullah, I. W. Lenggoro, and F. Iskandar. 2006. Preparation of functional nanostructured particles by spray drying. *Advanced Powder Technology* 17(6): 587–611.

Olejnik, M., B. Krajnik, D. Kowalska, G. Lin, and S., Mackowski. 2012. Spectroscopic studies of plasmon coupling between photosynthetic complexes and metallic quantum dots. *Journal of Physics: Condensed Matter* 25: 194103.

Ovenden, C., and H. Xiao. 2002. Flocculation behaviour and mechanisms of cationic inorganic microparticle/polymer systems. *Colloids and Surfaces A: Physicochemical and Engineering Aspects* 197(1–3): 225–234.

Pandey, A. C., S. S. Sanjay, and R. S. Yadav. 2010. Application of ZnO nanoparticles in influencing the growth rate of *Cicer arietinum*. *Journal of Experimental Nanoscience* 5(6): 488–497. doi: 10.1080/17458081003649648.

Pandey, G. 2018. Challenges and future prospects of agri-nanotechnology for sustainable agriculture in India. *Environmental Technology and Innovation* 11: 299–307.

Perrin, T. S., D. T. Drost, J. Boettinger, and J. M. Norton. 1998. Ammonium-loaded clinoptilolite: A slow-release nitrogen fertilizer for sweet corn. *Journal of Plant Nutrition* 21(3): 515–530. doi: 10.1080/01904169809365421.

Porter, A. E., M. Gass, K. Muller, J. N. Skepper, P. A. Midgley, and M. Welland. 2007. Direct imaging of single-walled carbon nanotubes in cells. *Nature Nanotechnology* 2(11): 713–717. doi: 10.1038/nnano.2007.347.

Prasad, R., V. Kumar, and K. S. Prasad. 2014. Nanotechnology in sustainable agriculture: Present concerns and future aspects. *African Journal of Biotechnology* 13(6): 705–713. doi: 10.5897/AJBX2013.13554.

Prasad, R., R. Pandey, and I. Barman. 2016. Engineering tailored nanoparticles with microbes: Quo vadis? *WIREs Nanomedicine and Nanobiotechnology* 8(2): 316–330. doi: 10.1002/wnan.1363.

Qian, K., T. Shi, T. Tang, S. Zhang, X. Liu, and Y. Cao. 2011. Preparation and characterization of nano-sized calcium carbonate as controlled release pesticide carrier for validamycin against Rhizoctonia solani. *Microchimica Acta* 173(1–2): 51–57. doi: 10.1007/s00604-010-0523-x.

Qiua, Z., E. Egidi, H. Liu, S. Kaur, and B. K. Singh. 2019. New frontiers in agriculture productivity: Optimised microbial inoculants and in situ microbiome engineering. *Biotechnology Advances* 37(6):107371. doi: 10.1016/j.biotechadv.2019.03.010.

Quiterio-Gutierrez, T., H. Ortega-Ortiz, G. Cadenas-Pliego et al. 2019. The application of selenium and copper nanoparticles modifies the biochemical responses of tomato plants under stress by Alternaria solani. *International Journal of Molecular Sciences* 20(8): 1950. doi: 10.3390/ijms20081950.

Rahman, P., and M. Green. 2009. The synthesis of rare earth fluoride based nanoparticles. *Nanoscale* 1(2): 214–224.

Rathna, R., A. Kalaiselvi, and E. Nakkeeran. 2018. Potential applications of nanotechnology in agriculture: Current status and future aspects. In: *Bioorganic Phase in Natural Food: An Overview*, Roopan, S., Madhumitha, G., Eds., Springer, Cham, pp. 187–209.

Ryman-Rasmussen, J. P., J. E. Riviere, and N. A. MonteiroRiviere. 2006. Penetration of intact skin by quantum dots with diverse physicochemical properties. *Toxicological Sciences* 91(1): 159–165. doi: 10.1093/toxsci/kfj122.

Sandhya, G., S. Kumar, D. Kumar, and N. Dilbaghi. 2017. Preparation, characterization, and bio-efficacy evaluation of controlled release carbendazim loaded polymeric nanoparticles. *Environmental Science and Pollution Research* 24(1): 926–937. doi: 10.1007/s11356-016-7774-y.

Sarlak, N., A. Taherifar, and F. Salehi. 2014. Synthesis of nanopesticides by encapsulating pesticide nanoparticles using functionalized carbon nanotubes and application of new nanocomposite for plant disease treatment. *Journal of Agricultural and Food Chemistry* 62(21): 4833–4838. doi: 10.1021/jf404720d.

Sathiyabama, M., and A. Manikandan. 2018. Application of copper-chitosan nanoparticles stimulate growth and induce resistance in finger millet (Eleusine coracana Gaertn.) plants against blast disease. *Journal of Agricultural and Food Chemistry* 66(8): 1784–1790. doi: 10.1021/acs.jafc.7b05921.

Sathiyabama, M., I. Murugan, and M. Selvaraj. 2019. Chitosan nanoparticles loaded with thiamine stimulate growth and enhances protection against wilt disease in Chickpea. *Carbohydrate Polymers* 212: 169–177. doi: 10.1016/j.carbpol.2019.02.037.

Shang, Y., M. D. Hasan, G. J. Ahammed, M. Li, H. Yin, and J. Zhou. 2019. Applications of nanotechnology in plant growth and crop protection: A review. *Molecules* 24(14): 2558.

Sharma, D., A. Nagpal, Y. B. Pakade, and J. K. Katnoria. 2010. Analytical methods for estimation of organophosphorus pesticide residues in fruits and vegetables: A review. *Talanta* 82(4): 1077–1089. doi: 10.1016/j.talanta.2010.06.043.

Shojaei, T. R., M. A. M. Salleh, M. Tabatabaei et al. 2019. Applications of nanotechnology and carbon nanoparticles in agriculture. In: *Synthesis, Technology and Applications of Carbon Nanomaterials*, Suraya, A. R., Raja, N. I. R. O., Mohd, Z. H., Eds., Elsevier, Amsterdam, Netherlands, pp. 247–277.

Singh, S., B. K. Singh, S. M. Yadav, and A. K. Gupta. 2015. Applications of nanotechnology in agricultural and their role in disease management. *Research Journal of Nanoscience and Nanotechnology* 5(1): 1–5. doi: 10.3923/rjnn.2015.1.1.

Solanki, J. N., and Z. V. P. Murthy. 2011. Controlled size silver nanoparticles synthesis with water-in-oil microemulsion method: A topical review. *Industrial and Engineering Chemistry Research* 50(22): 12311–12323. doi: 10.1021/ie201649x.

Song, Y., J. Chen, M. Sun, C. Gong, Y. Shen, Y. Song, and L. Wang. 2016. A simple electrochemical biosensor based on AuNPs/MPS/Au electrode sensing layer for monitoring carbamate pesticides in real samples. *Journal of Hazardous Materials* 304: 103–109. doi: 10.1016/j.jhazmat.2015.10.058.

Songa, E. A., and J. O. Okonkwo. 2016. Recent approaches to improving selectivity and sensitivity of enzyme-based biosensors for organophosphorus pesticides: A review. *Talanta* 155: 289–304. doi: 10.1016/j.talanta.2016.04.046.

Swaminathan, M. S. (ed). 1993. *Wheat Revolution: A Dialogue*. MacMillan India Ltd, Chennai, p. 166.

Tan, X., Q. Hu, J. Wu et al. 2015. Electrochemical sensor based on molecularly imprinted polymer reduced graphene oxide and gold nanoparticles modified electrode for detection of carbofuran. *Sensors and Actuators. Part B: Chemical* 220: 216–221. doi: 10.1016/j.snb.2015.05.048.

Tang, J., J. R. Durrant, and D. R. Klug. 2008. Mechanism of photocatalytic water splitting in TiO2. Reaction of water with photoholes, importance of charge carrier dynamics, and evidence for four-hole chemistry. *Journal of the American Chemical Society* 130(42): 13885–13891. doi: 10.1021/ja8034637.

TechNavio, T. Report. 2015. Hybrid seeds market in India. 2015–2019 SKU: IRTNTR5497.

Tomer, V., J. K. Sangha, and H. G. Ramya. 2015. Pesticide: An appraisal on human health implications. *Proceedings of the National Academy of Sciences, India Section B: (Biological Sciences)* 85(2): 451–463.

Torney, F., B. G. Trewyn, V. S. Y. Lin, and K. Wang. 2007. Mesoporous silica nanoparticles deliver DNA and chemicals into plants. *Nature Nanotechnology* 2(5): 295–300.

Tran, C. L., L. D. Buchanan, R. T. Cullen, A. Searl, A. D. Jones, and K. Donaldson. 2000. Inhalation of poorly soluble particles. II. Influence of particle surface area on inflammation and clearance. *Inhalation Toxicology* 12(12): 1113–1126. doi: 10.1080/08958370050166796.

Tran, C. L., R. T. Cullen, D. Buchanan et al. 1999. Investigation and prediction of pulmonary responses to dust, part II. In: Investigations into the pulmonary effects of low toxicity dusts. *Contract Research Report* 216/1999. Health and Safety Executive, Suffolk, UK.

Tripathi, M., S. Kumar, A. Kumar, P. Tripathi, and S. Kumar. 2018. Agro-nanotechnology: A future technology for sustainable agriculture. *International Journal of Current Microbiology Applied Sciences* 7: 196–200.

Tripati, S., S. K. Sonkar, and S. Sarkar. 2011. Growth stimulation of gram (Cicer arietinum) plants by water soluble carbon nanotubes. *Nanoscale* 3(3): 1176–1181. doi: 10.1039/C0NR00722F.

Vaghela, C., M. Kulkarni, S. Haram, R. Aiyer, and M. Karve. 2018. A novel inhibition based biosensor using urease nanoconjugate entrapped biocomposite membrane for potentiometric glyphosate detection. *International Journal of Biological Macromolecules* 108: 32–40. doi: 10.1016/j.ijbiomac.2017.11.136.

Vass, I. 2012. Molecular mechanisms of photodamage in the photosystem II complex. *Biochimica et Biophysica Acta (BBA): Bioenergetics* 1817(1): 209–217. doi: 10.1016/j.bbabio.2011.04.014.

Wanyika, H. 2013. Sustained release of fungicide metalaxyl by mesoporous silica nanospheres. *Journal of Nanoparticle Research* 15(8): 321–329. doi: 10.1007/s11051-013-1831-y.

Watanabe, T., Y. Murata, T. Nakamura, Y. Sakai, and M. Osaki. 2009. Effect of zero-valent iron application on cadmium uptake in rice plants grown in cadmium-contaminated soils. *Journal of Plant Nutrition* 32(7): 1164–1172. doi: 10.1080/01904160902943189.

Wu, M. 2013. Effects of incorporation of nano-carbon into slow-released fertilizer on rice yield and nitrogen loss in surface water of paddy soil. *2013 Third International Conference on Intelligent System Design and Engineering Applications*, Hong Kong, pp. 676–681.

Yarizade, K., and H. Ramin. 2015. Expression analysis of ADS, DBR2, ALDH1 and SQS genes in Artemisia vulgaris hairy root culture under nano cobalt and nano zinc elicitation. *Ext Journal of Applied Sciences* 3: 69–76.

Zhang, B., L. P. Zheng, W. Yi Li, and J. W. Wang. 2013. Stimulation of artemisinin production in Artemesia annua hairy roots by Ag-SiO2 core shell nanoparticles. *Current Nanoscience* 9(3): 363–370. doi: 10.2174/1573413711309030012.

Zhang, X., J. Zhang, and Y. Zhu. 2010. Chitosan/double-stranded RNA nanoparticles-mediated RNA interference to silence chitin synthase genes through larval feeding in the African malaria mosquito (Anopheles gambiae). *Insect Molecular Biology* 19(5): 683–693.

Zhou, H., Y. Qu, T. Zeid, and X. Duan. 2012. Towards highly efficient photocatalysts using semiconductor nanoarchitectures. *Energy and Environmental Science* 5(5): 6732–6743. doi: 10.1039/C2EE03447F.

Zinchenko, A., Y. Miwa, L. I. Lopatina, V. G. Sergeyev, and S. Murata. 2014. DNA hydrogel as a template for synthesis of ultrasmall gold nanoparticles for catalytic applications. *ACS Applied Materials and Interfaces* 6(5): 3226–3323. doi: 10.1021/am5008886.

15 Metallic Nanoparticles at the Biointerface
Current Perspectives and Future Challenges

Mousmee Sharma and Parteek Prasher
Uttaranchal University, Dehradun, India; and University
of Petroleum and Energy Studies, Dehradun, India

CONTENTS

15.1   INTRODUCTION

The biological applications of metallic nanoparticles emanate typically from their characteristic optical and magnetic properties (Prasher et al. 2020). The maneuverable physicochemical profile, ability to transit across the physiological barriers, and potency to conjugate to the biomacromolecules further endorse their applications in biosensing, theranostics, targeted drug delivery, and molecular medicine (Alavi and Rai 2019). The noble metal nanoparticles present a marked oligodynamic effect and hence display a robust candidature as next-generation antibiotics that effectively overcome the therapeutic limitations of representative antimicrobial pharmaceuticals, such as multidrug resistance (Prasher et al. 2018). The biocidal properties of the metallic nanoparticles principally arise due to the offset of intracellular oxidative stress that further triggers a cascade of deleterious events, hazardous to the microbial cells (Singh and Prasher 2018). Interestingly, the metallic nanoparticles exhibit characteristic surface plasmon resonance (SPR), arising due to the interactions between the conduction electrons of metal nanoparticles and the incident photons (Craciun et al. 2017). The phenomenon of SPR demonstrated by metallic nanoparticles manifests in marked applications in biosensing, biotagging, as diagnostic probes, and in the colorimetric detection of bioanalytes (Yang et al. 2016). Notably, the metallic nanoparticles, appended with an appropriate surface functionalization upon deliberated bioconjugation with biomacromolecules, function as targeted-drug-delivery vehicles. This occurs where the cargo drug detaches from the nano-bio-conjugated system under the physiologically distinct conditions at the target site, thereby assuring optimal performance of the payload drug (Pollok et al. 2019). Apart from this, the magnetic nanoparticles present applications as MRI contrast agents, bioimaging, stem cell tracking, and rapid DNA sequencing, as well as early detection of cellular morbidity (Petrov et al. 2020). Hence, the plethora of applications of the metallic nanoparticles at the biointerface justifies their position as materials of the future. Figure 15.1 demonstrates the potential applications of functionalized metal nanoparticles.

15.2 MICROBICIDAL METAL NANOPARTICLES

The oligodynamic effect of metal nanoparticles is a widely accepted phenomenon (Chakravarty and Kundu 2019). The metallic nanoparticles based on noble metals such as gold and silver exhibit a considerable biocidal profile; however, the associated physiological toxicity poses significant challenges for the development of nanometal-based antimicrobial therapeutics (Burdusel et al. 2018). While contacting the microbial cells, the metallic nanoparticles first attach to the cell membrane via transmembrane proteins, triggering the structural deformities and blocking the vital transport channels (Azharuddin et al. 2019). These events adversely influence the selective permeability of the microbial membrane, leading to the leakage of K$^+$ ions and internalization of metallic nanoparticles (Dakal et al. 2016). Upon entering the microbial cells, the metallic nanoparticles instigate oxidative stress with the decoupling of oxidative phosphorylation from the electron transport chain (ETC) in mitochondria, hence causing a splurge of intracellular reactive oxygen species (ROS) (Lopez et al. 2019). However, the therapeutic utilization of nanometal-based systems requires sensible consideration of the physiological toxicity. Decoration of the

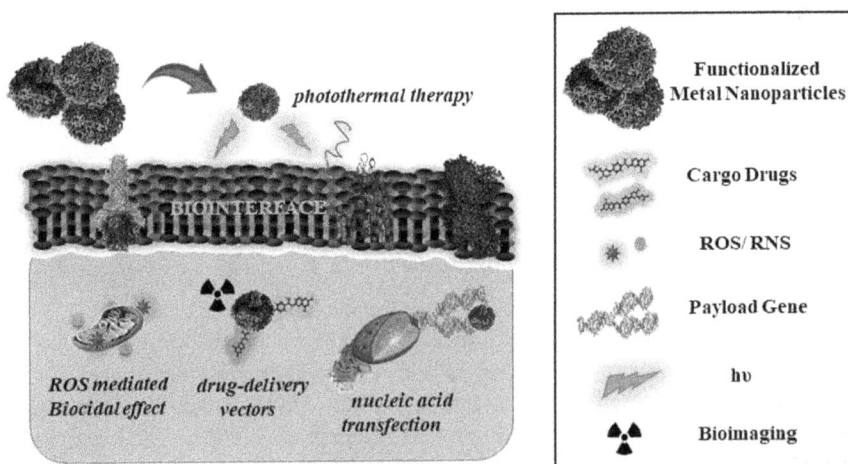

FIGURE 15.1 Applications of surface functionalized metal nanoparticles at the biointerface.

surface of nanoparticles with suitable molecules mitigates the associated toxicity, improving their physiological tolerance and provides supplementary functionalities that enable their further conjugation to the biomacromolecules for extended applications (Ravindran et al. 2013). The non-functionalized metal nanoparticles reportedly develop a proteinaceous corona upon internalization into the cells, which conceals their engineered surface and affords physiological stability (Abdelkhaliq et al. 2018). Nevertheless, the development of corona enhances the nanoparticle toxicity that impedes the efforts directed on the development of nanometal-based therapeutics (Oh et al. 2018). The biological activity of metallic nanoparticles depends on their size, surface charge, surface functionality, and interactions with biomacromolecules. The nanoparticles with an ultrafine size exhibit superior antimicrobial properties compared to their large-sized counterparts because of the larger effective surface area of the former (Raza et al. 2016). However, the large-sized biofabricated nanoparticles effectively evade the microbial efflux pumps due to the "size exclusion effect", that represents the inability of the efflux system to pump out macromolecules (Roy et al. 2019). Interestingly, the metallic nanoparticles demonstrate an excellent synergism with the representative antibiotics by complex formation, thereby creating a surge of the local concentration of metallic nanoparticles near target microbial cells, eventually resulting in its mortality (Shaikh et al. 2019). Khatoon et al. (2019) demonstrated the synergistic inhibitory effect of silver nanoparticles covalently coated with ampicillin against multi-drug-resistant bacteria. The amino group of ampicillin prompted the reduction of precursor silver salt to nano-metallic silver with an average size 9-20 nm that exhibited a prolonged inhibitory effect in combination with the antibiotic with IC50 in the range 3-28 μg/ mL. Adil et al. (2019) reported a similar biocidal synergism of silver nanoparticles-ciprofloxacin combination against drug-resistant microbial strains. Notably, the silver nanoparticles and the partner antibiotic display different mechanisms of inhibition against the test microbe, which possesses multiple regulatory pathways to demonstrate resistance against the customary antibiotics. The nanoparticle-antibiotic combination prompts the multi-targeting of these pathways, thereby augmenting the efficacy of the combination, by combining the biocidal potency of silver nanoparticles to the inhibition profile of the test antibiotic. Munoz et al. (2019) investigated the synergistic effect of silver nanoparticle in combination with antibiotics: chloramphenicol, kanamycin, ampicillin, aztreonam, and biapenem. The antibiotics that function inside the microbial cells displayed a synergistic effect with silver nanoparticles because the latter promoted the antibiotic internalization by disturbing the membrane permeability, which resulted in enhanced microbicidal activity. Contrarily, the synergistic effect diminished for the β-lactam class of antibiotics that merely affect the microbial cell wall integrity. Carrizales et al. (2018) validated the efficacy of silver nanoparticle-antibiotics combination for countering the multidrug-resistant pathogens. Reportedly, the test antibiotics ampicillin and amikacin stabilized the silver nanoparticles via carbonyl and amine functional groups respectively. *In vitro* fractional inhibitory concentration (FIC) assay confirmed the synergistic effect of the nanoparticle-antibiotic combination with a superior physiological tolerance. Interestingly, the combination of silver nanoparticles with amikacin and ampicillin led to a 16 to 32-fold reduction in minimal inhibitory concentration (MIC) required for the microbicidal effect. Yu et al. (2020) developed a novel, multi-responsive silver nanoparticles/gentamycin nanocarrier stabilized by hyaluronic acid (HA), which demonstrated a synergistic antimicrobial effect against drug-resistant bacterial strains. The metabolic activity of bacteria lowered the pH, triggering the release of silver nanoparticles and the antibiotic from the nanosystem. The released silver nanoparticles induced morphological changes in the microbial plasma membrane and altered its selective permeability. It further diminished the transmembrane proton electrochemical gradient, resulting in the deactivation of energy-dependent metabolic reactions. These events endorsed the internalization of the antibiotic in the microbial cell, which further inhibits the protein synthesis by interacting with the ribosomal subunits. In addition, the silver nanoparticles interacted with essential intracellular enzymes and nucleic acids, eventually affecting the microbial metabolism and development. Wadhwani et al. (2017) synthesized homogeneously distributed gold nanoparticles decorated with antimicrobial cationic peptides via the N-terminal of cysteine residues. The representative peptides, when anchored to gold nanoparticles, switched their conformation to α-helical conformation in the presence of model membranes, while retaining their antimicrobial activity. Interestingly, the conjugation of gold nanoparticles with antimicrobial peptides improved the stability of the latter against protease digestion, thereby upgrading its activity and enhancing its lifetime. Jalaei et al. (2018) further validated the enhanced potency of therapeutic peptides extracted from *Vespa orientalis* (wasp venom) in conjugation with gold nanoparticles. The peptide nanoparticle conjugate exhibited superior antimicrobial properties compared to tetracycline and the unconjugated peptide. The antimicrobial activity of the extracted peptide appeared due to the presence of alanine, asparagine, leucine, and valine residues, which further improved on conjugation with gold nanoparticles. Alteriis et al. (2018) investigated microbicidal and anti-biofilm properties of gold nanoparticles coated with the antimicrobial peptide (AMP) indolicidin. Importantly, the AMP-gold nanoparticle combination penetrated the biofilm matrix, reduced the membrane permeability, thereby sabotaging the microbial population in the biofilm. Apparently, the AMP-nanoparticle combination also displayed enhanced physiological stability and minimal vulnerability to protease activity. However, the combination did not affect the membrane surface and activated the microbial efflux pumps post-treatment. The mechanism of biofilm inhibition entirely depended on the cellular penetration of the AMP-nanoparticle complex, which after internalization effectively targets the intracellular metabolism in the host microbe.

15.3 METAL NANOPARTICLES IN CANCER THERAPY

Conventional anticancer therapeutics lack target specificity, thereby manifesting considerable cytotoxicity (Sharma et al. 2015). In addition, the abnormal expression of cellular efflux towards the representative pharmaceuticals necessitates the development of novel, target-specific therapeutic agents for a effective targeting of the invasive cancers (Pugazhendi et al. 2018). A multifactorial targeting approach focused on multiple regulatory pathways associated with the diseased condition, presenting desirable outcomes for countering cancer-induced cellular morbidity, and sparing the optimal functioning of healthy cells (Rai et al. 2016). The noble metal nanoparticles achieve passive targeting of the cancer cells by direct penetration through the defective vasculature (Cordani and Somoza 2018), leading to enhanced retention and accumulation at the diseased site (Attia et al. 2019). The presence of hydrophilic functionalities on the nanoparticle surface prevents their loss via systemic circulation, hence providing considerable stability against the enzymatic degradation (Bazak et al. 2015). For active targeting of the cancer cells, the nanosystems generated by functionalization of metallic nanoparticles with biomacromolecules such as DNA/RNA, peptides, and antibodies specifically modulate critical regulatory pathways and receptors associated with the diseased cell metabolism, thereby pushing the cancer cells to death (Conde et al. 2014). Reportedly, the metal nanoparticles hamper the proliferation of cancer cells by arresting cell cycle in G_o/G_1 phase, leading to failure of mitosis and ceasing of the central dogma in target cells (Jeyaraj et al. 2014). The internalization of metal nanoparticles in cancer cells disrupts the intracellular homeostasis and triggers oxidative stress that adversely affects the cellular apoptosis (George et al. 2018). Benyettou et al. (2015) developed a silver-nanoparticle-based drug delivery system for the intracellular delivery of doxorubicin and alendronate. Biphosphonate alendronate served as the template for nanoparticle formation and a site for drug anchoring through the unreacted, free primary ammonium group through pH-sensitive imine bond formation, enabling the intracellular release of the drug in the acidic pH of lysosomes and endosomes. The cellular uptake of the nanosystem occurred through clatharin-mediated endocytosis and macropinocytosis. Notably, the nanosystem demonstrated a restrained premature release of the drug in the bloodstream, thereby improving its efficacy and therapeutic index. Brown et al. (2010) conjugated thiolated poly(ethylene glycol)-decorated gold nanoparticles with free carboxylic groups with the active pharmacophore of anticancer drug oxaliplatin for improved drug delivery at the cancer site. Cancer cells cause an uptake of the nanoconjugate via endocytosis by generating momentary pores in the membrane. The nanoconjugates displayed enhanced plasma retention and uptake into the solid tumors, with 5-fold efficacy, compared to the non-conjugated oxaliplatin. Paciotti et al. (2016) conjugated thiolated gold nanoparticles to paclitaxel through different linkers, for targeted drug delivery to cancer cells. The nanosystem exhibited considerable stability in the circulatory system due to the hydrophobic nature of surface ligands, and released the cargo drug at acidic pH or underwent reductive cleavage at the tumor site, coupled with TNF-mediated disruption of the tumor vasculature, prompting drug release from the nanosystem. Spherical gold nanoparticles demonstrate considerable photothermal effect for the annihilation of cancer cells. Mendes et al. (2017) exhibited synergistic obliteration of cancer cells due to the photothermal characteristics of gold nanoparticles and cytotoxic property of doxorubicin. The cancer cells treated with gold nanoparticles and doxorubicin led to a 58% reduction in cell viability due to considerable DNA-damage owing to the slowdown of the replication fork, which induces breaks in the DNA double strands. Notably, the combination therapy of nanoparticles and doxorubicin, coupled with photoirradiation for inducing local hyperthermia, triggered denaturation of cytoplasmic proteins and nucleic acids in addition to damaging cellular and organelles membranes. Sengupta et al. (2018) investigated the anticancer potency of silver nanoparticles decorated with murine serum albumin (MSA). Reportedly, the nanoparticles instigated a localized redox-sensitive signaling cascade effect in the tumor microenvironment, in addition to elevating the level of reactive oxygen/ nitrogen species in the tumor-associated macrophages (TAM). The heightened oxidative and nitrosative stress in TAM prevents their overexpression in modulating the immunomodulatory activity of the cancer cells, resulting in depletion of TNF-α and IL-10 followed by an upsurge in the immunoregulatory cytokine IL-12. Dzwonek et al. (2018) reported the anti-cancer potency of the nanosystem, containing lipoic acid-functionalized ultrafine gold nanoparticles bioconjugated to doxorubicin via amide linkage. The nanosystem demonstrated a controlled drug release following enhanced uptake by the cancer cells, which manifests a marked nuclear accumulation. The amide bond involved in conjugation of drug and functionalized nanoparticle displayed marked cytoplasmic stability for 22 hours, hence affording minimal systemic toxicity and slower body clearance to the payload drug, compared to its unconjugated form. siRNA-based anticancer chemotherapy is of great therapeutic interest owing to remarkable target specificity. However, the fragility of siRNA in serum poses significant challenges for an effective *in vivo* delivery to the target site (Babu et al. 2016). Fitzgerald et al. (2016) designed poly(ethylenimine)-capped gold nanoparticles conjugated to anisamide, which prompted siRNA uptake into the PC3 prostate cancer cells by binding to the sigma receptor present on the surface of target cells. The nanosystem prevented the degradation of siRNA by serum nucleases for up to 24 hours; however, the shielding effect by negatively charged serum proteins limits the approach of the nanosystem to sigma receptors on target PC3 prostate cancer cells. Elbakry et al. (2009) overcome the limitations of effective siRNA delivery via layer-by-layer coating of poly(ethylenimine) for conjugation to siRNA. This approach prevents the aggregation of nanoparticles after conjugation with siRNA, while ensuring superior stability in serum, and thus enables effective cellular delivery of the cargo nucleic acid. Liu et al. (2019) developed flower-like gold nanoparticles that served as a core for layer-by-layer coating of biodegradable polymers, which further carry the negatively charged gene drugs. Further conjugation of the nanosystem to siRNA assists in photothermal and gene silencing anticancer therapy. The gold nanoparticle-siRNA complex possesses high-loading efficiency, and displayed considerable physiological tolerance, thereby prompting the tumor cell death following laser irradiation by decreasing BAG3 expression.

15.4   METAL NANOPARTICLES FOR BIOIMAGING

Nanosized metal nanoparticles exhibit extraordinary physicochemical and optical properties compared to their bulk analogs. Metallic nanoparticles such as nanosized iron oxide demonstrate superparamagnetism and behave as individual magnetic domains (Wahajuddin 2012). Similarly, the noble metal nanoparticles possess characteristic dielectric constants, which interact with incident electromagnetic radiation to induce collective oscillation of conduction electrons (Miller and Lazarides 2005). The phenomenon referred to as surface plasmon resonance (SPR) results in the localized enhancement of the electromagnetic field on excitation of conduction electrons of the irradiated nanoparticles (Jana et al. 2016). The SPR band, being highly localized and characteristic of the nanoparticle under investigation, supports their bioimaging applications. Gold nanoparticles, due to a high X-ray absorption coefficient, have considerable biocompatibility and a unique SPR; they present strong candidature as contrast agents for techniques such as computed tomography (CT) (Hayashi et al. 2013). Dong et al. (2019) investigated the relationship between the size of polyethylene glycol (PEG)-coated gold nanoparticles and CT contrast. The investigations with coupled plasma optical emission spectroscopy validated that gold nanoparticles with an average diameter of 15 nm possess an enhanced lifespan in systemic circulation, whereas large-sized gold nanoparticles rapidly accumulated in the liver and spleen, hence providing a strong contrast of the latter. Popovtzer et al. (2008) developed gold nanoprobes that induced a distinct contrast in CT imaging for highly sensitive and selective targeting of tumor-selective antigens. Reportedly, the gold nanoparticles specifically assembled on cancer cells, thereby manifesting a characteristic X-ray attenuation, differentiating them from the healthy cells. This event enables highly selective contrasting of target cancer cells. Reuveni et al. (2011) reported gold nanoparticles (30 nm diameter) conjugated to an anti-epidermal growth factor receptor, for the detection and contrasting of small tumors. The X-ray attenuation associated with the gold nanoprobe assembly on target cancer cells transforms the morbid cells to discrete and detectable features in CT imaging. Khademi et al. (2019) further reported *in vitro* targeted imaging of cancer cells by computed tomography, assisted by cysteamine linked folic acid gold nanoparticles with 13 nm average diameter, acting as a contrast agent. The nanosystem enabled the detection and imaging of small tumors via X-ray attenuation with high tolerability and stability in blood. The investigations on gold-nanoparticle-based nanoprobes supported high specificity of active tumor cells compared to their passive analogs. Liu et al. (2010) reported dendrimer-stabilized silver nanoparticles as image contrast agents in computed tomography. Reportedly, the nanosystem with diameter 16.1 nm displayed X-ray attenuation intensity comparable to a clinically utilized iodine-based contrast agent. The nanosystem exhibited considerable stability in pH range 5–8 at temperatures 25–50° C. Siddiqui et al. (2009) reported silver nanoparticles functionalized with lanthanide complexes as hyper-intensity magnetic resonance imaging (MRI) contrast agents. The average r_1 relaxivity of nano-complex at 9.4 T and 25°C appeared to be 10.7 and 9.7 s^{-1} mM^{-1} respectively, compared to 4.7 s^{-1} mM^{-1} for clinically used FDA-approved contrast agent Magnevist. The reported nano-complex holds potential application for targeted cellular labeling and for detecting lacerations available from the circulatory system. Superparamagnetic iron oxide nanoparticles (SPION) present robust candidature as MRI contrast agents, thereby extending their biomedical utility. Liao et al. (2010) reported biocompatible surfactin-stabilized superparamagnetic iron oxide nanoparticles (SPION) as sensitive MRI contrast agents. The observed r_1 and r_2 relaxivities and r_2/r_1 ratio suggest that surfactin stabilized silver nanoparticles act as T_2 contrast agents. Tsai et al. (2010) further extended the application of high relaxivity chitosan-coated SPIONS as MRI contrast agents. The rapid uptake of the nanoprobes by macrophages at pH 2-3 enabled for cellular tracking and labeling. Na et al. 2007; developed MnO nanoparticles as biocompatible T_1 contrast agent for MRI. Conjugation of MnO nanoparticles with tumor-specific antibodies prompts selective imaging of cancer cells in metastatic brain tumors and breast cancer. The MnO nanoparticles exhibited swift delivery, tolerable physiological toxicity, and rapid biological clearance, thereby validating their clinical applications. Wei et al. (2017) developed ultrafine SPION nanoparticles coated with zwitterions constituting of 3-nm inorganic cores surrounded with a 1-nm ultrathin hydrophilic shell. These nanosystems function as high T_1 contrast agents in preclinical MRI. These SPION nanosystems display satisfactory renal clearance, and their prolonged T_1 contrast power provides applications in magnetic resonance angiography. Further, the reported SPION nanosystems provide suitable pharmacokinetic characteristics that regulate the iron overdose in the body within the safe limit. Yin et al. (2018) demonstrated SPIONs (11–22 nm) with strong T_1 contrast enhancement in an ultra-low field MRI at 0.13 mT. The nanoparticles of size 18 nm displayed superior longitudinal relaxivity ($r_1 = 615 \, s^{-1} \, mM^{-1}$) compared to the representative gadolinium-based contrast agents operating at high magnetic fields. The improvement in r_1 at an ultra-low magnetic field appears due to the coupling of SPION magnetic fluctuations with proton spins. The SPION-based nanosystem provided an enhanced signal, superior biocompatibility, and shorter imaging time. Though SPIONs provide a desirable profile as MRI contrast agents, their application is limited due to poor clinical contrast performance. Hobson et al. (2019) developed advanced SPION-based organ-targeted MRI contrast agents. The encapsulation of SPIONs in amphiphilic chitosan polymer resulted in their accumulation and clustering in liver extravascular space, producing a superior contrast owing to a reduction in spin-lattice relaxation, and enhancement in spin-spin relaxation. The reported SPION-based nanosystem exhibited considerable blood half-life of 30 minutes, and displayed colloidal stability in a range of biomedical fluids.

15.5   METAL NANOPARTICLES FOR GENE THERAPY

Gene therapy incorporates the introduction of engineered gene coding for functional peptides that are useful for modulating the regulatory pathways contributing to disease progression, or alteration of endogenous gene expressions for the same purpose. The nanoparticle-mediated gene therapy prompts target specific delivery, considerable biocompatibility, and high stimulus responsiveness. Peng et al. (2014) reported cationic noble metal nanoparticles bioconjugated to TAT peptide as vehicles for stem cell gene delivery. The nanosystem possesses the advantages of nanoparticles for mitigating the toxicity of the cationic molecules, whereas the cell-penetrating TAT peptide confers high transfection efficiency, due to enhanced intracellular and intranuclear penetration. Thomas and Klibanov (2003) covalently linked branched-chain polyethylenimine (PEI) with gold nanoparticles for the vector delivery of plasmid DNA in the presence of serum. The transfection efficiency of the nanosystem varied as PEI/nanoparticle molar ratio, while the conjugation of PEI with N-dodecyl further amplified the DNA transfection by 100-fold. The hydrophobic interactions between dodecyl groups manifest self-assembly, and improves the effective molecular weight, resulting in highly efficient DNA condensation. Additionally, the presence of hydrophobic substituents improves the interactions of the nanosystem with plasma membrane and membranes of endosomes and lysosomes, thereby affecting the cellular uptake and endosomal escape. Ghosh et al. (2008) designed amino-acid-functionalized gold nanoparticles for effectively binding to DNA. While acting as transfection agents, the nanoparticles with lysine functionality displayed minimal physiological toxicity, and reportedly formed compact complexes with cargo gene for its efficient delivery. In addition, the nanosystem displayed high responsiveness to intracellular glutathione, thereby affording their controlled payload release *in vivo*. Du et al. (2017) reported folic acid functionalized, lipid-coated gold nanoparticles for targeted gene delivery. The nanosystem afforded superior transfection efficiency of the cargo gene, due to enhanced cellular uptake prompted by folic acid functionality. The reported nanosystem accumulated at the tumor site, thereby influencing the gene delivery *in vivo*. Kalimuthu et al. (2018) reported non-covalently loaded PEGylated-gold nanoparticles as vectors for enhancing the bioavailability of peptide drug conjugated (PDC). Interestingly, the gold nanoparticles improved the blood and plasma half-life of PDCs, which otherwise demonstrated high vulnerability to biological fluids and tissues. The conjugation to PEGylated-gold nanoparticles promotes the penetration of PDC to tumor cells, prompting their targeted delivery and a controlled release. Cheong et al. (2009) developed SPIONs fabricated with water-soluble chitosan and linoleic acid for hepatocyte-targeted gene delivery. The nanosystem reportedly accumulated in the liver as confirmed by MRI, and resulted in the considerable enhancement in the *in vivo* expression of green fluorescent protein in hepatocytes. The nanosystem provided for *in vivo* and *in vitro* targeted delivery of plasmid DNA in hepatocytes. Kievit et al. (2009) presented SPIONs decorated with PEI-PEG-chitosan copolymer for gene delivery. PEI in the nanosystem ensures effective DNA binding and its transfection in tumor cells, whereas chitosan mitigates the toxicity associated with PEI. The reported nanosystem effectively delivered intact DNA to tumors cells, and the SPIONs enabled its *in vivo* tracking. Lee et al. (2012) developed positively charged SPIONs fabricated with polymersomes for effective and targeted delivery of genes. MRI enabled the *in vivo* and *in vitro* detection of SPIONs, which accumulated in the tumor site. Tightly packed nanoparticles improved the image contrast considerably, mainly by increasing T2 relaxation, which resulted in the achievement of a tumor-selective high-signal drop with significant MRI sensitivity. Kim et al. (2015) developed SPIONs coated with PEI (polyethyleneimine) and PCL (polycaprolactone) as gene delivery vectors with high transfection efficiency and low toxicity. The SPIONs' encapsulation with amphiphilic polymers improves the nucleic acid condensation and endosomal/lysosomal escape through a proton sponge effect after entering the cells. Notably, the external magnetic field prompts the sedimentation of a reported SPION nanosystem that results in their accumulation on tumor cell surface. These results further validated the application of magnetic nanoparticles for effective gene delivery.

15.6   FUTURE CHALLENGES

The emergence of metal-resistant microbial strains marred the efforts to develop antimicrobials based on nanometals. Reportedly, the nanoparticles with a diameter less than 10 nm agglomerate in the extracellular matrix led to their deactivation (Martinez et al. 2019). The large-sized nanoparticles display a similar effect due to the overexpression of flagelin matrix, extracellular substances produced by microbial cells exposed to nanoparticles, and proteinaceous coronas which alters the morphology and zeta potential of the nanoparticles, thereby preventing their further contact with microbial cells (Hachicho et al. 2014). McHugh et al. (1975) identified pMG101-plasmid coding for resistance against silver ions in *Salmonella typhimurium*. Franke (2007) identified a SiIP efflux pump that facilitated the active transportation of silver ions from cytoplasm to periplasm, further supporting microbial resistance from metallic silver and silver ions. Similarly, Riggle and Kumamoto (2000) reported an ATP-dependent efflux of silver ions from *Candida albicans* via copper efflux protein. Importantly, the lack of clinically validated MIC value for nanometal-based therapeutics further enhances the proneness of nanometallic formulations to antimicrobial resistance. The nanoparticles as drug delivery vectors suffer limitations such as low-drug loading capacity, especially for hydrophilic drugs (Fratoddi et al. 2014). Moreover, the biodistribution of metal nanoparticles, pharmacokinetics, and associated physiological toxicity further pose significant challenges to the potential biological applications of metal-based nanoparticles.

15.7 CONCLUSION

Metal nanoparticles present numerous applications at the biointerface, which have revolutionized the contemporary molecular medicine paradigm. The functionalized noble metal nanoparticles present robust candidature in designing the next generation of antibiotics, which effectively evade the challenges faced by representative antimicrobial molecules, such as drug resistance. Similarly, the functionalized magnetic metal nanoparticles present excellent applications in bioimaging and as image contrast agents in bioanalytical instruments. Besides, the metal nanoparticles play an important role in drug delivery and intracellular nucleic acid delivery with high target specificity. Hence, the metal nanoparticles serve as an essential ingredient for exploring vital processes at the biointerface for designing state-of-the-art theranostics and bioprobes useful for the next generation.

REFERENCES

Abdelkhaliq, A., Zande, M., Punt, A., Helsdingen, R., Boeren, S., Vervoort, J.J.M., Rietjens, I.M.C.M., Bouwmeester, H. 2018. Impact of nanoparticle surface functionalization on the protein corona and cellular adhesion, uptake and transport. *J. Nanobiotechnol.* 16(1): 70.

Adil, M., Khan, T., Aasim, M., Khan, A.A., Ashraf, M. 2019. Evaluation of the antibacterial potential of silver nanoparticles synthesized through the interaction of antibiotic and aqueous callus extract of Fagonia indica. *AMB Express* 9(1): 75.

Alavi, M., Rai, M. 2019. Recent advances in antibacterial applications of metal nanoparticles (MNPs) and metal nanocomposites (MNCs) against multidrug-resistant (MDR) bacteria. *Expert Rev. Anti-Infect. Ther.* 17(6): 419–428.

Alteriis, E.D., Maselli, V., Falanga, A., Galdiero, S., Lella, F.M.D., Gesuele, R., Guida, M., Galdiero, E. 2018. Efficiency of gold nanoparticles coated with the antimicrobial peptide indolicidin against biofilm formation and development of Candida spp. clinical isolates. *Infect. Drug Resist.* 11: 915–925.

Attia, M.F., Anton, N., Wallyn, J., Omran, Z., Vandamme, T.F. 2019. An overview of active and passive targeting strategies to improve the nanocarrier efficiency to tumor sites. *J. Pharm. Pharmacol.* 71(8): 1185–1198.

Azharuddin, M., Zhu, G.H., Das, D., Ozgur, E., Uzun, L., Turner, A.P.F., Patra, H.K. 2019. A repertoire of biomedical applications of noble metal nanoparticles. *Chem. Commun. (Camb.)* 55(49): 6964–6996.

Babu, A., Muralidharan, R., Amreddy, N., Mehta, M., Munshi, A., Ramesh, R. 2016. Nanoparticles for siRNA-based gene silencing in tumor therapy. *IEEE Trans. Nanobiosci.* 15(8): 849–863.

Bazak, R., Houri, M., Achy, S.E., Kamel, S., Refaat, T. 2015. Cancer active targeting by nanoparticles: A comprehensive review of literature. *J. Cancer Res. Clin. Oncol.* 141(5): 769–784.

Benyettou, F., Rezgui, R., Ravaux, F., Jaber, T., Blumer, K., Jouiad, M., Motte, L., Olsen, J.-C., Iglesias, C.P., Magzoub, M., Trabolsi, A. 2015. Synthesis of silver nanoparticles for the dual delivery of doxorubicin and alendronate to cancer cells. *J. Mater. Chem. B* 3(36): 7237–7245.

Brown, S.D., Nativo, P., Smith, J.-A., Stirling, D., Edwards, P.R., Venugopal, B., Flint, D.J., Plumb, J.A., Graham, D., Wheate, N.J. 2010. Gold nanoparticles for the improved anticancer drug delivery of the active component of oxaliplatin. *J. Am. Chem. Soc.* 132(13): 4678–4684.

Burdusel, A.C., Gherasim, O., Grumezescu, A.M., Mogoanta, L., Ficai, A., Andronescu, E. 2018. Biomedical applications of silver nanoparticles: An up-to-date overview. *Nanomater* 8(9): 681.

Carrizales, M.L., Velasco, K.I., Castillo, C., Flores, A., Magana, M., Castanon, G.A.M., Gutierrez, F.M. 2018. In vitro synergism of silver nanoparticles with antibiotics as an alternative treatment in multiresistant uropathogens. *Antibiotics* 7: 50.

Chakravarty, I., Kundu, S. 2019. Oligodynamic boons of daptomycin and noble metal nanoparticles packaged in an anti-MRSA topical gel formulation. *Curr. Pharm. Biotechnol.* 20(9): 707–718.

Cheong, S.-J., Lee, C.-M., Kim, S.-L., Jeong, H.-J., Kim, E.-M., Park, E.-H., Kim, D.W., Lim, S.T., Sohn, M.-H. 2009. Superparamagnetic iron oxide nanoparticles-loaded chitosan-linoleic acid nanoparticles as an effective hepatocyte-targeted gene delivery system. *Int. J. Pharm.* 372(1–2): 169–176.

Conde, J., Dias, J.T., Grazu, V., Moros, M., Baptista, P.V., Fuente, J.M. 2014. Revisiting 30 years of biofunctionalization and surface chemistry of inorganic nanoparticles for nanomedicine. *Front. Chem.* 2: 48.

Cordani, M., Somoza, A. 2018. Targeting autophagy using metallic nanoparticles: A promising strategy for cancer treatment. *Cell. Mol. Life Sci.* 76(7): 1215–1242.

Craciun, A.M., Focsan, M., Magyari, K., Vulpoi, A., Pap, Z. 2017. Surface plasmon resonance or biocompatibility—Key properties for determining the applicability of noble metal nanoparticles. *Materials* 10(7): 836.

Dakal, T.C., Kumar, A., Majumdar, R.S., Yadav, V. 2016. Mechanistic basis of antimicrobial action of silver nanoparticles. *Front. Microbiol.* 7: 1831.

Dong, Y.C., Hajfathalian, M., Maidment, P.S.N., Hsu, J.C., Naha, P.C., Mohamed, S., Breuilly, M., Kim, J., Chhour, P., Douek, P., Litt, H.I., Cormode, D.P. 2019. Effect of gold nanoparticle size on their properties as contrast agents for computed tomography. *Sci. Rep.* 9(1): 14912.

Du, B., Gu, X., Han, X., Ding, G., Wang, Y., Li, D., Wang, E., Wang, J. 2017. Lipid-coated gold nanoparticles functionalized by folic acid as gene vector for targeted gene delivery in vitro and *in vivo*. *ChemMedChem* 12(21): 1768–1775.

Dzwonek, M., Zalubiniak, D., Piatek, P., Cichowicz, G., Wielgosz, S.M., Stepkowski, T., Kruszewski, M., Wieckowska, A., Bilewicz, R. 2018. Towards potent but less toxic nanopharmaceuticals – Lipoic acid bioconjugates of ultrasmall gold nanoparticles with an anticancer drug and addressing unit. *RSC Adv.* 8(27): 14947–14957.

Elbakry, A., Zaky, A., Liebl, R., Rachel, R., Goepferich, A., Breunig, M. 2009. Layer-by-layer assembled gold nanoparticles for siRNA delivery. *Nano Lett.* 9(5): 2059–2064.

Fitzgerald, K.A., Rahme, K., Guo, J., Holmes, J.D., O'Driscoll, C.M. 2016. Anisamide-targeted gold nanoparticles for siRNA delivery in prostate cancer-synthesis, physico-chemical characterisation and *in vitro* evaluation. *J. Mater. Chem. B* 4(13): 2242–2252.

Franke, S. 2007. Microbiology of the toxic noble metal silver. In: Nies D, Silver S (eds), *Molecular Biology of Heavy Metals. Microbiology Monographs*. Springer, Berlin.

Fratoddi, I., Venditti, I., Cametti, C., Russo, M.V. 2014. Gold nanoparticles and gold nanopar-ticle-conjugates for delivery of therapeutic molecules. Progress and challenges. *J. Mater. Chem. B* 2(27): 4204–4220.

George, B.P.A., Kumar, N., Abrahmse, H., Ray, S.S. 2018. Apoptotic efficacy of multifaceted biosynthesized silver nanoparticles on human adenocarcinoma cells. *Sci. Rep.* 8: 14368.

Ghosh, P.S., Kim, C.-K., Han, G., Forbes, N.S., Rotello, V.M. 2008. Efficient gene deliv-ery vectors by tuning the surface charge density of amino acid-functionalized gold nanoparticles. *ACS Nano* 2(11): 2213–2218.

Hachicho, N., Hoffmann, P., Ahlert, K., Heipieper, H.J. 2014. Effect of silver nanoparticles and silver ions on growth and adaptive response mechanisms of *Pseudomonas putida* mt-2. *FEMS Microbiol. Lett.* 355(1): 71–77.

Hayashi, K., Nakamura, M., Miki, H., Ozaki, S., Abe, M., Matsumoto, T., Ishimura, K. 2013. Gold nanoparticle cluster–plasmon-enhanced fluorescent silica core–shell nanopar-ticles for X-ray computed tomography–fluorescence dual-mode imaging of tumors. *Chem. Commun.* 49(46): 5334–5336.

Hobson, N.J., Weng, X., Siow, B., Veiga, C., Ashford, M., Thanh, N.T.K., Schatzlein, A.G., Uchegbu, I.F. 2019. Clustering superparamagnetic iron oxide nanoparticles pro-duces organ-targeted high-contrast magnetic resonance images. *Nanomedicine* 14(9): 1135–1152.

Jalaei, J., Ghalehsoukhteh, S.L., Hosseini, A., Fazeli, M. 2018. Antibacterial effects of gold nanoparticles functionalized with the extracted peptide from Vespa orientalis wasp venom. *J. Pep Sci.* 24(12): e3124.

Jana, J., Ganguly, M., Pal, T. 2016. Enlightening surface plasmon resonance effect of metal nanoparticles for practical spectroscopic application. *RSC Adv.* 6(89): 86174–86211.

Jeyaraj, M., Arun, R., Sathishkumar, G., Rajesh, M., Sivanandan, G., Kapildev, G., Manickavasagam, M., Thajuddin, N., Ganapathi, A. 2014. An evidence on G2/M arrest, DNA damage and caspase mediated apoptotic effect of biosynthesized gold nanopar-ticles on human cervical carcinoma cells (HeLa). *Mater. Res. Bull.* 52: 15–24.

Kalimuthu, K., Lubin, B.-C., Bazylevich, A., Gallerman, G., Shpilberg, O., Luboshits, G., Firer, M.A. 2018. Gold nanoparticles stabilize peptide-drug-conjugates for sustained targeted drug delivery to cancer cells. *J. Nanobiotechnol.* 16(1): 34.

Khademi, S., Sarkar, S., Zadeh, A.S., Attaran, N., Kharrazi, S., Ay, M.R., Azimian, H., Ghadiri, H. 2019. Targeted gold nanoparticles enable molecular CT imaging of head and neck cancer: An in vivo study. *Int. J. Biochem. Cell Biol.* 114: 105554.

Khatoon, N., Alam, H., Khan, A., Raza, K., Sardar, M. 2019. Ampicillin silver nanoformula-tions against multidrug resistant bacteria. *Sci. Rep.* 9(1): 6848.

Kievit, F.M., Veiseh, O., Bhattarai, N., Fang, C., Gunn, J.W., Lee, D., Ellenbogen, R.G., Olson, J.M., Zhang, M. 2009. PEI–PEG–chitosan-copolymer-coated iron oxide nanoparticles for safe gene delivery: Synthesis, complexation, and transfection. *Adv. Funct. Mater.* 19(14): 2244–2251.

Kim, M.-C., Lin, M.M., Sohn, Y., Kim, J.-J., Kang, B.S., Kim, D.K. 2015. Polyethyleneimine-associated polycaprolactone – Superparamagnetic iron oxide nanoparticles as a gene delivery vector. *J. Biomed. Mater. Res. Part B.* 105(1): 145–154.

Lee, S.J., Muthiah, M., Lee, H.J., Lee, H.-J., Moon, M.-J., Che, H.-L., Heo, S.U., Lee, H.C., Jeong, Y.Y., Park, I.-K. 2012. Synthesis and characterization of magnetic nanoparti-cle-embedded multi-functional polymeric micelles for MRI-guided gene delivery. *Macromol. Res.* 20(2): 188–196.

Liao, Z., Wang, H., Wang, X., Wang, C., Hu, X., Cao, X., Chang, J. 2010. Biocompatible surfactin-stabilized superparamagnetic iron oxide nanoparticles as contrast agents for magnetic resonance imaging. *Colloids Surf. A* 370(1–3): 1–5.

Liu, H., Wang, H., Guo, R., Cao, X., Zhao, J., Luo, Y., Shen, M., Zhang, G., Shi, X. 2010. Size-controlled synthesis of dendrimer-stabilized silver nanoparticles for X-ray computed tomography-imaging applications. *Polym. Chem.* 1(10): 1677–1683.

Liu, Y., Xu, M., Zhao, Y., Chen, X., Zhu, X., Wei, C., Zhao, S., Liu, L., Qin, X. 2019. Flower-like gold nanoparticles for enhanced photothermal anticancer therapy by the delivery of pooled siRNA to inhibit heat shock stress response. *J. Mater. Chem. B* 7(4): 586–597.

Lopez, L.Z.F., Gomez, H.E., Somanathan, R. 2019. Silver nanoparticles: Electron transfer, reactive oxygen species, oxidative stress, beneficial and toxicological effects. Mini review. *J. Appl. Toxicol.* 39(1): 16–26.

Martinez, N.N., Mendez, M.F.S., Ruiz, F. 2019. Molecular mechanisms of bacterial resistance to metal and metal oxide nanoparticles. *Int. J. Mol. Sci.* 20: 2808.

McHugh, G.L., Moellering, R.C., Hopkins, C.C., Swartz, M.N. 1975. Salmonella typhimurium resistant to silver nitrate, chloramphenicol, and ampicillin. *Lancet* 305(7901): 235–240.

Mendes, R., Pedrosa, P., Lima, J.C., Fernandes, A.R., Baptista, P.V. 2017. Photothermal enhancement of chemotherapy in breast cancer by visible irradiation of gold nanoparticles. *Sci. Rep.* 7(1): 10872.

Miller, M.M., Lazarides, A.A. 2005. Sensitivity of metal nanoparticle surface plasmon resonance to the dielectric environment. *J. Phys. Chem. B* 109(46): 21556–21565.

Munoz, R.V., Villezcas, A.M., Fournier, P.G.J., Castro, E.S., Moreno, K.J., Hernandez, A.L.G., Bogdanchikova, N., Duhalt, R.V., Saquero, A.H. 2019. Enhancement of antibiotics antimicrobial activity due to the silver nanoparticles impact on the cell membrane. *PLOS ONE* 14: 0224904.

Na, H.B., Lee, J.H., An, K., Park, Y., Park, M., et al. 2007. Development of a T_I contrast agent for magnetic resonance imaging using MnO nanoparticles. *Angew. Chem. Int. Ed. Engl.* 46(28): 5397–5401.

Oh, J.Y., Kim, H.S., Palanikumar, L., Go, E.M., Jana, B., Park, S.A., Kim, H.Y., Kim, K., Seo, J.K., Kwak, S.K., Kim, C., Kang, S., Ryu, J.-H. 2018. Cloaking nanoparticles with protein corona shield for targeted drug delivery. *Nat. Commun.* 9(1): 4548.

Paciotti, G.F., Zhao, J., Cao, S., Brodie, P.J., Tamarkin, L., Huhta, M., Myer, L.D., Friedman, J., Kingston, D.G.I. 2016. Synthesis and evaluation of paclitaxel-loaded gold nanoparticles for tumor-targeted drug delivery. *Bioconjug. Chem.* 27(11): 2646–2657.

Peng, L.-H., Niu, J., Zhang, C.-Z., Yu, W., Wu, J.-H., Shan, Y.-H., Wang, X.-R., Shen, Y.-Q., Mao, Z.-W., Liang, W.-Q. 2014. TAT conjugated cationic noble metal nanoparticles for gene delivery to epidermal stem cells. *Biomaterials* 35(21): 5605–5618.

Petrov, D.A., Ivantsov, R.D., Zharkov, S.M., Velikanov, D.A., Molokeev, M.S., Lin, C.R., Tso, C.T., Hsu, H.S., Tseng, Y.T., Lin, E.S., Edelman, I.S. 2020. Magnetic and magneto-optical properties of Fe_3O_4 nanoparticles modified with Ag. *J. Magn. Magn. Mater.* 493: 165692.

Pollok, N.E., Rabin, C., Smith, L., Crooks, R.M. 2019. Orientation controlled bioconjugation of antibodies to silver nanoparticles. *Bioconjug. Chem.* 30(12): 3078–3086.

Popovtzer, R., Agrawal, A., Kotov, N.A., Popovtzer, A., Balter, J., Carey, T.E., Kopelman, R. 2008. Targeted gold nanoparticles enable molecular CT imaging of cancer. *Nano Lett.* 8(12): 4593–4596.

Prasher, P., Sharma, M., Mudila, H., Gupta, G., Sharma, A.K., Kumar, D., Bakshi, H.A., Negi, P., Kapoor, D.N., Chellappan, D.K., Tambuwala, M.M., Dua, K. 2020. Emerging trends in clinical implications of bio-conjugated silver nanoparticles in drug delivery. *Colloids Interface Sci. Commun.* 35: 100244.

Prasher, P., Singh, M., Mudila, H. 2018. Oligodynamic effect of silver nanoparticles. *BioNanoScience* 8(4): 951–962.

Pugazhendi, A., Edison, T.N.J.I., Karuppusamy, I., Kathirvel, B. 2018. Inorganic nanoparticles: A potential cancer therapy for human welfare. *Int. J. Pharm.* 539(1–2): 104–111.

Rai, M., Ingle, A.P., Birla, S., Yadav, A., Santos, C.A.D. 2016. Strategic role of selected noble metal nanoparticles in medicine. *Crit. Rev. Microbiol.* 42(5): 696–719.

Ravindran, A., Chandran, P., Khan, S.S. 2013. Biofunctionalized silver nanoparticles: Advances and prospects. *Colloids Surf. B* 105: 342–352.

Raza, M.A., Kanwal, Z., Rauf, A., Sabri, A.N., Riaz, S., Naseem, S. 2016. Size- and shape-dependent antibacterial studies of silver nanoparticles synthesized by wet chemical routes. *Nanomaterials* 6(4): 74.

Reuveni, T., Motiei, M., Romman, Z., Popovtzer, A., Popovtzer, R. 2011. Targeted gold nanoparticles enable molecular CT imaging of cancer: An in vivo study. *Int. J. Nanomed.* 6: 2859–2864.

Riggle, P.J., Kumamoto, C.A. 2000. Role of a Candida albicans P1-type ATPase in resistance to copper and silver ion toxicity. *J. Bacteriol.* 182(17): 4899–4905.

Roy, A., Bulut, O., Some, S., Mandal, A.K., Yilmaz, M.D. 2019. Green synthesis of silver nanoparticles: Biomolecule-nanoparticle organizations targeting antimicrobial activity. *RSC Adv.* 9(5): 2673–2702.

Sengupta, M., Pal, R., Nath, A., Chakraborty, B., Singh, L.M., Das, B., Ghosh, S.K. 2018. Anticancer efficacy of noble metal nanoparticles relies on reprogramming tumor-associated macrophages through redox pathways and pro-inflammatory cytokine cascades. *Cell. Mol. Immunol.* 15(12): 1088–1090.

Shaikh, S., Nazam, N., Rizvi, S.M.D., Ahmad, K., Baig, M.H., Lee, E.J., Choi, I. 2019. Mechanistic insights into the antimicrobial actions of metallic nanoparticles and their implications for multidrug resistance. *Int. J. Mol. Sci.* 20(10): 2468.

Sharma, H., Mishra, P.K., Talegaonkar, S., Vaidhya, B. 2015. Metal nanoparticles: A theranostics nanotool against cancer. *Drug Disc. Today* 20(9): 1143–1151.

Siddiqui, T.S., Jani, A., Williams, F., Muller, R.N., Elst, L.V., Laurent, S., Yao, F., Wadghiri, Y.Z., Walters, M.A. 2009. Lanthanide complexes on Ag nanoparticles: Designing contrast agents for magnetic resonance imaging. *J. Colloid. Interface Sci.* 337(1): 88–96.

Singh, M., Prasher, P. 2018. Ultrafine silver nanoparticles: Synthesis and biocidal studies. *BioNanoScience* 8(3): 735–741.

Thomas, M., Klibanov, A.M. 2003. Conjugation to gold nanoparticles enhances polyethylenimine's transfer of plasmid DNA into mammalian cells. *Proc. Nat'l. Acad. Sci* 100(16): 9138–9143.

Tsai, Z.-T., Wang, J.-F., Kuo, H.-Y., Shen, C.-R., Wang, J.-J., Yen, T.-C. 2010. In situ preparation of high relaxivity iron oxide nanoparticles by coating with chitosan: A potential MRI contrast agent useful for cell tracking. *J. Magn. Mag. Mater.* 322(2): 208–213.

Wadhwani, P., Heidenreich, N., Podeyn, B., Burck, J., Ulrich, A.S. 2017. Antibiotic gold: Tethering of antimicrobial peptides to gold nanoparticles maintains conformational flexibility of peptides and improves trypsin susceptibility. *Biomater. Sci.* 5(4): 817–827.

Wahajuddin, Arora, S. 2012. Superparamagnetic iron oxide nanoparticles: Magnetic nanoplatforms as drug carriers. *Int. J. Nanomed.* 7: 3445–3471.

Wei, H., Burns, O.T., Kaul, M.G., Hansen, E.C., Barch, M., et al. 2017. Exceedingly small iron oxide nanoparticles as positive MRI contrast agents. *Proc. Nat'l. Acad. Sci.* 114(9): 2325–2330.

Yang, P., Zheng, J., Xu, Y., Zhang, Q., Jiang, L. 2016. Colloidal synthesis and applications of plasmonic metal nanoparticles. *Adv. Mater.* 28(47): 10508–10517.

Yin, X., Russek, S.E., Zabow, G., Sun, F., Mohapatra, J., Keenan, K.E., Boss, M.A., Zeng, H., Liu, J.P., Viert, A., Liou, S.-H., Moreland, J. 2018. Large T_1 contrast enhancement using superparamagnetic nanoparticles in ultra-low field MRI. *Sci. Rep.* 8(1): 11863.

Yu, N., Wang, X., Qiu, L., Cai, T., Jiang, C., Sun, Y., Li, Y., Peng, H., Xiong, H. 2020. Bacteria-triggered hyaluronan/AgNPs/gentamicin nanocarrier for synergistic bacteria disinfection and wound healing application. *Chem. Engg. J.* 380: 122582.

16 Synthesis, Characterization, and Activity of Maghemite (γ-Fe$_2$O$_3$) Nanoparticles through a Facile Solvent Hydrothermal Phase Transformation of Fe$_2$O$_3$

Rafia Azmat, Iqbal Altaf, Amina Pervaiz, and Summyia Masood
University of Karachi, Karachi, Pakistan

CONTENTS

16.1   INTRODUCTION

The advancements in science and technology at nanoscale has attained a significant position in the field of research globally, due to the elimination of pollutants and toxins during all biological and chemical processing (Chen et al. 2016, Jarzębski, Kościński, and Białopiotrowicz 2017, Maity and Agrawal 2007, Abkenar et al. 2014, Iram et al. 2010, Azmat, Altaf, and Asma 2020). The recent progress in nanoscience and nanotechnology, involving the synthesis of magnetic nanoparticles (MNPs), has attracted intense experimental activity because of their potential applications. This is due to their size, composition, and growth morphology, with tremendous physical, chemical, thermal and mechanical properties. Iron oxide nanoparticles, with magnetite (Fe_3O_4) and maghemite (Fe_2O_3), are useful in nanomedicine in pharmaceuticals where it is used in cell separation, drug carriers for targeted cancer therapy, microfluidic sensors for pathogens diagnosis as Nano markers (Liu et al. 2016, Liu et al. 2003, Cheng et al. 1998, Togashi et al. 2011, Aivazoglou, Metaxa, and Hristoforou 2018, Qadri, et al., 2009). The processes of the synthesis of Iron NPs are very significant for diverse applications in the environment (Jeong et al. 2004). Several processes developed for the synthesis of MNP, including the microwave-assisted process, electrochemical methods, microemulsions, polyol, and sol-gel process, and the decomposition of organic precursors at high temperature according to their biomedical applications (Lin, Chang, and Chuang 2008, Gui et al. 2012). Literature in the field has revealed that the reactive metallic element iron at nanoscale has high potential to be used as a catalyst. Studies have shown its reactivity against the degradation of dyes but usually, the synthetic procedures are quite complicated with the involvement of various chemicals (Hoag et al. 2009). Therefore, the need has arisen to formulate a cost-effective method of the synthesis of iron NPs that is much more straight forward to prepare, and that can also be used at bulk scale. The synthesis of iron NPs depends on their applications in a diverse field. For instance, the removal of acridine orange dye was carried out by Fe_2O_3 NPs and the reaction kinetics were studied (Gao et al. 2008). Ti and Fe-mediated nanocomposites have been synthesized by the seed-mediated method and used for the degradation of methylene blue under Hg light in a photocatalytic reactor (Wu et al. 2011). In another report, green tea- synthesized nZVI (Fe0) NPs was employed for the catalytic degradation of methylene blue (MB) and methyl orange (MO) dyes. The almost complete removal of the dyes was achieved after 200 min for MB and 350 min for MO, under the studied conditions (Shahwan et al. 2011, Sun and Zeng 2002, Zhang et al. 2006, Njagi et al. 2010, Ramasubbu, Saravanan, and Vasanthkumar 2011). Fe_2O_3 NPs prepared by hydrothermal phase transfer, was used for the degradation of paracetamol in aqueous solution, and monitored by infrared spectroscopy (Altaf and Azmat 2020). The maghemite (γ-Fe_2O_3) is a significant material having several applications in science and technology, including the synthesis of nanomedicine, spin electronic devices, biosensors, and high-density magnetic recording. Keeping the significance, related to the diverse application of iron nanomaterials in the scientific field, an effort has been made to develop an economical, cost-effective and direct process for controlling biological and chemical organic contaminants. Current research reports a unified approach for the synthesis of γ-Fe$_2$O$_3$ NPs through a facile solvent hydrothermal method, followed by the calcination process. The synthesis is different from reported work, as it based on the direct use of a Fe$_2$O$_3$ compound without adjusting or maintaining pH, and reaction complexity involving several reagents. The characterizations were carried out using advanced technologies with chemical applications on the degradation of Congo red (Figure 16.1), a carcinogenic, mutagenic, and toxic azo dye, used extensively. It is vulnerable for many living organisms due to its stability with aromaticity which is difficult to remove from wastewater.

16.2 EXPERIMENTAL SECTION

16.2.1 PREPARATION OF FE NANOPARTICLES

All chemicals purchased were of analytical grade and used as received without further purification. The hydrothermal method for synthesis of iron nanoparticles (NPs) as described by Liu and co-workers (Liu et al. 2003), modified to synthesize Fe NPs, in order to get the required morphology and particle size. 5 g Fe$_2$O$_3$ (Merck), 20 ml CH$_2$CH$_3$OH (Merck), and 280 mL of distilled water were added in a 500 ml conical flask (the opening covered with aluminum foil with holes), and placed in an autoclave at 120–130 kPa pressure for 12 hours at 120° C. The obtained red precipitates were filtered through Whattman no. 40 filter paper, then washed with distilled water and dried in an oven at 100° C for 4 hours. The obtained product was calcined in a furnace at 350° C for 4 hours.

16.2.2 CHARACTERIZATION OF SYNTHETIC NPS

The morphology of the synthesized nanomaterial was investigated using a Quanta Inspect F scanning electron microscope (SEM), operating at 25 kV in a vacuum in conjunction with a surface chemical elemental analysis energy dispersive spectroscopy

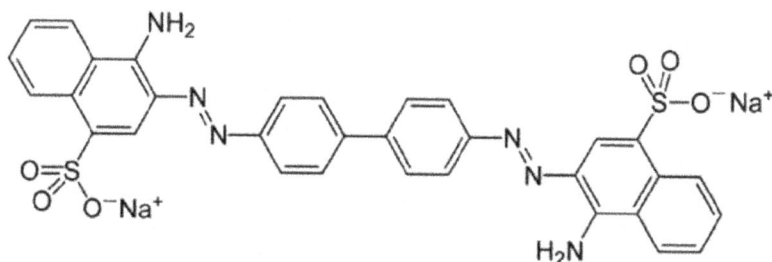

FIGURE 16.1 Congo Red (Disodium salt of 3,3'-(1E,1'E)-biphenyl-4,4'-diylbis(diazene-2,1-diyl)bis(4-aminonaphthalene-1-sulfonate).

(EDS) detector from EDAX. The operating conditions were an accelerating voltage between 2 and 25 keV (depending on the signal/noise ratio) for samples tilted at 25° in order to get the optimal take-off angle (30°), allowing a dead time around 20–30% and a collecting time of 90–120. The optical absorption spectra recorded with UV–Vis 1800 A Shimadzu spectrophotometer. Distilled water was used as a reflectance standard in the experiment. Fourier transform infrared spectrograph (FTIR) analysis was carried out by using Perkin-Elmer Inc. Samples prepared by mixing ~2 mg of Hap powder and ~200 mg of spectroscopic-grade KBr (Merck) and pressing them to a disk at 15 MPa. Infrared spectra were recorded in the region 4000–400 cm^{-1}, with a resolution of 4.00 cm^{-1}.

16.2.3 SORPTION STUDIES

The Congo red dye was selected for studying the catalytic properties of the transformed NPs (Figure 16.1).

16.2.3.1 pH Optimization

Batch sorption studies were monitored on aqueous solutions of CR to determine the optimum pH for dye wastewater removal through synthesized Iron NPs. The required concentration of dye (100 mg/L) solution prepared via dilution of the stock solution and pH adjusted using NaOH and H$_2$SO$_4$ in the range of 2–9. The absorbance of the solution was recorded after filtration using a spectrophotometer (UV-1800 Shimadzu).

16.2.3.2 Optimization of Contact Time

The contact time was determined by diluting the stock solution to an optimum concentration of 100 ppm at a pH of 2.5 followed by adding 0.5 g of iron NP with 0.4 ml of 1 M of H$_2$SO$_4$. Initially, the 5 ml of reaction mixture was withdrawn before the addition of NPs powder (t = 0) to record initial absorbance. Subsequent withdrawals were conducted between 2 and 20 minutes for recording the spectral changes. The mixture was stirred continuously with a magnetic stirrer. The 5 ml portions of this solution were withdrawn at regular time intervals of 2 min, and each portion was immediately centrifuged through an electronic centrifuge at 2500 rpm.

16.2.3.3 Optimization of Initial Dye Concentration and Adsorption Capacity

100 mL of standard Congo red dye solutions of concentrations between 10–250 mg/L at their respective optimum pH were equilibrated with 0.5 g of iron NPs powder for 20 min. 5 mL aliquot was withdrawn and centrifuged through electronic centrifuge at 2500 rpm to record spectral change. The data obtained was fitted to two adsorption isotherms models. The initial dye concentration in the reaction mixture and the adsorbent amount varied, allowing for the exploration of their consequences on the adsorption kinetics.

16.2.3.4 Recovery of NPs

The NPs were recovered after the reaction to check catalytic activity by simple washing and drying at a temperature of 80°C, then subjected to FTIR, SEM, and EDS.

16.3 RESULTS AND DISCUSSIONS

Magnetic maghemite, γ-Fe$_2$O$_3$ NPs were synthesized by the phase transferred solvent hydrothermal method, using a Fe$_2$O$_3$ precursor, and characterized by following advanced technologies.

16.3.1 CHARACTERIZATION OF γ-FE$_2$O$_3$ THROUGH SEM AND EDS

The image of synthesized NPs obtained through an emission-scanning electron microscopy (FESEM), presented in Figure 16.2a, b, and c. These are in comparison to the original bulk compound (Fe$_2$O$_3$). It revealed the morphologies of synthesized nanostructures of the phase transformation of Fe$_2$O$_3$ into the γ-Fe$_2$O$_3$ revealed small honeycomb-like structures and nanorods. Surface scan and structure distribution can be observed in Figure 16.2a to 16.2c for the identification and understanding of the structural science. The different phase structure and morphology of prepared NPs suggested that Fe$_2$O$_3$ can be transformed into active NPs (Figure 16.2b) which showed that NPs somehow aggregated, i.e. were found in clusters, which is the characteristic property of γ-Fe$_2$O$_3$ NPs. Figure 16.2a showed original iron precursor with

(a)

(b)

(c)

FIGURE 16.2 (a) Scanning electron microscopy images of bulk Fe$_2$O$_3$ showing particle size. (b) Scanning electron microscopy images of synthesized Fe$_2$O$_3$ nanoparticles showing particle size. (c) Scanning electron microscopy images of recovered Fe$_2$O$_3$ nanoparticles showing particle size.

a size of 1.5 to 2 μm, while as-prepared γ-Fe_2O_3 structures have diameters of 74 to 97 nm with an average diameter of 83 nm (Figure 16.2b).

The electron dispersive spectrum reveals a strong signal in the iron region, confirming the formation of NPs. The percentage of iron in the precursor is 76.69; oxygen is 23.31 (Table 16.1) which displays a pure particle. The EDS spectrum (Figure 16.3b) of Fe_2O_3 nanocomposites has Fe (67.19%), O (28.59%), and C (4.22%) (Table 16.1). It showed that the EDS spectra are free from any contaminations and interfering species which confirmed the chemical composition of the target particles that it contains iron and oxygen only. It verified the integrity and simplicity of the method of synthesis. Figure 16.2c showed recovered NPs in which particle size is 78.8 and 99.4 nm whereas Fe (68.63%), O (25.14%), and C (6.24%) (Table 16.1). The results of the present investigation are comparable with the earlier work (Maity and Agrawal 2007), who reported the synthesis of maghemite NPs in an oxidizing environment. They observed the size through TEM micrographs as 14, 10, and 9 nm for the water, kerosene, and dodecane-based ferrofluids, respectively. Jarzębski and co-workers (2017) synthesized iron oxide-NPs (Fe_3O_4) by three methods through modification, using dextran and gelatin as a shell for iron oxide particles. Furthermore, they confirmed that the synthesized magnetic nanoparticles (MNP) have a diameter in a range from 5 to 50 nm (Jarzębski, Kościński, and Białopiotrowicz 2017). Abkenar and co-workers (2014) observed spherical shapes and a nearly uniform distribution of particle size of less than 50 nm through the TEM technique (Abkenar et al. 2014). Vélez and co-workers (2016) observed the SEM images of Iron NPs, it showed particle sizes less than 100 nm which shows the agreement of current results with the previously reported work.

16.3.2 CHARACTERIZATION OF γ-Fe_2O_3 THROUGH FTIR SPECTRA

The FT-IR spectra of phase transferred γ-Fe_2O_3 NPs are presented in Figure 16.4. It shows good agreement with those presented in earlier published work. Figure 16.4b shows the IR spectrum of phase transferred γ-Fe_2O_3 NPs which is comparable with bulk Fe_2O_3 (Figure 16.4a. Both spectrums reflect three main areas, 3000–3800 cm^{-1}, 1300–2200 cm^{-1}, and 400–1250 cm^{-1}. The spectrum of Fe_2O_3 and γ-Fe_2O_3 NPs showed a fixed broad band at 3437 and 3439.08 cm^{-1} respectively, which is due to the stretching vibration of H_2O molecules, or bonded H, or absorbed water molecule, or OH groups found on the surface, which was similar to the previous reports

TABLE 16.1

ZAF Method Standard Less Quantitative Analysis Fitting Coefficient: 0.3368

Element	(keV)	Fe Compound (mass%)	Original Np (mass%)	Recovered Nanoparticles (mass%)
C K	0.277	–	4.22	6.24
O K	0.525	23.31	28.59	25.14
Fe K	6.398	76.69	67.19	68.63

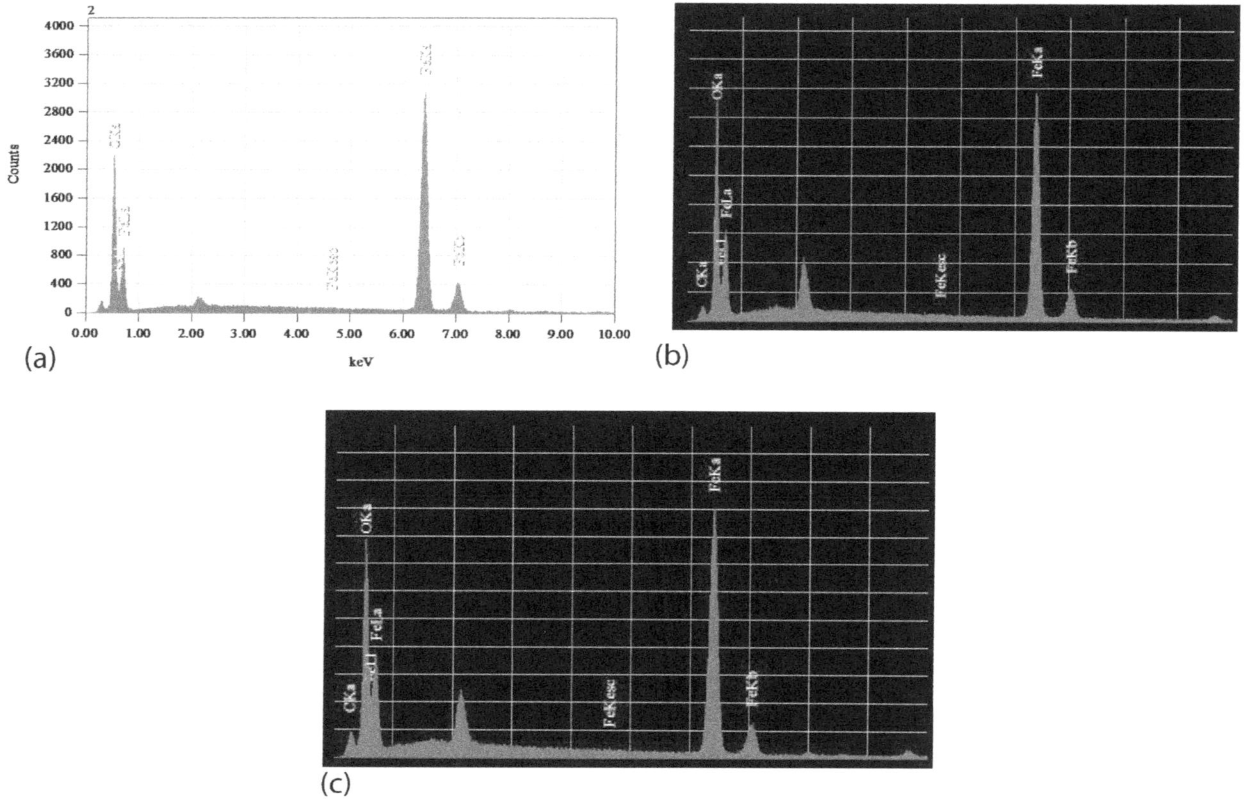

FIGURE 16.3 (a) Energy dispersive X-ray spectroscopy of bulk Fe₂O₃. (b) Energy dispersive X-ray spectroscopy of synthesized Fe₂O₃ nanoparticles. (c) Energy dispersive X-ray spectroscopy of recovered Fe₂O₃ NPs.

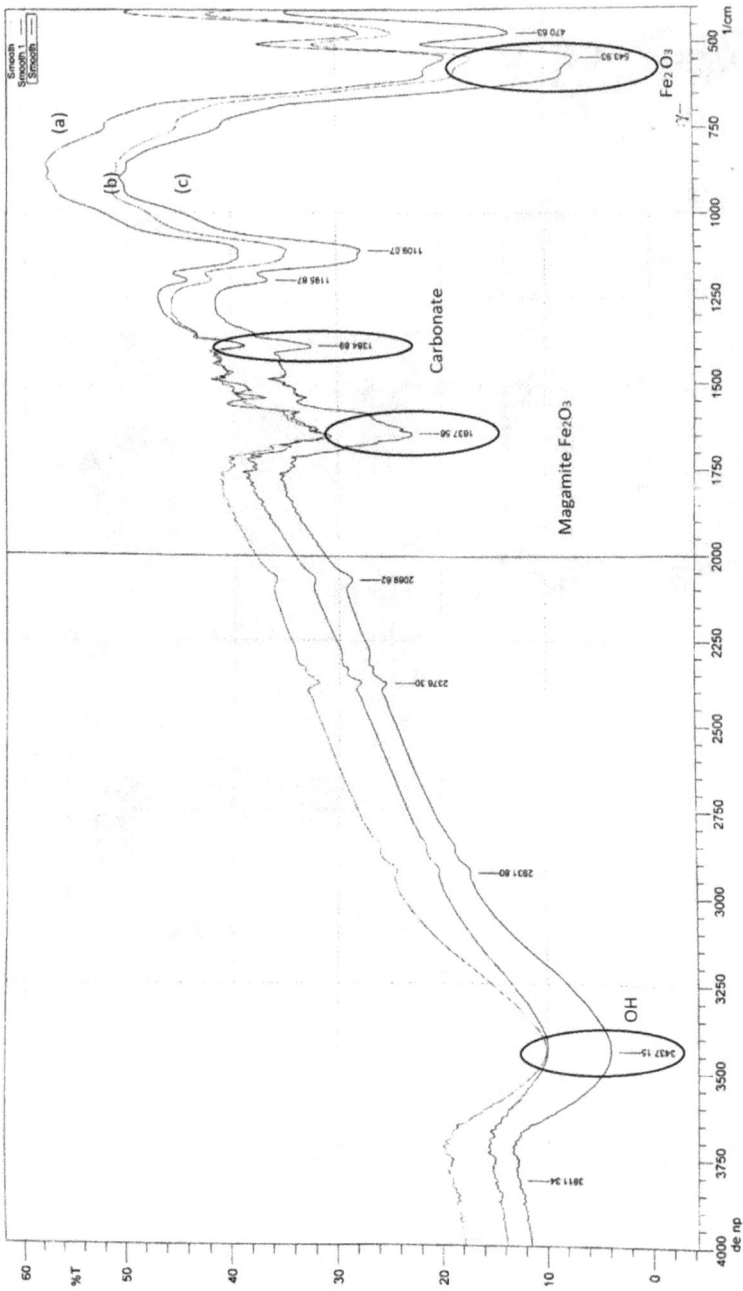

FIGURE 16.4   Fourier transform infrared analysis of: (a) bulk Fe_2O_3; (b) synthesized Fe_2O_3 nanoparticles; (c) recovered Fe_2O_3 nanoparticles.

related to the Fe NPs (Cheng et al. 1998, Yu and Chow 2004, Jeong et al. 2004). The band corresponding to the bending vibrations of H_2O molecules positioned at 1624 cm^{-1} in γ-Fe$_2$O$_3$ while it was absent in bulk compound Fe$_2$O$_3$. The IR bands at 1512, 1354, and 1195 cm^{-1} in the spectrum of γ-Fe$_2$O$_3$ NPs showed the presence of carbonate groups which is comparable with earlier work (Aivazoglou, Metaxa, and Hristoforou 2018). The presence of a carbonate group is due to the adsorption of atmospheric carbon dioxide by γ-Fe$_2$O$_3$ NPs, while these peaks were absent in bulk Fe$_2$O$_3$ (Figure 16.4(a)), which was according to the reports in literature (Liu et al. 2016). Two strong IR bands at 594 and 470 cm^{-1} are typical for ferrihydrite or γ-Fe$_2$O$_3$ NPs represent the bands of Fe–O stretching vibrations of γ-Fe$_2$O$_3$ and the bands at 914 and 746 cm^{-1} can be assigned to Fe–OH···H bending vibrations, showed good agreement with the reported work (Chahkandi 2017). Darezereshki (2010) also reported an FTIR analysis of γ-Fe$_2$O$_3$ where absorption peak at 587 cm^{-1} was related to the vibration of Fe–O, the other peaks at 454, 632, 795, and 892 cm^{-1} are pure maghemite, which was approximately similar to current work. Jarlbring et al. 2005 observed two broad bands with shoulders at 724, 694,638, 584, 558, 442, and 396 cm^{-1} which were characteristic for well-ordered maghemite. The band at 580 cm^{-1} assigned to Fe–O deformation in the octahedral and tetrahedral sites. These IR bands of Fe$_2$O$_3$ and Fe$_3$O$_4$ showed that bulk iron oxide changes into maghemite and magnetite NPs as Fe$_2$O$_3$ is a simple oxide in which the oxidation state of Fe is +3 only, while Fe$_3$O$_4$ is a mixed oxide in which Fe is present in both +2 and +3 oxidation states. It may also be written as FeO.Fe$_2$O$_3$, while Fe$_2$O$_3$ is written as iron (III) oxide, while in Fe$_3$O$_4$ iron (II,III) oxide (Figure 16.4b).

The IR spectra of recovered particle showed more or less the same bands as in original NPs, according to which Fe-O is at 594 cm^{-1} and 590 cm^{-1} while the other same bands at 1195 and 1195, 1388, and 1390 cm^{-1} etc. in synthesized and recycled NPs respectively (Figure 16.4c).

16.3.3 UV/VISIBLE ANALYSIS OF CONGO RED DEGRADATION

The activity of synthesized NPs checked on CR degradation in batch experiments by keeping one parameter as variable and other constant (including contact time, pH, the effect of dye, and NPs dosage). The reaction was monitored by adding several dosages of adsorbent (0.05–0.6 g) into the flask containing 100 mL of the dye solution. The initial dye concentrations and the pH of the solutions were fixed (acidic) at 100 mg/L respectively for all batch experiments. The suspension stirred and underwent centrifuge after a regular time interval for recording the absorbance of the remaining dye. The slow kinetics observed at neutral pH where reaction completed in 24 h. On the other hand, the rapid kinetics were pragmatic at pH 2.5 in 20 min. The spectral changes related to the degradation of CR with respect to time presented in Figure 16.5, while a visual illustration is shown in Figure 16.6. Furthermore, no change in the reaction was recorded at the alkaline medium.

The amount of NPs varied from 0.05 to 0.6 g where it was observed that the degradation capacity reached a maximum from 64–93 % respectively at 0.5 g when the dye concentration varied from 10 ppm to 250 ppm. It was observed that 64% of the dye was removed at the initial dosage of 0.05 g. The removal of the dye increased

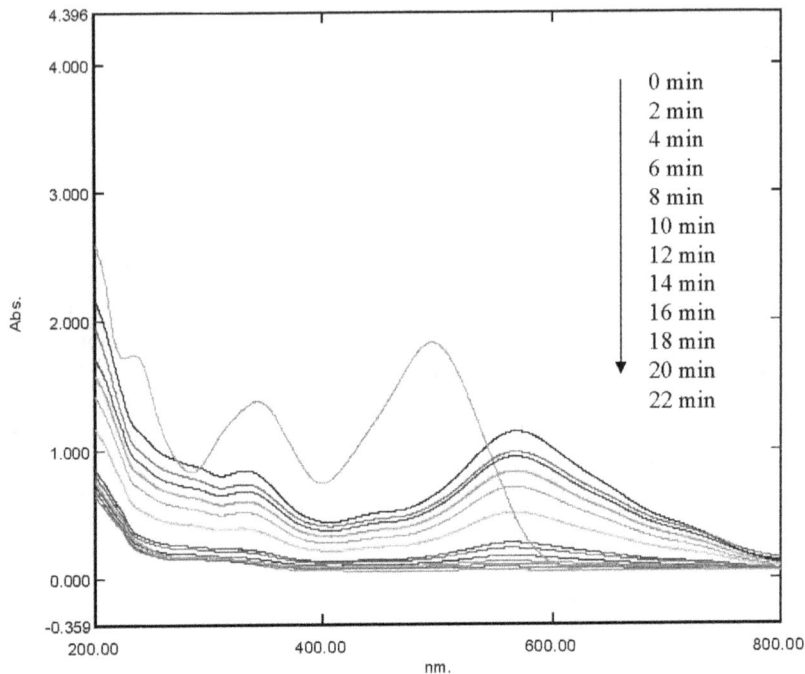

FIGURE 16.5 Structural change of Congo Red 100 (mg/L) during catalytic degradation by Fe$_2$O$_3$ nanoparticles.

FIGURE 16.6 Visual change of CR solution. (a) Congo Red solution; (b) Congo Red solution after reaction with Fe$_2$O$_3$ nanoparticles.

with raising the γ-Fe$_2$O$_3$ dosage to 0.5 g where it reached over 95%. The removal of the CR with such dosages can be explained by the higher number of adsorption sites available for dye molecules. Furthermore, the amount of NPs greater than 0.5 g did not affect the removal of any of the dye. Hence, the optimum dosage of γ-Fe$_2$O$_3$ powder for removing CR dye was found to be 0.5 g. It is noteworthy that the sorption capability for CR on the γ-Fe$_2$O$_3$ in degradation can reach a high adsorption capacity with 200 ppm, which is similar to previous reports (Chen et al. 2016). Therefore, when γ-Fe$_2$O$_3$ is applied for dye wastewater treatment, the amount of adsorbent can play a significant role in adsorption and reduction of the time of adsorption.

16.3.4 INVESTIGATION OF KINETICS AND REACTION MECHANISM

The adsorption studies monitored using Langmuir adsorption isotherm (Equation 16.1) and Freundlich adsorption isotherm (Equation 16.2) are given below.

16.3.4.1 Langmuir Adsorption Isotherm

$$C_e/q_e = C_e/c_m + 1/k_L C_m \qquad (16.1)$$

Where q_e is the amount of dye adsorbed (mg/g), C_e is the equilibrium concentration of the adsorbate (mgL^{-1}), and c_m and k_L are Langmuir constants related to the maximum adsorption capacity (mg g^{-1}) and energy of adsorption (L mg^{-1}).

16.3.4.2 Freundlich Adsorption Isotherm

$$\ln q_e = \ln C_e/n + \ln k_f \qquad (16.2)$$

Where q_e is the amount of dye adsorbed onto the adsorbent at equilibrium (mg/g), C_e is the equilibrium concentration of the adsorbate (mgL^{-1}), and k_f (min^{-1}) is the rate constant of the Freundlich isotherm.

The adsorption data is used for modeling and for kinetics of first- and second-order reactions. The Langmuir and Freundlich models were applied to analyze the experimental sorption equilibrium parameters. According to the Langmuir isotherm model, there is a finite number of active sites available which are homogeneously distributed, over the surface where molecules are adsorbed in a single layer with no interaction between adsorbed molecules. The adsorption equilibrium data at a high and low concentration of dye fitted very closely to the Langmuir adsorption isotherm model, confirming monolayer adsorption (Figure 16.7a and b, Table 16.2). According to the equation (Equation 16.1), when the adsorption obeys the Langmuir equation, a plot of C_e/q_e versus C_e is a straight line with a slope of $1/c_m$ and intercepts $1/k_L$. This important characteristic of the Langmuir isotherm can be expressed in terms of a dimensionless factor, R_L (Chahkandi 2017), which is defined as that the R_L values indicate the type of adsorption as either unfavorable ($R_L > 1$), linear ($R_L = 1$), favorable ($0 < R_L < 1$), or irreversible ($R_L = 0$). The value of R_L in current research is 0.99, which is less than 1 and indicates that adsorption is favorable and irreversible (Table 16.2). The complete degradation of dye during sorption supports the value

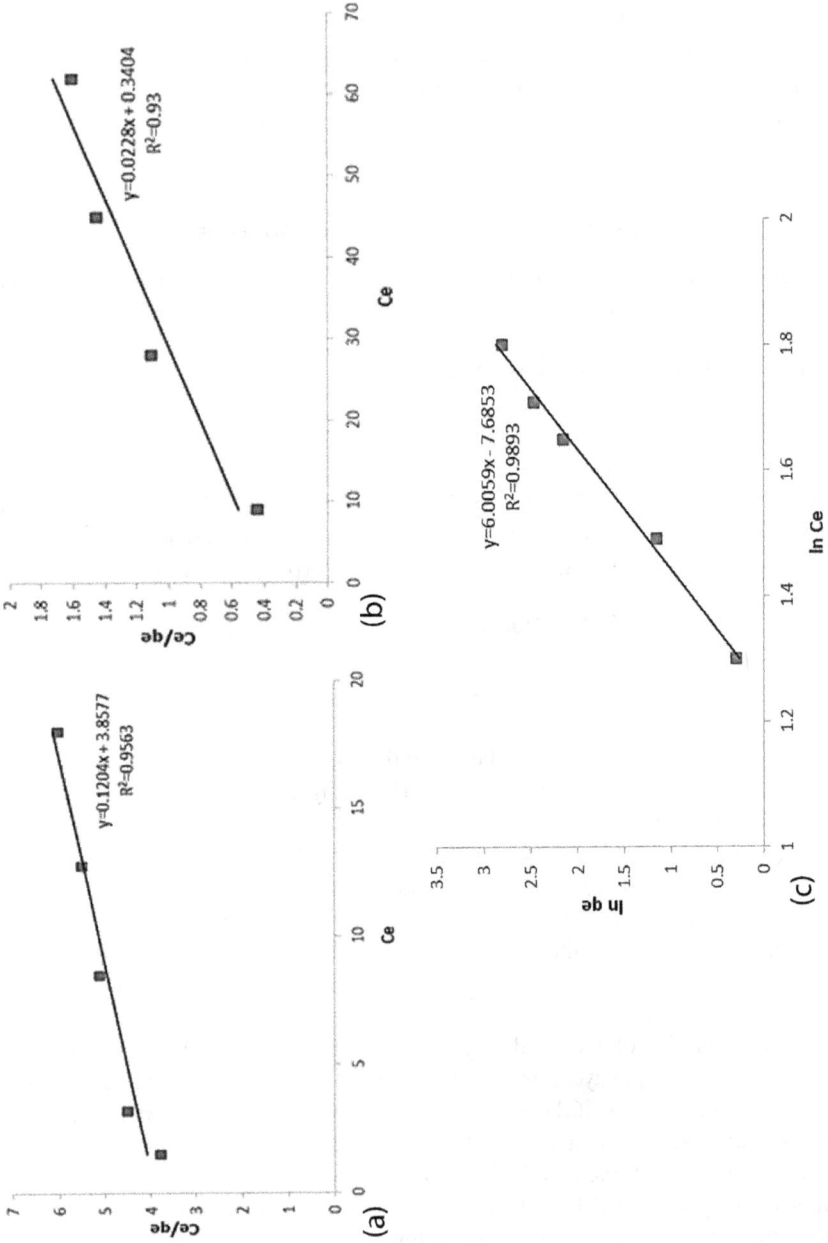

FIGURE 16.7 (a) Adsorption isotherm Langmuir plot at low concentration. (b) Adsorption isotherm Langmuir plot at high concentration. (c) Adsorption isotherm Freundlich plot at high concentration.

TABLE 16.2
The Various Constants Appearing in Langmuir and Freundlich Adsorption Isotherm

	Langmuir Adsorption Isotherm				Freundlich Adsorption Isotherm		
	C_m	K_l	R_L	R^2	K_f	N	R^2
Low conc	8.30	0.03	1	0.95	4.59×10^{-4}	0.16	0.98
High conc	43.85	0.06	1	0.93	11.18	3.66	0.84
Literature Values	105,20.74	0.04,0.05	–	–	4.79,1.21	7.69,1.82	–
Reference	[5],[30]	[5],[30]	–	–	[25]	[25]	–

of R^2 0.9563 (Table 16.2). Table 16.2 shows the various constant appearing in the Langmuir and Freundlich adsorption isotherm models.

The Freundlich isotherm model (Equation 16.2) was tested in adsorption studies of CR on NP surfaces. It showed interaction between the adsorbed molecules, and was not restricted to the formation of a monolayer (Afkhami and Moosavi 2010). This model assumes that as the adsorbate concentration increases, the concentration of adsorbate on the adsorbent surface also increases and, correspondingly, the sorption energy exponentially decreases on completion of the sorption centers of the adsorbent. It is observed that the Freundlich isotherm model fitted at a high concentration where the value of n reaches 4 indicates that at a high concentration surface, energies exponentially decrease after complete adsorption of the dye, followed by degradation of the CR (Figure 16.7c). Figure 16.7c shows adsorption of an isotherm Freundlich plot at high concentration.

The kinetics and order of reaction were monitored at different operational parameters to optimize the catalytic degradation of CR. The Lagergren-pseudo-first-order and pseudo-second-order model were applied to ascertain the type of reaction, either physi-sorption or chemisorption (Shah et al. 2015). These models were used to fit the experimental data to study the kinetics of heterogeneous catalytic degradation of the CR in the Lagergren-first-order model expressed as the equation:

$$Ln(q_e\text{-}q_t) = \ln q_e\text{-}k_1t \qquad (16.3)$$

Where q_t and q_e are the amounts of products (mg g^{-1}) at a time 't' and at equilibrium, k_1 is the rate constant of the pseudo-first-order process (min^{-1}). Straight line plots of $\ln(q_e - q_t)$ against t were used to determine the rate constant k_1, and adsorption capacity (Figure 16.8a and b). From the results, it was concluded that the heterogeneous catalytic process on magnetite did not follow pseudo-first-order kinetics, because the pseudo-first-order model data did not fall on the straight line. The pseudo-second-order equation is expressed as:

$$t/q_t = t/q_e + 1/k_2q_e \qquad (16.4)$$

$$h = kq_{e2}$$

Where h (mg g^{-1}min^{-1}) is the initial degradation rate as t' and k$_2$ is the rate constant of pseudo-second-order degradation (g mg^{-1} min^{-1}). The plot of t/qt versus t for pseudo-second-order model (Figure 16.8b) yields good straight lines (R^2 = 0.9881) as compared to the plot of pseudo-first-order, where the values of q$_e$, k, and h are 23.98 mg/g, 0.388 g/mg/s, and 0.1073 mg/g/s respectively, and were determined from the slope and intercept of the plot. Therefore, it is possible to suggest that the degradation of CR by maghemite NPs followed pseudo-second-order reaction (Table 16.3). The Lagergren-pseudo-first-order and pseudo-second-order kinetics models are shown in Table 16.3.

The results of the present investigation are in accordance with the work of Shah and co-workers (Shah et al. 2015), who reported the successful catalytic hydrogenation of acetone to alcohol by NaBH$_4$ in the presence of microwave radiation, followed by second-order kinetics and the Langmuir-Hinshelwood kinetic mechanism.

16.3.5 RECOVERY OF USED NANOPARTICLES

The adsorption ability of used NPs was checked for its surface activity by a simple washing and heating strategy. The recovered used Fe NPs were washed with distilled water and then with ethanol. After that, heating was done at 250° C for 3 hours for the combustion of dye molecules from the surface of used NPs, and allowed to cool in a desiccator to avoid moisture. These NPs were again subjected to SEM, EDS, FTIR (Figure 16.2c, Figure 16.3c, and Figure 16.4c), and UV/Visible spectroscopy for size, and tested again for CR degradation respectively. It was found that these particles were active at least four times with a non-significant loss of its efficiency. These results are in accordance with the reported work of Das and co-workers (2017) who observed that the Humi Fe$_3$O$_4$ nanocomposite material established an excellent source of magnetically recoverable catalyst for the chemical reduction of toxic p-nitrophenol to p-aminophenol where 90–92% p-nitrophenol converted into p-aminophenol after 6 h of reaction (Das et al. 2017). The used magnetic Humi Fe$_3$O$_4$ was reused three times with a non-significant loss of its catalytic activity. The recovered particle was tested at least three times with more or less the same efficacy. Therefore, it is suggested on the basis of the catalytic activity of recycled NPs that the developed method is more efficient and cost-effective for recovering and managing wastewater.

16.4   CONCLUSIONS

This study provides a simple, modified hydrothermal method for the phase transfer of direct Fe$_2$O$_3$ compound into NPs for their effective use in various fields of science and technology. The method is simple, requiring no special reaction for the conversion of the Fe-containing compound into Fe$_2$O$_3$ except to produce at bulk scale for the industry. The IR and SEM studies showed that the magnetic γ-Fe$_2$O$_3$ NPs at nanoscale could be prepared by using the simple bulk compound for its huge applications with an average diameter (74 to 99 nm) of NPs. The results showed that γ-Fe$_2$O$_3$ is an active adsorbent catalyst for the degradation of CR from aqueous solutions, as 93.6% removal was achieved using 250 ppm of the dye at pH 2.5 and an adsorbent amount of 0.5 g/L. The pseudo-second-order kinetic model fitted the kinetic adsorption process of CR. The NPs were recoverable with all active sites.

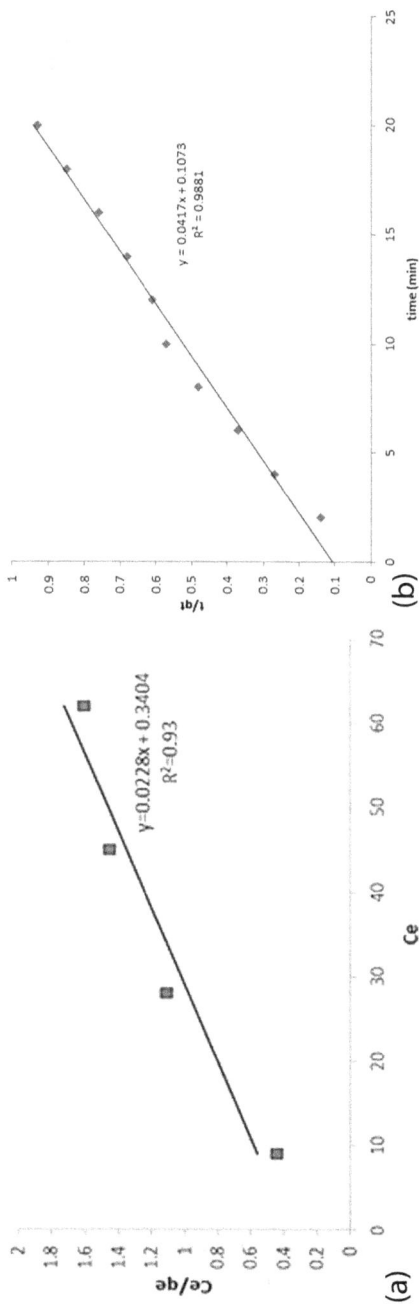

FIGURE 16.8 (a) Pseudo-first-order plot for adsorption kinetic of Congo Red on Fe$_2$O$_3$ nanoparticles. (b) Pseudo-second-order plot for adsorption kinetic of Congo Red on Fe$_2$O$_3$ nanoparticles.

TABLE 16.3
Lagergren Pseudo-First-Order and Pseudo-Second-Order Kinetics Models

Kinetic Model	Parameters	100 ppm Congo Red	Literature Values	References
Pseudo-first-order	K1	0.18	0.0299	[32]
	qe	15.74	36.01	
	R^2	0.82	0.83	
Pseudo-second-order	K2	0.388	0.001	[32]
	qe	23.9	72.99	
	R^2	0.9881	0.99	

ACKNOWLEDGMENTS

The author is thankful to the Dean, Faculty of Science, the University of Karachi for financial support and HEC Pakistan for permanent equipment under Project No. No.20-2282/NRPU/R&D/HEC/12/5014.

CONFLICT OF INTEREST

The authors declare there is no conflict of interest between the authors.

REFERENCES

Abkenar, Shiva Dehghan, Mehdi Khoobi, Roghayeh Tarasi, Morteza Hosseini, Abbas Shafiee, and Mohammad R Ganjali. 2014. "Fast removal of methylene blue from aqueous solution using magnetic-modified Fe_3O_4 nanoparticles." *Journal of Environmental Engineering* 141(1):04014049.
Afkhami, Abbas, and Razieh Moosavi. 2010. "Adsorptive removal of Congo red, a carcinogenic textile dye, from aqueous solutions by maghemite nanoparticles." *Journal of Hazardous Materials* 174(1–3):398–403.
Aivazoglou, E, E Metaxa, and E Hristoforou. 2018. "Microwave-assisted synthesis of iron oxide nanoparticles in biocompatible organic environment." *AIP Advances* 8(4):048201.
Altaf, Iqbal, and Rafia Azmat. 2020. "IR study of degradation of acetaminophen by iron nano-structured catalyst." *Pakistan Journal of Pharmaceutical Sciences* (Accepted, In press article).
Azmat, Rafia, Iqbal, Altaf, and Asma . 2020. "The preparation of TiO2 nanoparticles through hydrothermal phase transformation and its activity in water chemistry." *Pakistan Journal of Pharmaceutical Sciences* no. (Accepted, In press. article).
Chahkandi, Mohammad. 2017. "Mechanism of Congo red adsorption on new sol-gel-derived hydroxyapatite nano-particle." *Materials Chemistry and Physics* 202:340–351.
Chen, Dong, Ziyang Zeng, Yubin Zeng, Fan Zhang, and Mian Wang. 2016. "Removal of methylene blue and mechanism on magnetic γ-Fe2O3/SiO2 nanocomposite from aqueous solution." *Water Resources and Industry* 15:1–13.
Cheng, Zhi Hua, Akemi Yasukawa, Kazuhiko Kandori, and Tatsuo Ishikawa. 1998. "FTIR study on incorporation of CO_2 into calcium hydroxyapatite." *Journal of the Chemical Society, Faraday Transactions* 94(10):1501–1505.

Darezereshki, Esmaeel. 2010. "Synthesis of maghemite (γ-Fe2O3) nanoparticles by wet chemical method at room temperature." *Materials Letters* 64(13):1471–1472.

Das, Tonkeswar, Gayatri Kalita, Priyam Jyoti Bora, Dipak Prajapati, Gakul Baishya, and Binoy K Saikia. 2017. "Humi-Fe3O4 nanocomposites from low-quality coal with amazing catalytic performance in reduction of nitrophenols." *Journal of Environmental Chemical Engineering* 5(2):1855–1865.

Gao, Shuyan, Youguo Shi, Shuxia Zhang, Kai Jiang, Shuxia Yang, Zhengdao Li, and Eiji Takayama-Muromachi. 2008. "Biopolymer-assisted green synthesis of iron oxide nanoparticles and their magnetic properties." *The Journal of Physical Chemistry C* 112(28):10398–10401.

Gui, Minghui, Vasile Smuleac, Lindell E Ormsbee, David L Sedlak, and Dibakar Bhattacharyya. 2012. "Iron oxide nanoparticle synthesis in aqueous and membrane systems for oxidative degradation of trichloroethylene from water." *Journal of Nanoparticle Research: An Interdisciplinary Forum for Nanoscale Science and Technology* 14(5):861.

Hoag, George E, John B Collins, Jennifer L Holcomb, Jessica R Hoag, Mallikarjuna N Nadagouda, and Rajender S Varma. 2009. "Degradation of bromothymol blue by greener nano-scale zero-valent iron synthesized using tea polyphenols." *Journal of Materials Chemistry* 19(45):8671–8677.

Iram, Mahmood, Chen Guo, Yueping Guan, Ahmad Ishfaq, and Huizhou Liu. 2010. "Adsorption and magnetic removal of neutral red dye from aqueous solution using Fe3O4 hollow nanospheres." *Journal of Hazardous Materials* 181(1–3):1039–1050.

Jarlbring, Mathias, Lars Gunneriusson, Björn Hussmann, and Willis Forsling. 2005. "Surface complex characteristics of synthetic maghemite and hematite in aqueous suspensions." *Journal of Colloid and Interface Science* 285(1):212–217.

Jarzębski, Maciej, Mikołaj Kościński, and Tomasz Białopiotrowicz. 2017. "Determining the size of nanoparticles in the example of magnetic iron oxide core-shell systems." *Journal of Physics: Conference Series* 885(1):012007. IOP Publishing.

Jeong, Sang-Hun, Jae-Keun Kim, Bong-Soo Kim, Seok-Ho Shim, and Byung-Teak Lee. 2004. "Characterization of SiO2 and TiO2 films prepared by rf magnetron sputtering and their application to anti-reflection coating." *Vacuum* 76(4):507–515.

Lin, Kuen-Song, Ni-Bin Chang, and Tien-Deng Chuang. 2008. "Fine structure characterization of zero-valent iron nanoparticles for decontamination of nitrites and nitrates in wastewater and groundwater." *Science and Technology of Advanced Materials* 9(2):025015.

Liu, Jingbing, Xiaoyue Ye, Hao Wang, Mankang Zhu, Bo Wang, and Hui Yan. 2003. "The influence of pH and temperature on the morphology of hydroxyapatite synthesized by hydrothermal method." *Ceramics International* 29(6):629–633.

Liu, Shan, Ke Yao, Lian-Hua Fu, and Ming-Guo Ma. 2016. "Selective synthesis of Fe₃O₄, γ-Fe₂O₃, and α-Fe₂O₃ using cellulose-based composites as precursors." *RSC Advances* 6(3):2135–2140.

Maity, D, and DC Agrawal. 2007. "Synthesis of iron oxide nanoparticles under oxidizing environment and their stabilization in aqueous and non-aqueous media." *Journal of Magnetism and Magnetic Materials* 308(1):46–55.

Njagi, Eric C, Hui Huang, Lisa Stafford, Homer Genuino, Hugo M Galindo, John B Collins, George E Hoag, and Steven L Suib. 2010. "Biosynthesis of iron and silver nanoparticles at room temperature using aqueous sorghum bran extracts." *Langmuir* 27(1):264–271.

Qadri, Shahnaz, Ashley Ganoe, and Yousef Haik. 2009. "Removal and recovery of acridine orange from solutions by use of magnetic nanoparticles." *Journal of Hazardous Materials* 169(1–3):318–323.

Ramasubbu, A, S Saravanan, and S Vasanthkumar. 2011. "One-pot synthesis and characterization of biopolymer–iron oxide nanocomposite." *International Journal of Nano Dimension* 2(2):105–110.

Shah, Muhammad Tariq, Aamna Balouch, Kausar Rajar, Imdad Ali Brohi, and Akrajas Ali Umar. 2015. "Selective heterogeneous catalytic hydrogenation of ketone (C==O) to alcohol (OH) by magnetite nanoparticles following Langmuir–Hinshelwood kinetic approach." *ACS Applied Materials & Interfaces* 7(12):6480–6489.

Shahwan, Talal, S Abu Sirriah, Muath Nairat, Ezel Boyacı, Ahmet E Eroğlu, Thomas B Scott, and Keith R Hallam. 2011. "Green synthesis of iron nanoparticles and their application as a Fenton-like catalyst for the degradation of aqueous cationic and anionic dyes." *Chemical Engineering Journal* 172(1):258–266.

Sun, Shouheng, and Hao Zeng. 2002. "Size-controlled synthesis of magnetite nanoparticles." *Journal of the American Chemical Society* 124(28):8204–8205.

Togashi, Takanari, Takashi Naka, Shunsuke Asahina, Koichi Sato, Seiichi Takami, and Tadafumi Adschiri. 2011. "Surfactant-assisted one-pot synthesis of superparamagnetic magnetite nanoparticle clusters with tunable cluster size and magnetic field sensitivity." *Dalton Transactions* 40(5):1073–1078.

Vélez, E., Campillo, G.E., Morales, G., Hincapié, C., Osorio, J., Arnache, O., Uribe, J.I., and Jaramillo, F. 2016. "Mercury removal in wastewater by iron oxide nanoparticles." *Journal of Physics: Conference Series* 687(1): 012050. IOP Publishing.

Wu, Wei, Xiangheng Xiao, Shaofeng Zhang, Feng Ren, and Changzhong Jiang. 2011. "Facile method to synthesize magnetic iron oxides/TiO 2 hybrid nanoparticles and their photodegradation application of methylene blue." *Nanoscale Research Letters* 6(1):533.

Yu, Shi, and Gan Moog Chow. 2004. "Carboxyl group (–CO 2 H) functionalized ferrimagnetic iron oxide nanoparticles for potential bio-applications." *Journal of Materials Chemistry* 14(18):2781–2786.

Zhang, Weixin, Zeheng Yang, Xue Wang, Yuancheng Zhang, Xiaogang Wen, and Shihe Yang. 2006. "Large-scale synthesis of β-MnO2 nanorods and their rapid and efficient catalytic oxidation of methylene blue dye." *Catalysis Communications* 7(6):408–412.

17 Programmable Delay for Nanodevices

Abhishek Kumar
Lovely Professional University, Phagwara, India

CONTENTS

17.1   INTRODUCTION

Three types of Delay Locked Loop (DLL) based on a controlled delay line are (a) the analog/voltage-controlled delay line (VCDL); (b) the digitally controlled delay line (DDL); and (c) hybrid control delay lines (HDL). The DLL is used to postpone the clock signal by the required value with respect to the circuit in the design [1, 2]. In usual emphasis, the clock-timing properties thereby increase the output produced to an expected level and are controlled. DLLs are also integrated for clock and data recovery (CDR) systems. To define the functioning of a DLL, we can say that it is utilized as a functional block, as a negative edge gate, being used as a reference clock input [3–8]. There is a kind of PLL circuit, the only difference being the usage of a delay element in place of an oscillator circuit. The internal component of the DLL circuit contains a phase detector (PD), which compares the phase of the reference clock and feedback clock phase. Depending on the UP or DOWN signal of the reference clock signal, the charge pump is either charged upward or downward [9–16]. The charge pump output is filtered out through a low pass filter (LPF), which is

295

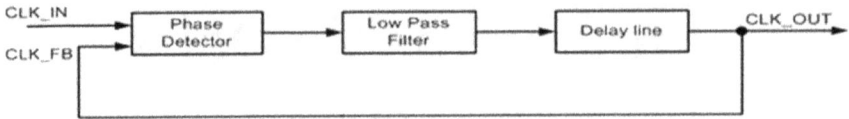

FIGURE 17.1 DLL architecture.

finally received at the far end by the delay line. The phase detector is at input, the delay line postpones the clock frequency. The delay line of cascaded connection of logic cell. The delay gates are in the form of gates with a combinational cell. The architecture is as shown in Figure 17.1.

A voltage-controlled analog delay line is the delay line whose loiter is dependent upon the analog source variations, resulting in a delayed output. A binary-controlled delay line not only embeds complex cells to accumulate delay, but is also determined by binary "1" and binary "0" for the result at output. [17–20]. Delay lines can be controlled simply when the output of each stage of the delay propagates to the next stage.

17.2 DIFFERENCE BETWEEN DLL AND PLL

The main difference between PLL and DLL is highlighted below [21–24]:

1. The DLL depends on the variation into phase but PLL work on the variation is the frequency.
2. A DLL compares the last end phase with the reference input clock. DLL produces an impulse signal which is first included and feedback as the control element. This procedure results in zero error which maintains the control signal from up to down.
3. A PLL is based on the consideration of the stage of the oscillator to produce an error signal. The generated error signal is then amplified to create a control input for the VCO. This control signal reflects their effect on the oscillator's frequency.
4. PLL uses VCO in the feedback, whereas DLL uses a controlled delay line.
5. A clean version of the clock is a major source of difference between PLL and DLL. PLL blocks the jitter in the source whence can affect the VCO output, while DLL propagates the jitter.
6. PLL is a viable choice over DLL where the main sampling point needs to pull from the signal that is arriving or ignoring the jitter.
7. DLL is used for multiphase sampling delay time to reduce the jitter between two signals.
8. DLL is preferred over PLL for high-speed on-chip communication, like communication between DDR SDRAM and memory controller.
9. PLL is a hardware-based frequency control device, while DLL is based on a software dynamic-linked library.

17.3 APPLICATIONS OF DLL

DLL is found in applications in the following areas [25–28]:

1. Delay compensation – DLL compensate the delay in complex circuits with the issue of hardware delay. Systems and circuits where delay elements are available to synchronize the clock signal from input are probably completely forfeited. To achieve it, DLL is utilized as a buffer chain in the feedback path.
2. Multiphase clock generation – A multiphase clock generation circuit needs to originate a clock with different phases of the same reference clock. Spacing of phase is given by DLL. The ability to generate multiple phase shift replicates the same clock signal, and can be suggested for use in clock data recovery (CDR). A CDR system functions as binary ON/OFF state conversion into RF modulation
3. Frequency synthesis – The expected frequency at which output is required and received is monitored. DLLs control the clock pulse phases efficiently. A number of the clock pulse phases can be included, to determine a higher modulated frequency at output.
4. Clock and data recovery (CDR) systems – With the ability of DDL to synchronize and observe clock reference signals.
5. Military and biomedical applications – All the applications like military code encoding and decoding, biomedical sensors, as well as Patient Monitoring Systems (PMS), require high precision with a high resolution. Hence, the clock generation circuitry in these circuits employs a DLL/PLL with the efficient use of delay lines to provide the desirable synchronization.

This chapter focuses on the design of the voltage-controlled delay line, where delay can be controlled by the analog voltage generated by the analog MOS-based circuit. The delay lines that are designed to date either use the logic gates (preferably an inverter) or the buffer. The power consumption of delay lines is large. The alternative method to the delay line is a delay due to the combinational logic cells like multiplexers. Apart from the above-stated applications, delay lines are entirely independent of the DLL. Some of these applications are [29]:

1. Low-power devices
2. Wireless transmission architecture
3. Memory elements, like SRAM
4. Large processor circuits, like microprocessors
5. Handshake signal architecture

All three kinds of delay lines were designed namely: voltage-controlled, digitally controlled, and hybrid delay line. The delay can be controlled by a single DC voltage, or with the help of multiple binary bits of data. In this chapter, the inverter-based delay and buffer-based delay line have uses as an inverter and buffer or as a basic delay unit, respectively. In the inverter-based delay lines, a minimum delay in the transition of the 50% output level from the 50% input level is offered. The total delay in the circuit is not only due to the delay of the not gates but also includes the delay offered by the interconnect and wires during the transmission of the input to the output. The voltage-controlled delay line (VCDL) is an integral component of the DLL. The main objective of DLL-circuit synchronization of other peripheral available onto the circuit. Clock synchronization has a vital role in circuit designs with clock referenced to be one of the stable inputs in any circuit originating from a crystal oscillator. The DLL locks the circuit to a stable synchronous state. Un-synchronization into the circuit leads to uncontrolled output, which is disturbing as the functionality of the circuit might fail the purpose of usage.

The limitations of conventional CMOS delay lines are (a) jitter performance, which is in the range of several picoseconds (ps) or femto seconds (fs). Extensive work to produce a jitter-less circuit or sub-picosecond jitter performance CMOS delay is actively undertaken by many parties. IC-based delay lines are robust in terms of system integration and cost reduction when compared to their counterpart. Jitter deviates the clock from its ideal position and the active edge at which data transfer is uncertain; (b) observing a long delay value that follows a linear relation with high-resolution delay steps. Fine-resolution CMOS delay lines obtained by the cascaded structure of delay lines; due to delay increases, there is mainly a non-linear element. This is due to the cumbersome attribute of the parasitic capacitance. The methodology involves a complex PCB fabrication [30–32]. A single-chip solution should be developed to overcome these shortcomings. Two basic categories of delay elements for VCDL are inverter-based VCDL and buffer-based VCDL. A jitter problem in a delay-locked loop (DLL) arises because of uncertainties in the controlled delay line. The jitter at the internal stages of the delay line and the jitter in the conventional delay cell is analyzed.

17.4 CONTROLLED DELAY LINE

17.4.1 INVERTER-BASED VCDL

In the inverter-based voltage control delay line, the research device design started fore mostly using Cadence Virtuoso CMOS 90 nm technology. The inverter and the buffer-based delay lines work in the following manner: When an input voltage is supplied as the reference clock signal to the first input terminal of the inverter or the buffer, the input voltage passes through all of the transistors after experiencing some propagation delay, and is transmitted to the next inverter stage. There can be many inverter stages. The amount of delay achieved increases with the number of stages. This chapter uses, as the basis of the thesis work, a five-stage inverter-based delay line which was made using the Cadence tool. To increase the delay in the delay line, we need to exploit the RC delay offered in the circuit. To accomplish this in many designs, a cascaded stage of other devices is applied in connection to the main delay line. These devices are called "delay elements." Examples of these delay elements include NAND gates, 3T XOR gates, etc.

Then, another delay line was made using the 3T XOR transistor as the delay element. In this delay line, the 3T XOR was connected in a cascaded manner to the inverter and 5 of such inverter stages were interconnected to get the delay line. When the voltage is applied across the inverter at the first stage, the inverter output is then passed on to the 3T XOR gate connected, the 3T XOR acts as a capacitive path. Hence, with the method of Variable Capacitive Loading, the 3T XOR passes on a significantly delayed output to the next stage and hence, a large delay is achieved in this manner at the output. Also, a delay line was created using a Modified 3T XOR gate. The delay line works in the same manner as stated above except for the Modified 3T XOR gate working. Working criteria are described as follows: out of three PMOS, two PMOS P1 and P2 were connected in cascade to each other, and then the PMOS P3 was connected in cascade to the whole pair. The PMOS P1 and P2 were given the input terminal supply, each being connected to a voltage source. However, the PMOS P3 was connected to a separate constant voltage source. This made the PMOS P3 act as always ON. So, when a strong 1 was needed as at the output, the input from the PMOS P1 and P2 was passed on to the PMOS P3, which acted as a resistive path, hence passing on a strong 1 value. The aim of this work is to develop programmable delay lines which can be easily configured to achieve the maximum delay. This will be accomplished by adding different kinds of delay units to the basic delay lines, or designing delay lines based upon the usage of different sequential and combinational logic elements.

17.4.1.1   NOT Gate

NOT gate has been used to offer the basic delay unit in the delay line designed with CMOS 90 technology. A pulsating input signal provided an input to the inverter and a determined preset V_{DD} and ground in the circuit [33]. The PMOS was taken on the top, as the width of the PMOS was considered to be twice the width of the NMOS. Thus output derives high and output derives low, offering equal energy to the output node. The reason for these values of the width is the fact that the majority carriers – which are holes in the case of the PMOS and electrons in the case of NMOS – have a difference in the mobility of approximately two times, i.e., the mobility of the holes is two times slower than that of the electrons. Thereby, the rising and the falling edges of the NMOS and the PMOS will be symmetrical. The width of the NMOS is set at 120 nm, and the PMOS at 240 nm.

17.4.1.2   Five-Stage Inverter-Based Delay Line

A delay stage is a cascaded connection of the odd number of delay stages. Figure 17.2 shows the five-stage delay line using an inverter, where all five inverters were in series. The reference clock input was applied to the first stage inverter as the reference clock voltage (CLK_{REF}); controlling voltage (V_{CTRL}) on the next stage, using V_{DD} given to the device which controlled the device. The output of each inverter was analyzed with respect to the input provided [34, 35] and the loiter produced by each circuit element (inverter) at each stage. The response on the fifth stage is identical to the NOT gate, due to an odd number of compliments of the delay measured by substituting an output pin at each of the inverter outputs. Figure 17.2 shows the delay-line-based inverter.

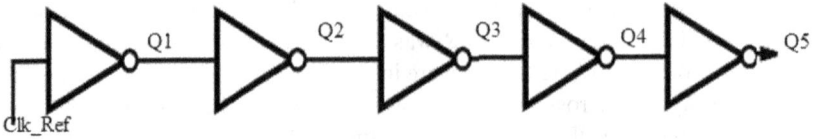

FIGURE 17.2   Inverter-based five-stage delay line.

The delay of VCDL are calculated using delay offered by cell and wires used for the interconnection:

Delay of a NOT gate $= 23.08\ ps$

Delay of 5 NOT gates $= 5 \times 23.08\,ps = 115.4\ ps$

Output trigger at the output *of the last NOT gate* $= 180.6\ ps$

Delay due to wire $=$ *Total Delay* in *the circuit* $-$ *The delay of* 5 NOT *gate*

$$= 180.6 - 1154.4 = 65.2\ ps$$

17.4.1.3   Seven-Stage Inverter-Based Delay Line

Similarly, delay of the seven-stage inverter-based delay was designed, and can be calculated as follows:

Delay of a NOT gate $= 23.1$ ps

Delay of 7 NOT gates $= 7 \times 23.1$ ps $= 161.7$ ps

Output trigger at the output $= 212.5$ ps

Total delay in the circuit $= 212.5$ ps

Hence, the delay offered by the wire

$=$ Total Delay in the circuit $-$ The delay of VCDL

$= 212.5$ ps $- 7 \times 23.1$ ps

$= 50.8$ ps

17.4.2   Buffer-Based Delay Line

A CMOS-based buffer is a reverse structure of the inverter design, using NMOS and a PMOS. As shown in Figure 17.3, the NMOS was kept on the top and the PMOS on the bottom; connection to the V_{DD} with NMOS and the PMOS to GND respectively. The structure was designed such that they should only pass the voltage as a buffering unit without any error in signal – the transient simulation result is presented in Figure 17.4 of the buffer circuit using cadence specter simulator. The schematic of buffer design is shown in Figure 17.3, and its transient response is shown in Figure 17.4.

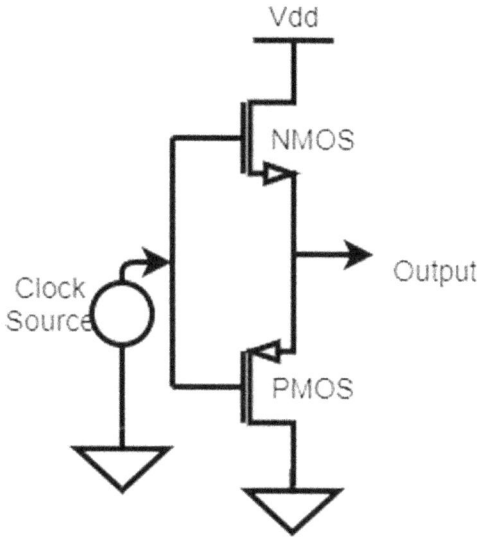

FIGURE 17.3   Buffer design schematic.

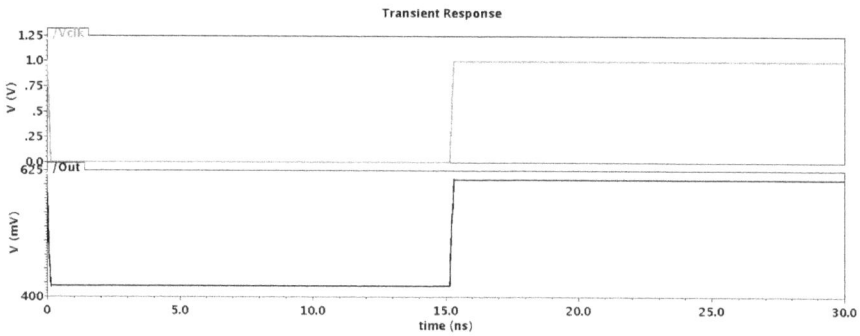

FIGURE 17.4   Transient response of buffer.

A delay line is a series connection of a delay stage; a five-stage buffer unit as a delay element. The total delay offer by VCDL is the cumulative effect of the cell and RC delay offered due to interconnection wire. The delay of five- and seven-stage delay line is calculated as shown below.

Delay calculation for five-stage voltage control delay line:

Delay of one buffer = 3.285 ps

Delay due to 5 buffers = $5 \times 3.285 \, \text{ps} = 16.425 \, \text{ps}$

Output triggers at output terminal = 168.46 ps

Total delay in the circuit = Delay at last buffer = 168.46 ps

Delay offered by the wire = $168.46 \, \text{ps} - 5 \times 3.285 \, \text{ps} = 152.035 \, \text{ps}$

Delay calculation for seven-stage voltage control delay line:

> Delay of one buffer $= 3.285$ ps
>
> Delay due to 5 buffers $= 7 \times 3.285$ ps $= 22.995$ ps
>
> Output triggers at output terminal $= 176.67$ ps
>
> Total delay in the circuit $=$ Delay at last buffer $= 176.67$ ps
>
> Delay offered by the wire $= 176.67$ ps $- 5 \times 3.285$ ps
>
> $= 152.035$ ps

17.5 DELAY LINE WITH A DELAY ELEMENT

A delay line is composed of a cascaded connection of delay stage, either that of inverter or buffer. Each internal terminal consists of unique delay elements. This delay element adds up to an additional delay in the computation and additional control input, as shown in Figure 17.5. Delay elements are the combinatorial two-input logic gate. Each input acts as a control input. For each input delay, elements find unique changing and discharging current results in a unique delay. In this work, a two-input XOR gate is used as a delay element. Inclusion of additional elements increases the hardware resources. Advances in technology can save the area. The static structure of the XOR gate requires 12 MOS transistors. But the same functionality of XOR is achieved with three transistors. Here, conventional and modified 3T XOR add the delay in the internal stage.

17.5.1 XOR as Delay Element

In this work, the 3T XOR-based delay element creates a controllable delay based on (i) conventional 3-T XOR and (ii) modified 3-T XOR.

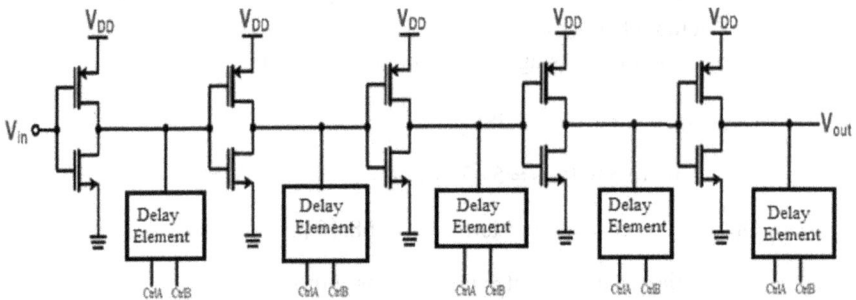

FIGURE 17.5   Delay line with a controlled delay element.

17.5.1.1 Conventional 3T XOR

Figure 17.6 shows [35] a 3T XOR required two PMOS and one NMOS. Aspect ratio $(W/L)_{P1}$ parametrized as 5 um/100 nm, $(W/L)_{P2}$ was kept at 20 um/100 nm and $(W/L)_{N1}$ was kept at 1 um/100 nm to achieve output. During the working of the 3T XOR gate, the value of the input set '01' depends on the PMOS P2 and for the input set '10', the value depends on the PMOS P1. However, when the total ratio of the PMOS pair (W/L) exceeds 1:4, a significant fall is seen in the value of the '01' input set. The NMOS (W/L) ratio should be kept in a 1:4 ratio to the PMOS P2. We already know that the PMOS is responsible for passing the strong 1 and NMOS for passing strong 0. The modified 3T XOR gate achieved the following set of results for each set of input. The conventional 3T XOR gate is shown in Figure 17.6.

17.5.1.2 Modified 3T XOR

A modified version of the 3T XOR gate is presented in Figure 17.7, consisting of three PMOS transistors. The PMOS are powered by a stable voltage of 440mV, which acts as resistance in the complete circuitry design. The aspect ratio $(W/L)_{P1, P2}$ was placed at 360 nm/120 nm and the $(W/L)_{P3}$ was kept at 120 nm/120 nm for the output. The modified 3T XOR gate is shown in Figure 17.7.

17.5.1.3 Five-Stage Inverter-Based Delay Lines

A five-stage inverter-based VCDL is implemented with CMOS 90 nm technology, designed with a cadence virtuoso schematic composer with CMOS 180 nm technology node and the simulation result is obtained with a cadence specter simulator [36, 37]. The waveform analyzed is the integrated plotting tool; it plots a drawn simulated result of the transient and dc response. The simulation results generate the jitter of the DLL, due to variations of the delay cells, it is zero at the start and end of the VCDL, and is highest at the center of the VCDL. All stages of the control input of

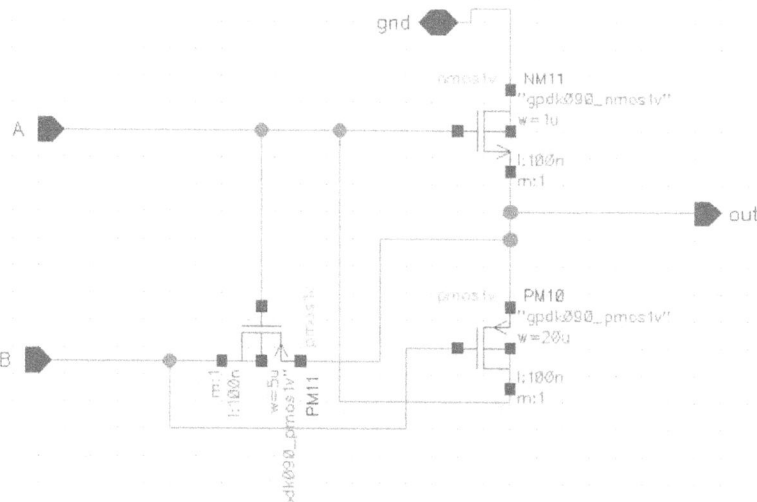

FIGURE 17.6 Conventional 3T XOR gate.

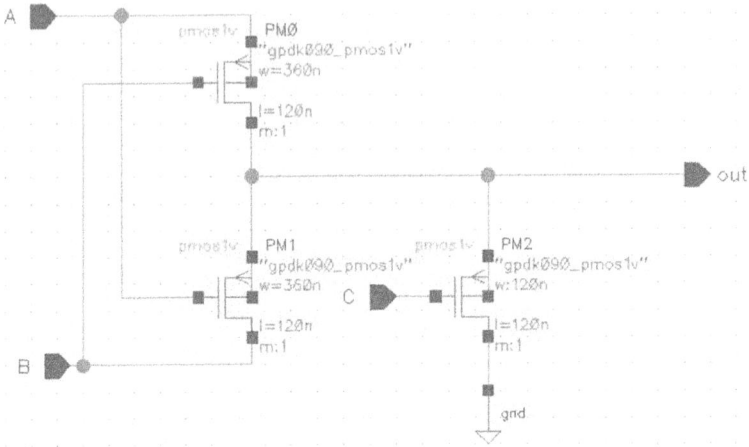

FIGURE 17.7   Modified 3T XOR gate.

the delay element are connected to the same input. A delay line using a modified 3T XOR is a cascaded connection of five inverting stages controlled by the XOR gate. The maximum delay results, while a control input "00" and minimum delay for control input "10" are shown in Table 17.1.

The modified 3T XOR gate is dependent on the working of three PMOS. The PMOS P3 is connected to a constant current source of 440 mV. Hence, for the voltage that is passed across the input terminals, the output for the input sets "01" and "10" depend on the P3 working as resistance, and because of this a logic1 is passed. A comparative analysis of delay five- and seven-stage inverter-based VCDL with conventional 3T XOR and modified 3T XOR is presented in Table 17.2.

VLSI trade-off says delay and power are inversely related, as the delay value reducing power consumption increases. Table 17.3 presents the comparative analysis of power delay, showing how it can easily be manipulated as the complexity of the attached delay chain increases. The overall average power is given by the table.

The total delay of a delay line is dependent upon the delay of an individual cell in the delay element, as well a number of delay stages into a forward path. Figure 17.8 and Figure 17.9 show the graphical representation delay and power of the various

TABLE 17.1
Delay Achieved for 3T Modified XOR Five-Stage Inverter-Based Delay Line

S. No.	Set of Control Input Voltages (V)	Delay Achieved (ps)
1.	00	117.2
2.	01	132.3
3.	10	72.3
4.	11	90.4

TABLE 17.2
Overall Delay Achieved for Different Delay Lines

S. No.	Different Delay Lines	Delay Achieved (ps)
1.	Inverter-based five-stage delay line	180.6
2.	Inverter-based seven-stage delay line	212.5
3.	Buffer-based five-stage delay line	168.46
4.	Buffer-based seven-stage delay line	176.67
5.	3T Conventional XOR five-stage inverter-based delay line	1674
6.	3T Conventional XOR seven-stage inverter-based delay line	2493.5
7.	3T Modified XOR five-stage inverter-based delay line	103.05

TABLE 17.3
Average Power for Different Delay Lines

S. No.	Delay Line	Average Power (uW)
1.	Inverter-based five-stage delay line	0.2997
2.	Inverter-based seven-stage delay line	0.4203
3.	Buffer-based five-stage delay line	6.199
4.	Buffer-based seven-stage delay line	6.1477
5.	3T Conventional XOR five-stage inverter-based delay line	400.9
6.	3T Conventional XOR seven-stage inverter-based delay line	585.5
7.	3T Modified XOR five-stage inverter-based delay line	151.5

stages of the inverter-based VCDL, with conventional and modified 3T XOR as the delay element.

Delay increases with the increase in delay stage and delay elements in the intermediate stage. Lower delay arises due to lowering supply voltage, which further decreases the average power consumption. 7T XOR shows the reverse characteristics; total delay and average power consumption reduce as the delay stage and delay element increase. Maximum delay is achieved for 3T conventional XOR with a seven--stage inverter-based delay line and minimum for inverter-based five-stage delay line. Similarly, power consumption is maximum for 3T conventional XOR with a seven-stage inverter-based delay line and minimum for an inverter-based five-stage delay line.

17.6 CONCLUSION

The MOSFETs being used offer an added advantage in the respect that their switching characteristics show a propagation delay time that is offered for the high to low and low to high ends of the input pulse. The MOSFET – both NMOS and PMOS

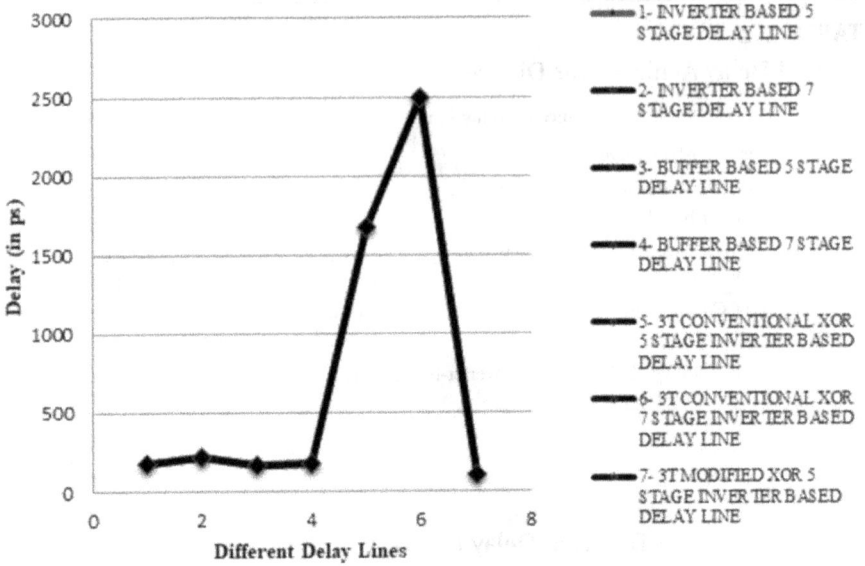

FIGURE 17.8 Delay comparison between the various delay lines.

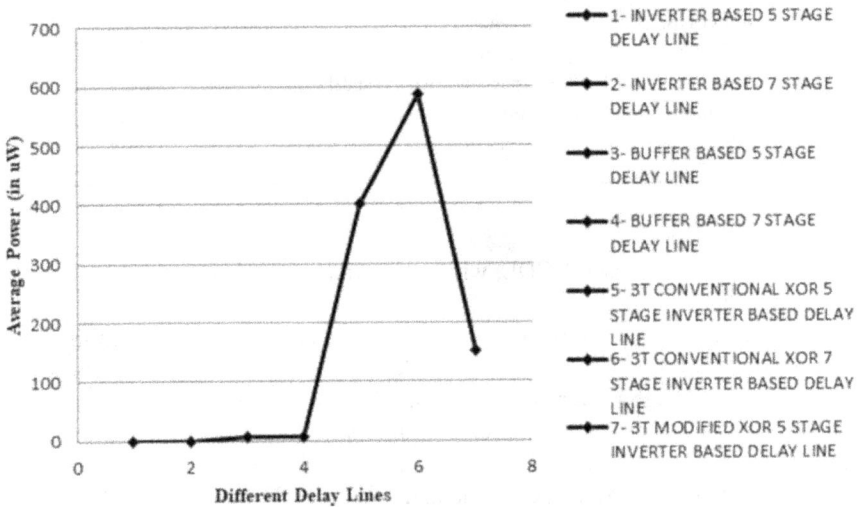

FIGURE 17.9 Average power comparison between the different delay lines.

– can be exploited for the delay because they act as an RC delay unit, i.e., they both offer a resistive and capacitive role. Therefore when the capacitor is charging, it exhibits a delay in transmission of the input signal to the output. The delay lines which are voltage-dependent, VCDL, are exploiting the above-stated principle only. The most efficient RC Model is of an inverter. Hence, most of the delay lines employ an inverter as a basic delay line element. Thereby, as the number of stages increases, the delay through each inverter increases. Also, when another delay element is connected in the cascade to the inverter, at each stage the RC delay increases for the parasitic capacitance charging and discharging. So, this variation is easily exploited. Also, as the voltage varies, so does the offered delay. The (W/L) ratio of the MOSFET also acts as an agent in the determination of the threshold voltage loss at each stage and consequently, the overall delay of the circuit can be modified in accordance. The delay achieved in the thesis work is in the range of 180.6 ps to 2.4935 ns. The delay achieved by using the Conventional 3T XOR is very much larger than the Modified 3T XOR. The delay can be further increased by increasing the complexity of the cascade stage.

REFERENCES

1. Jovanović, Goran, and Mile K. Stojčev. "Voltage controlled delay line for digital signal." *Facta Universitatis-Series: Electronics and Energetics* 16(2) (2003): 215–232.
2. Johnson, Mark G., and Edwin L. Hudson. "Variable delay line phase-locked loop circuit synchronization system." U.S. Patent 5,101,117, issued March 31, 1992.
3. Woo, Ann K. "Cmos digital-controlled delay gate." U.S. Patent 5,227,679, issued July 13, 1993.
4. Morichetti, Francesco, Andrea Melloni, Carlo Ferrari, and Mario Martinelli. "Error-free continuously-tunable delay at 10 Gbit/s in a reconfigurable on-chip delay-line." *Optics Express* 16(12) (2008): 8395–8405.
5. Morales, Lou. "Active delay line circuit." U.S. Patent 4,899,071, issued February 6, 1990.
6. Bult, Klaas, and Hans Wallinga. "A CMOS analog continuous-time delay line with adaptive delay-time control." *IEEE Journal of Solid-State Circuits* 23(3) (1988): 759–766.
7. Santos, D. M., S. F. Dow, J. M. Flasck, and M. E. Levi. "A CMOS delay locked loop and sub-nanosecond time-to-digital converter chip." *IEEE Transactions on Nuclear Science* 43(3) (1996): 1717–1719.
8. Santos, D. M., S. F. Dow, J. M. Flasck, and M. E. Levi. "A CMOS delay locked loop and sub-nanosecond time-to-digital converter chip." *IEEE Transactions on Nuclear Science* 43(3) (1996): 1717–1719.
9. Dehng, Guang-Kaai, June-Ming Hsu, Ching-Yuan Yang, and Shen-Iuan Liu. "Clock-deskew buffer using a SAR-controlled delay-locked loop." *IEEE Journal of Solid-State Circuits* 35(8) (2000): 1128–1136.
10. Cheng, Kuo-Hsing, and Yu-Lung Lo. "A fast-lock wide-range delay-locked loop using frequency-range selector for multiphase clock generator." *IEEE Transactions on Circuits and Systems. Part II: Express Briefs* 54(7) (2007): 561–565.
11. Cheng, Kuo-Hsing, and Yu-Lung Lo. "A fast-lock wide-range delay-locked loop using frequency-range selector for multiphase clock generator." *IEEE Transactions on Circuits and Systems. Part II: Express Briefs* 54(7) (2007): 561–565.
12. Priya, M. Geetha, and K. Baskaran. "Low power full adder with reduced transistor count." *International Journal of Engineering Trends and Technology (IJETT)* 4(5) (2013): 1755–1759.
13. Veeramachaneni, Sreehari, and M. B. Srinivas. "New improved 1-bit full adder cells." In: *2008 Canadian Conference on Electrical and Computer Engineering*, pp. 000735–000738. IEEE, 2008.
14. Kang, Sung-Mo, and Yusuf Leblebici. *CMOS Digital Integrated Circuits*. Tata McGraw-Hill Education, New Delhi, 2003, pp. 218–267.

15. Weste, Neil H. E., and Kamran Eshraghian. *CMOS VLSI Design: A Circuits and Systems Perspective*. Pearson Education India (1985), pp. 141–175.

16. Pahlevan, Maryam, Balakrishna Balakrishna, and Roman Obermaisser. "Simulation framework for clock synchronization in time sensitive networking." In: *2019 IEEE 22nd International Symposium on Real-Time Distributed Computing (ISORC)*, pp. 213–220. IEEE, 2019.

17. Prakash, S. R. Jaya, and Sujatha S. Hiremath. "Dual loop clock duty cycle corrector for high speed serial interface." In: *2017 International Conference on Smart Technologies for Smart Nation (SmartTechCon)*, pp. 935–939. IEEE, 2017.

18. Ross, Brady, Timothy Carbino, and Michael Temple. "Home automation simulcasted power line communications network (SPN) discrimination using wired signal distinct native attribute (WS-DNA)." In: *Proceedings of the Twelfth International Conference on Cyber Warfare and Security*, pp. 313–322. 2017.

19. Kazemier, Jan J., Georgios K. Ouzounis, and Michael H. F. Wilkinson. "Connected morphological attribute filters on distributed memory parallel machines." In: *International Symposium on Mathematical Morphology and Its Applications to Signal and Image Processing*, pp. 357–368. Springer, Cham, 2017.

20. Casto, Matthew James. "Multi-attribute design for authentication and reliability (MADAR)." PhD dissertation. The Ohio State University, 2018.

21. Gozzelino, Michele, Salvatore Micalizio, Filippo Levi, Aldo Godone, and Claudio Eligio Calosso. "Reducing cavity-pulling shift in Ramsey-operated compact clocks." *IEEE Transactions on Ultrasonics, Ferroelectrics, and Frequency Control* 65(7) (2018): 1294–1301.

22. Peng, Bin. "Building a memory reading circuit." *Journal of Computing Sciences in Colleges* 34(4) (2019): 114–116.

23. Johnson, Anju P., Sikhar Patranabis, Rajat Subhra Chakraborty, and Debdeep Mukhopadhyay. "Remote dynamic clock reconfiguration based attacks on internet of things applications." In: *2016 Euromicro Conference on Digital System Design (DSD)*, pp. 431–438. IEEE, 2016.

24. Deng, Zhongwen, Zhigang Liu, Shaowei Gu, Xingyu Jia, and Wen Deng. "Frequency-scanning interferometry for depth mapping using the Fabry–Perot cavity as a reference with compensation for nonlinear optical frequency scanning." *Optics Communications* 455 (2020): 124556.

25. Kim, Jaeha, Mark A. Horowitz, and Gu-Yeon Wei. "Design of CMOS adaptive-bandwidth PLL/DLLs: A general approach." *IEEE Transactions on Circuits and Systems. Part II: Analog and Digital Signal Processing* 50(11) (2003): 860–869.

26. Moon, Yongsam, Jongsang Choi, Kyeongho Lee, Deog-Kyoon Jeong, and Min-Kyu Kim. "An all-analog multiphase delay-locked loop using a replica delay line for wide-range operation and low-jitter performance." *IEEE Journal of Solid-State Circuits* 35(3) (2000): 377–384.

27. Zhang, Renyuan, and Mineo Kaneko. "Robust and low-power digitally programmable delay element designs employing neuron-MOS mechanism." *ACM Transactions on Design Automation of Electronic Systems (TODAES)* 20(4) (2015): 1–19.

28. Zhang, C. W., X. Y. Wang, and L. Forbes. "Simulation technique for noise and timing jitter in electronic oscillators." *IEE Proceedings-Circuits, Devices and Systems* 151(2) (2004): 184–189.

29. Xanthopoulos, Thucydides, Daniel W. Bailey, Atul K. Gangwar, Michael K. Gowan, Anil K. Jain, and Brian K. Prewitt. "The design and analysis of the clock distribution network for a 1.2 GHz alpha microprocessor." In: *2001 IEEE International Solid-State Circuits Conference. Digest of Technical Papers. ISSCC (Cat. No. 01CH37177)*, pp. 402–403. IEEE, 2001.

30. Andreani, Pietro, Franco Bigongiari, Roberto Roncella, Roberto Saletti, and Pierangelo Terreni. "A digitally controlled shunt capacitor CMOS delay line." *Analog Integrated Circuits and Signal Processing* 18(1) (1999): 89–96.
31. Saint-Laurent, Martin, and Gabriel Patrick Muyshondt. "A digitally controlled oscillator constructed using adjustable resistors." In: *2001 Southwest Symposium on Mixed-Signal Design (Cat. No. 01EX475)*, pp. 80–82. IEEE, 2001.
32. Yousefzadeh, Vahid, Toru Takayama, and Dragan Maksimovi. "Hybrid DPWM with digital delay-locked loop." In: *2006 IEEE Workshops on Computers in Power Electronics*, pp. 142–148. IEEE, 2006.
33. De Caro, Davide. "Glitch-free NAND-based digitally controlled delay-lines." *IEEE Transactions on Very Large Scale Integration (VLSI) Systems* 21(1) (2012): 55–66.
34. Sheng, Duo, Ching-Che Chung, and Chen-Yi Lee. "An ultra-low-power and portable digitally controlled oscillator for SoC applications." *IEEE Transactions on Circuits and Systems. Part II: Express Briefs* 54(11) (2007): 954–958.
35. Gao, Yanxia, Shuibao Guo, Yanping Xu, Shi Xuefang Lin, and Bruno Allard. "FPGA-based DPWM for digitally controlled high-frequency DC-DC SMPS." In: *2009 3rd International Conference on Power Electronics Systems and Applications (PESA)*, pp. 1–7. IEEE, 2009.
36. Xiao, Jinwen, Angel Peterchev, Jianhui Zhang, and Seth Sanders. "An ultra-low-power digitally-controlled buck converter IC for cellular phone applications." In: *Nineteenth Annual IEEE Applied Power Electronics Conference and Exposition, 2004. APEC'04.*, vol. 1, pp. 383–391. IEEE, 2004.
37. Kabiri, Ali, Qing He, Mohammed H. Kermani, and Omar M. Ramahi. "Design of a controllable delay line." *IEEE Transactions on Advanced Packaging* 33(4) (2010): 1080–1087.

18 MANET Routing Optimization using Nanotechnology

Sandeep Gupta, Vibha Aggarwal,
and Virinder Kumar Singla
Punjabi University Neighborhood
Campus, Rampura Phul, India

CONTENTS

18.1   INTRODUCTION AND BACKGROUND

Mobile ad-hoc networks (MANETs) are self-assembling systems of nodes associated through remote without the need for unified organization. In MANETs, every node demonstration both as host and switch, so it must be fit for sending/accepting information to/from a different node. As the clients in a remote system can move anyplace and anytime, this results in a system topology which is ever-changing and somewhat irregular. This is the motivation to develop new directing conventions to find, keep up, and deal with the courses, in light of the fact that customary routing conventions for wired systems cannot work proficiently in MANETs. A typical mobile ad-hoc network is shown in Figure 18.1.

An ordinary MANET structure is shown in Figure 18.1 (Murthy and Manoj 2004). The correspondence scope of any individual node is spoken to by the circle. In reality, the range could conceivably follow the specific roundabout shape and the connections in certainty can even be unidirectional much of the time. For instance, node "X" can arrive at node "Y" on interface 1. However, node "Y" will most likely be unable to utilize this connection to arrive at node "X". It might happen in light of the fact that the sign qualities of the two transmitters are inconsistent or can even be found on the transmission path.

For MANETs, directing conventions should work so that with minimal time, maximum data must be transferred from source to destination, with a least possible estimation of overheads, as both transmission capacity and power are imperative for MANETs.

Numerous MANET routing conventions have been projected so far. Precedent concentrated on structuring novel conventions, contrasting offered conventions, or improving conventions prior to standard MANET directing conventions are characterized. The major focus of research is on rebuilding an exploration of the directing conventions of enthusiasm for ad-hoc systems with certain traffic designs. However, rebuilding resulting from various research is not steady on account of the lack of steadiness in MANET-directing convention models and functional conditions,

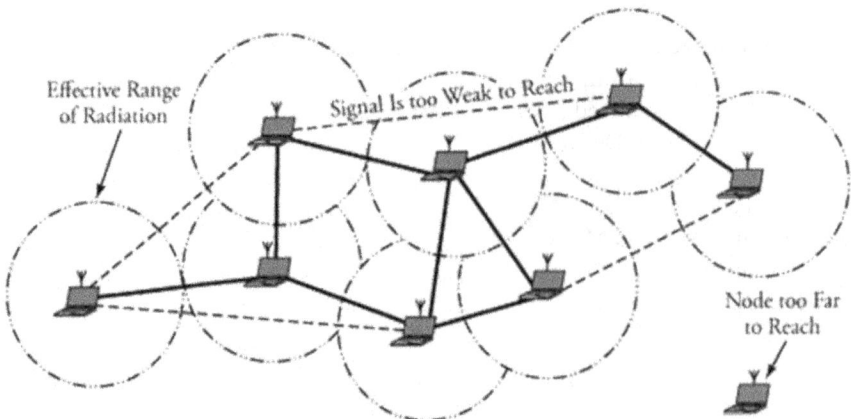

FIGURE 18.1   A typical mobile ad-hoc network.

together with traffic designs. For this reason, reproduction conditions utilized in past inspections are not impartial for all conventions, and their decisions may not be relied upon. At the same time, it is not viable to single out a suitable routing convention for a MANET to use.

18.2 PROPERTIES OF MANETS

MANETs possess various properties such as dynamic topology, unpredictable link quality, exposed terminal problems, and hidden terminal problems, etc.

18.2.1 DYNAMIC TOPOLOGY

In MANETs, the taking of an interest node may move indiscriminately, which prompts an evolving topology. This never occurs in conventional wired systems. Normally in high versatility applications like vehicular correspondence, topology changes quickly. On the one hand, dynamic topology may build the expense of keeping up courses because of connection failure. As the transmission scope of a node is restricted and node development is discretionary, the remote connection is both unreliable and insecure, prompting troubles in keeping up the course. Thus, the MANET-directing convention ought to be sufficient in managing the course break rather than the course revelation.

18.2.2 UNPREDICTABLE LINK QUALITY

Remote system administrations are time and area subordinate. Signal engendering from source to destination experiences channel debilitations like blurring, impedance, and multipath scratch-off during transmission. Also, remote connections have a lower limit than their wired frameworks, expanding the probability of system blockage. Since MANETs are treated as an augmentation of the current wired system in numerous applications, the data transfer capacity issue ought to be considered. Irregular connection quality, together with a constrained transfer speed, makes it a truly testing assignment.

18.2.3 EXPOSED TERMINAL PROBLEM AND HIDDEN TERMINAL PROBLEM

For MANETs, two typical problems are hidden terminals and exposed terminals. A hidden terminal problem happens when a node is noticeable from a remote passageway however not shapes different nodes speaking with that passageway. Right now, result from impacts between transmissions began by any node. For example, A which cannot listen to the progressing communication to its compare node B. The chance of such a strain is relative to the all-out number of terminals which have escaped from A. An uncovered node problem happens when a node is spared from transporting packets to different nodes because of an adjoining transmitter. Fruitless transmission happens from nodes. For example, A is spared from transmitting, in light of the fact that their relating node cannot send a CTS. Again, such ineffective transmissions are relative to the quantity of uncovered terminals. These dealings cause deprivation of a node's throughput (Jayasuriya et al. 2004).

18.3   ROUTING PROTOCOLS

The routing conventions can be named: location-aware routing, hierarchical routing, and flat routing. In the case of flat routing, each node is equivalent and assumes a similar job. Nevertheless, in hierarchical routing, hardly any nodes appreciate the advantages of bunch heads and some assume an alternate job in the system. In the case of location-aware routing, node positions may be utilized to course information in the system. As shown by the routing approach, the routing can be arranged as table-driven/proactive and source started/reactive, while as per the system structure, these delegates are: location-aware routing, hierarchical routing, and flat routing (Panisson et al. 2011). Figure 18.2 shows the flat routing protocols under every classification. Proactive routing protocols keep routers persistently refreshed, whereas reactive routing protocols respond on request. Routing protocols can likewise be classified as link state protocols or distance vector protocols. Routers using a link state routing protocol keep up a full or halfway copy of the system topology and costs for every single known connection. Routers utilizing a distance vector protocol only keep data about the next hop to adjoining neighbors and cost for ways to every single known destination (Kaur and Kumar 2013). The classification of routing protocols is shown in Figure 18.2.

18.3.1   PROACTIVE PROTOCOLS

In the case of proactive routing protocol, routing tables store at least one route to any conceivable destination. At the point when information is prepared for transmission, in any event at least one route to the target is as of now in the routing table. Proactive routing protocols show low stillness, yet high routing overheads, as the nodes intermittently trade control messages and directing table data so as to stay up with the latest routes to any dynamic node in the system. Proactive routing protocols can likewise notice better security vulnerabilities, in light of the occasional trade of routing table data and control messages.

18.3.1.1   OLSR (Optimized Link State Routing)

In OLSR to decrease the control overheads, multipoint relays (MPRs) are utilized. For the flooding procedure, only MPRs forward the broadcast message. To decrease the message size, each router proclaims just a little subset of the entirety of its neighbors. The principal hindrance of OLSR is that the routing way may not be the shortest one.

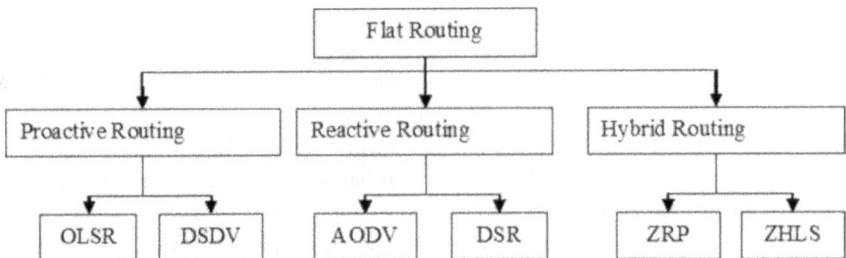

FIGURE 18.2   Classification of routing protocols.

The working of OLSR can be separated into three sections: packet sending, neighbor detecting, and topology revelation. Bundle sending and neighbor-detecting components give routers data about neighbors, and offer an improved method to flood messages in the OLSR organization by utilizing MPRs. The neighbor-detecting activity allows routers to spread nearby data to the all-out system. To compute and to refresh the directing tables, topology revelation is utilized. OLSR utilizes four message types: Hello message, Topology Control (TC) message, Multiple Interface Declaration (MID) message, and Host and Network Association (HNA) message. "Hello messages" are utilized for neighbor detecting. Topology affirmations depend on TC messages. MID messages contain different interface addresses and play out the undertaking of numerous interface affirmations. Since there are numerous interfaces associated with various subnets, HNA messages are utilized to proclaim related system data.

OLSR routers occasionally communicate hello packets to one hop neighbors. Every router sets up a rundown of its neighbors and a rundown of two hop neighbors dependent on which received hello messages. Every router likewise makes one MPR set and one MPR selector set. Routers that have non-void MPRS records communicate their MPRS sets to neighbors by means of TC packets. This is to lessen the size of control messages contrasted, and broadcast a rundown of all neighbor routers. A router rebroadcasts bundles if the sender of that packet is in its MPR selector set. It assists in decreasing the recurrence of flooding. Routers construct routing tables dependent on receive TC packets (Clausen et al. 2003).

18.3.1.2   DSDV (Destination Sequenced Distance Vector)

DSDV is a loop-free proactive routing protocol where the shortest-path calculation depends on the Bellman-Ford calculation. Information bundles are transmitted between the source node to the destination node through routing nodes utilizing routing tables put away at every node. Each routing table contains all the potential destinations from a node to some other node in the system, and the number of hops to every destination.

The protocol has three fundamental angles: to avoid loops, to determine the "count to infinity" problem, and to lessen a high routing overhead. A succession number is given by every node, and is connected to each new routing table update message. It uses two unique sorts of routing table updates, namely "full" and "steady dumps". Their role is to limit the number of control messages. Every node keeps measurable information, in order to lessen the number of rebroadcasts of conceivable routing sections that may land at a node from various paths. However, this is achieved with a similar arrangement number. DSDV only considers bidirectional connections between nodes.

Every node in the system intermittently shares control messages to set the multi-hop ways. As the routes are labelled with a destination succession number, so any route to a destination with a higher destination sequence number replaces a similar route with a smaller destination arrangement number in the node's routing table. This is regardless of the number of hops to this destination. Each node promptly promotes any noteworthy change in its routing table. For example, this could be the inability of a connection to link to its neighboring node(s). Indeed, it remains firmly in place for a specific measure of time known as "settling time" in order to publicize different changes. It occurs in order to limit the quantity of course refreshes transmitted by a node. In this way, when a node receives a course update for a destination from one of its neighboring nodes, and a couple of moments later, it is notified of a second update from an alternate neighboring node for a similar destination with a similar destination arrangement number. Despite a lower number of hops, the node does not quickly communicate the adjustment in its routing table.

As a node's position is ever-changing, DSDV is highlighted with the idea of full and incomplete routing table promotion, in order to keep away from the high overheads of control messages. With this, every node that experiences an adjustment in the way to any destination may decide to only notify this specific change, rather than its total routing table. This **fractional commercial** of a node's routing table is called a "steady update". If numerous progressions happen in a node's routing table, the node may decide to promote full routing table sections, called a "full update" (Narra et al. 2011).

18.3.2 REACTIVE PROTOCOLS

In reactive routing protocols, when the upper transport layer has information to send, and in the unlikely event that that route is not now in existence, at that point the protocol starts a route discovery process. Reactive protocols do not keep up exceptional routes to any destination in the system and do not generally trade any intermittent control messages. In this way, they present a low routing overhead, yet high inertness when contrasted with proactive protocols. Receptive protocols are increasingly defenseless against security threats, as any misfortune or adjustment of route revelation and upkeep messages may have serious ramifications for network performance.

18.3.2.1 AODV (Ad-Hoc On-Demand Distance Vector)

The AODV-routing protocol is a reactive-routing protocol in nature. AODV communicates a route request to find a route in a reactive mode. It is significant that in the case of AODV, hops are included in the route record. Each middle node sets up a transitory reverse link interface during the time spent in a route revelation. This connection focuses on the node that is sent via solicitation. It assists in finding the back way to the source node. At the point when the intermediate nodes get the reply, they are likewise able to set up relating forward-routing passages. To keep away from the stale routing information being utilized as an answer to the most recent solicitation, a destination sequence number is utilized in the route disclosure packet and the route reply bundle. A higher succession number demonstrates a fresher route request.

Route preservation in AODV is the same as in DSR. Some margin of variation is possible with AODV in that it is loop-free because of the destination sequence numbers related to the routes. Like DSR, poor versatility is one drawback with the use of AODV (Royer and Perkins 2000).

18.3.2.2 DSR (Dynamic Source Routing)

The DSR protocol is a distance vector routing protocol. It is reactive by nature: this implies that when a source node wishes to transmit a packet to a specific destination, and if there is no known route to that destination, this node begins a route revelation methodology. Some flexibility is allowed with the use of DSR, since no periodic routing packets are required. DSR, likewise, has the ability to deal with unidirectional connections. Since DSR finds routes on request, it might have terrible regarding control overhead in systems with high versatility and substantial traffic loads. Adaptability is said to be another hindrance of DSR, in light of the fact that DSR depends on daze communicates to trace routes.

DSR performs two primary tasks: Route discovery and route maintenance. For the period of the route discovery method, routers keep up ID arrangements of the crisp solicitations to stay away from the preparing of copy route demand and also loop.

Because of the movement of nodes, on the off chance that connection disruption happens, at that point it is taken care of with the use of route upkeep methodology. Every router keeps an eye on the connections that it uses to advance bundles. At the point when a connection breaks, a route blunder packet is promptly sent to the initiator of the related route. Hence, the invalid route is immediately disposed of (Usop et al. 2009).

18.3.2.3   Hybrid Protocols

In hybrid routing protocols, every node acts reactively in the locale and proactively outside of that area, or zone. Hybrid protocols appreciate some leeway of both reactive and proactive protocols, but may require extra equipment, for example, GPS, or be isolated or incorporated into the specialized gadget.

18.3.2.4   ZRP (Zone Routing Protocol)

The ZRP shares the benefits of both reactive and proactive protocols with a hybrid design, exploiting proactive discovery inside nodes in the vicinity, and exploit a reactive protocol for correspondence between these areas. In a MANET, it can safely be expected that most communication happens among nodes near one another. The partition of a node's neighborhood from the worldwide topology of the whole system takes into consideration applying various methodologies – and consequently maintains a favorable position in every system's **highlights** for a given circumstance. These nearby neighborhoods are called zones; every node may be inside various covering zones, and each zone might be of an alternate size. The "size" of a zone is not dictated by a topographical estimation. In fact, it is given by a range of lengths, where the number of hops to the border of the zone is the determiner. By partitioning the system into covering and variable-size zones, the Zone Routing Protocol is comprised of a few segments, which together give a full routing advantage to ZRP. Each segment works autonomously of the other, and they may utilize various advances so as to amplify effectiveness in their specific area.

18.3.2.5   ZHLS (Zone-Based Hierarchal Link State)

In ZHLS, a communication area is separated into zones in which versatile nodes move haphazardly. A node determines which zone it lives in by utilizing a GPS or signal. Every node floods its Link State Packet (LSP). There are two sorts of LSPs: node LSP for making routing tables to route packets inside a zone, and zone LSP for making between zone routing tables to route bundles between zones. Since ZHLS divides the area into zones which can be additionally isolated into sub-zones, it is considered an appropriate topology for enormous scale MANETs. Besides, because of the progression in GPS innovation and decrease in its cost, a GPS beneficiary can be effortlessly connected to any cell phone, and the area of the gadget can be precisely determined inside a separation of one meter.

18.4 PERFORMANCE METRICS

The job of the routing protocols is especially clear and it is to improve the capacity of the system to transmit the information in the given remote condition. But the decision of determining which is best, depends upon the structure and parameters of the network, such as the speed of nodes that are participating in the network, i.e. whether they are moving quickly or slowly. It additionally relies upon nodes which are packed in an extremely small region, or spread generally all through the system. As there are such a significant number of unanswered inquiries that are tended to by the producer of the protocol, the single protocol could conceivably cause the arrangement to have a considerable number of issues. In order to quantify the difference between the performances of protocols, the MANET's performance can be analyzed using standard performance metrics.

18.4.1 MAC COLLISION RATE

Mac collision rate is the quantity of information bundle impacts happening at the MAC layer in a system over a predetermined timeframe. It shows the rate at which information bundles impact or are missing in collisions. It is estimated as a level of the information packets effectively conveyed.

18.4.2 NORMALIZED ROUTING OVERHEADS

These refer to the fraction of the control packet size (including the RREQ, RREP, RERR, and Hello) and the data packet size delivered to the target.

$$\text{Normalized Routing Overheads} = \frac{\text{Number of Routing packet Sent}}{\text{Number of Data packet Received}} \quad (18.1)$$

18.4.3 PACKET DELIVER RATIO (PDR)

This is the fraction of number of data packets successfully received by the CBR destination to the number of data packets generated by the CBR source. It gauges the misfortune rate by transport protocols. Ideally, the packet delivery ratio should be at a high to minimize the losses. Mathematically, it can be expressed as:

$$\text{PDR} = \frac{\sum \left(\text{all the packets received by destination} \right)}{\sum \left(\text{all the packets sent by source} \right)} \quad (18.2)$$

18.4.4 Error Rate

This is the rate at which error may happen in the transmitted information bundles. More mistakes imply higher misfortunes in information bundles, and more retransmissions are required, which increase the overheads, and diminish the throughput.

18.4.5 Throughput

This is the amount of digital data per time unit delivered over a physical or logical link. It is measured in packets per second, and mathematically can be expressed as:

$$\text{Throughput} = \frac{\text{Data packets sent sucessfully}}{\text{Total Time}} \qquad (18.3)$$

The average of the total throughput is called an average throughput.

18.5 OPTIMIZED NANO MANET

A nanomachine is a mechanical or electromechanical tool whose measurements are estimated in nanometers (units of 10^{-9} meter). The performance of a MANET is governed by the data collision rate, which is estimated by some portability models; for example, random mobility and then random waypoint. There is collision on handling and arranging information in the versality of nanaomachines. Utilizing the nanotechnology, performance is expanded at a pace of 10^{-9}/sec. In this manner, collision speed and transmission of information are improved. On assessing the speed of the current innovation, nanotechnology performs quicker without loss of information bundles and postponement of throughput. Utilizing the nanoinnovation execution is determined for the collision speed and transmission of each nanoarranged topology. These nanoinnovation executions assess the current speed and execution determined for proficiency with this strategy. The minute size of nanomachines converts into high operational speed. This is a consequence of the regular propensity of all things considered, and frameworks to work quicker as their size decreases. Nanomachines could be modified to repeat themselves, or to work at fabricating bigger machines, or to develop nanochips. Specific nanomachines called nanorobots may be structured not exclusively to analyze, but also to improve the system execution. The nanomachines utilized here utilize the earth, assemble the data, and convey them to the infostation. The infostation makes a choice for an appropriate move and it likewise acts as a portal, or indeed, as a cushion. As the structure of MANET is huge, the likelihood of meeting the purpose of the nanomachine and the infostation is slight. Therefore, only a few optional nanomachines are used, which act as relay nodes.

18.6 SUMMARY

The routing of information packets in MANET systems is the key to achieving a proficient nature with the administration parameters. To build the system throughput, the routing convention must be extremely productive. In MANETs, the portable nodes perform agreeable transmission to convey the packet to the destination. In MANET, nodes are effortlessly moved and organized arbitrarily. This mobility causes the topology of the MANET system to be changed. The nodes' mobility puts an additional load on the congestion control mechanism. It cannot find the system elements of specially appointed systems. All nodes can join, connect, and leave forms in the network. The node group has incredibly dynamic encompassing, and property of information course. Subsequently, productive information directing is one of the fundamental issues in MANETs.

The enhanced technique incorporates nanotechnology with standard accessible proactive, reactive, and hybrid routing protocols. These changed frameworks improve the packet delivery ratio and throughput, by diminishing the errors and collisions which eventually upgrade the system performance.

REFERENCES

Clausen, T., Jacquet, P., Adjih, C., Laouiti, A., Minet, P., Muhlethaler, P., and Viennot, L. 2003. Optimized link state routing protocol (OLSR). *Project Hipercom*, INRIA, France, Network Working Group.

Jayasuriya, A., Perreau, S., Dadej, A., and Gordon, S. 2004. Hidden vs. exposed terminal problem in ad hoc networks. In: *Proceedings of the Australian Telecommunication Networks and Applications Conference*, Sydney, Australia, December 2004.

Kaur, D., and Kumar, N. 2013. Comparative analysis of AODV, OLSR, TORA, DSR and DSDV routing protocols in mobile ad-hoc networks. *International Journal of Computer Network and information Security* 3:39–46.

Murthy, C.S.R., and Manoj, B.S. 2004 *Ad Hoc Wireless Networks: Architectures and Protocols*. Prentice Hall PTR, Upper Saddle River, New Jersey.

Narra, H., Cheng, Y., Cetinkaya, E.K., Rohrer, J.P., and Sterbenz, J.P. 2011. Destination-sequenced distance vector (DSDV) routing protocol implementation in ns-3. In: *Proceedings of the 4th International ICST Conference on Simulation Tools and Techniques*, 439–446, 25 March, Barcelona, Spain.

Panisson, A., Barrat, A., Cattuto, C., Broeck, W.V.D., Ruffo, G., and Schifanella, R. 2011. On the dynamics of human proximity for data diffusion in ad-hoc networks. *Ad-Hoc Network* 10(8):1532–1543.

Royer, E.M., and Perkins, C.E. 2000. An implementation study of the AODV routing protocol. In: *2000 IEEE Wireless Communications and Networking Conference. Conference Record (Cat No. 00TH8540)*, 3, 1003–1008. 23–28 September, Chicago.

Usop, N.S.M., Abdullah, A., and Abidin, A.F.A. 2009. Performance evaluation of AODV, DSDV & DSR routing protocol in grid environment. *IJCSNS International Journal of Computer Science and Network Security* 9(7):261–268.

Index

A

Abraxane, 48
Absorption properties, 107
Active solar, 73
Ad-hoc on-demand distance vector (AODV), 316
Adsorption capacity, 280
Aerogel, 29
Agriculture, 27–29
    broad nanotechnology applications in, 28
    challenges, nanobiotechnology, 236
        postharvest losses (PHL), 239
        proper irrigation facility, 239
        seed availability, 238–239
        small and fragmented land-holdings, 238
        technology to overcome, 239–241
    nanotechnology area/types in, 28
Agroforestry systems, 240
Air Plasma Spraying (APS) technique, 98–100
Air quality, 39
Alzheimer's disease, 50
AMP, *see* Antimicrobial peptide
Ampholytic surfactants, 173
Anionic nutrient ions, 205
Anionic surfactants, 173
Anode graphite rod, 13
Anthocyanin, 106, 107
Antibacterial textiles, 52
Antibiotic resistance, 52
Antimicrobial peptide (AMP), 264
Antimicrobial property, 184
AODV, *see* Ad-hoc on-demand distance vector
Apatite, 243
APS, *see* Air Plasma Spraying technique
Arc discharge method, 13–14
Atrazine, 208
Autoclave, 21

B

Ball milling method, 15
Band-to-band tunneling phenomenon, 111
BBB, *see* Blood brain barrier
BCS, *see* Biopharmaceutical Classification System
Bellman-Ford calculation, 315
Bi-continuous system, 170
Binary controlled delay line, 296
Bioenergy, 74
Bioimaging, 222, 266–268
Biologically derived nanoparticles, 142

Biomass, 72–73
Biomedical applications, 34–38
Biomimicry, 51
Biomolecules integration, nanoscale range, 220
Biopharmaceutical Classification System (BCS), 170
Biopharmaceuticals, 225–230
Biosensors, 63
Biphosphonate alendronate, 265
Bladder transplant, 224–225
Blood brain barrier (BBB), 150, 151
Bond coat, 95, 96
Bone marrow, 149
Bone transplant, 224
Bottom-up fabrication methods, 11, 17–18
    chemical vapor deposition, 18
    hydrothermal synthesis, 20–21
    microwave method, 21
    molecular beam epitaxy (MBE), 19
    sol-gel synthesis, 19–20
Buckyballs, 219
Buffer
    schematic design, 301
    transient response, 301
Buffer-based delay line, 300–302
Bulk limited group, 122
Bulk nanomaterials, 4–5
Business applications, 65–66

C

Cadence Virtuoso CMOS 90 nm technology, 298
Cancer, 49
Cancer cells, 265
Cancer therapy, 264–266
Capping, 213
Carbendazime, 247
Carbon nanofibers (CNFs), 141
Carbon nanomaterials, 139
Carbon nanotube FETs (CNTFETs), 115
    conduction operation, 117–118
    device geometry, 116
    performance metrics, 118
Carbon nanotubes (CNTs), 13, 14, 31, 47, 55, 67, 100, 115, 139, 141
Carbon nanowires, 47
β-Catenin, 137
Cationic nutrient ions, nano-clays and zeolites, 205
Cationic surfactants, 173
CDR, *see* Clock and data recovery systems

321

For Product Safety Concerns and Information please contact our EU
representative GPSR@taylorandfrancis.com
Taylor & Francis Verlag GmbH, Kaufingerstraße 24, 80331 München, Germany

www.ingramcontent.com/pod-product-compliance
Lightning Source LLC
Chambersburg PA
CBHW060806220326
41598CB00022B/2552

* 9 7 8 0 3 6 7 5 5 8 3 6 9 *